Space Group Representations

Nikolai B. Melnikov · Boris I. Reser

Space Group Representations

Theory, Tables and Applications

Springer

Nikolai B. Melnikov
Moscow State University
Moscow, Russia

Boris I. Reser
Institute of Metal Physics
Russian Academy of Sciences
Ekaterinburg, Russia

ISBN 978-3-031-13990-1 ISBN 978-3-031-13991-8 (eBook)
https://doi.org/10.1007/978-3-031-13991-8

This Springer imprint is published by the registered company Springer Nature Switzerland AG
The registered company address is: Gewerbestrasse 11, 6330 Cham, Switzerland

To the memory of Rudolf Egorov

Preface

Hardly any other field of natural science is so well studied and at the same time is so difficult to master, as the representation theory of space groups.

This book delivers exactly what it promises in the title:

1. clear, compact and self-contained description of the representation theory of space groups;
2. complete tables of the irreducible projective representations of all 32 crystallographic point groups, which allow one to construct the irreducible representations, including double-valued, of all 230 space groups;
3. worked out examples of the use of the tables in practical problems and applications of the theory to energy band calculations, some of which have made it into textbooks[1].

Since the book is for sale in whole and in parts (as individual chapters), we did our best to make the chapters autonomous by reducing the number of cross-references to the minimum. We hope that this will promote mass appeal and wider use of our approach to space group representations.

A good book, just as good music, should be reminiscent of something familiar. Therefore, in our outline of introductory material we used the textbooks by Wigner and Streitwolf[2]. On the other hand, we avoided excessive mathematical details and being prolix as much as we could.

The book will be useful to everyone using symmetry analysis in solid-state physics and chemistry. We tried to make it assertible and valuable to researchers and graduate students alike. The reader is assumed to have only a modest knowledge of linear algebra and quantum mechanics.

[1] G.C. Fletcher, *The Electron Band Theory of Solids* (North Holland: Amsterdam, 1971), J. Callaway, *Quantum Theory of the Solid State*, 2nd edn (Academic, New York, 1991), W. Hergert, R.M. Geilhufe, *Group Theory in Solid State Physics and Photonics* (Wiley, New York, 2018).

[2] E.P. Wigner, *Group Theory and its Application to the Quantum Mechanics of Atomic Spectra*, 2nd edn (Academic, New York, 1959), H.W. Streitwolf, *Group Theory in Solid State Physics* (MacDonald, London, 1971).

One of us (B. Reser) started studying group theory at the lectures of A. Sokolov[3] and is grateful to him for the lifelong interest in symmetry of solids. For several years, B.R. developed group theoretical methods for energy band calculations together with R. Egorov, and we dedicate this book to his memory. It is a pleasure to thank V. Bobrovskii, A. Granovsky and B. Johansson for useful discussions. We are grateful to the editor Z. Evenson for his energetic support during the writing process. We thank I. Melnikova for her patience and encouragement. No one has read the manuscript, and all responsibility for the content of this book and typos lies solely with us.

Corrections and suggestions will be gratefully received and may be addressed by email to melnikov@cs.msu.ru and reser@imp.uran.ru.

Moscow, Russia
Ekaterinburg, Russia
May 2022

Nikolai B. Melnikov
Boris I. Reser

[3] A.V. Sokolov, V.P. Shirokovskii, Group-theoretical methods in the quantum physics of solids (spatial symmetry), *Sov. Phys. Usp.*, **3**, 551 (1961).

Contents

Symbols

As is customary in physics, we often omit the infinite limits of integration. Bold letters denote vectors; calligraphic letters are used for operators and group representations.

$a \equiv \{\alpha\|\mathbf{t}\}$	Linear spatial transformation, Sect. 7.2
$a_{i,j}^{\mathbf{k},\mathbf{k}+\kappa}$	Multiplicity coefficient, Sect. 5.8
A, B, C	Constants in the **kp**-approximation, (14.12)
$\mathbf{a}_1, \mathbf{a}_2, \mathbf{a}_3$	Basic vectors of the crystal lattice, (5.4)
$\mathbf{b}_1, \mathbf{b}_2, \mathbf{b}_3$	Basic vectors of the reciprocal lattice, (5.9)
C_n	Rotation through an angle $2\pi/n$, Sect. 2.1
C_n^2, C_n^3	Successive applications of the rotation C_n, Sect. 2.1
$\mathcal{D}$	Representation, Sect. 2.3
D	Matrix of the representation $\mathcal{D}$, Sect. 2.3
$\tilde{D}$	Transpose of D, Sect. 2.2
D^*	Complex conjugate of D, Sect. 2.2
$D^\dagger = \tilde{D}^*$	Adjoint (or Hermitian conjugate) of D, Sect. 2.2
$\mathcal{D}^i$	Regular continuation of the representation d^i, Sect. 3.1
$D^i(g)$	Matrix of the regular continuation of d^i, Sect. 3.1
$\mathcal{D}^{\mathbf{k}j}$	Irreducible representation of the space group, Sect. 5.7
$\mathcal{D}^{l\pm}$	Representation of the full rotation group, Sect. 12.5.1
$D^{l\pm}(\alpha)$	Matrix of the irreducible representation of the full rotation group, (17.47)
$\mathcal{D}^r$	Regular representation, Sect. 2.3
$D^r(a)$	Matrix of the regular representation, Sect. 2.3
$D^l_{m'm}$	Matrix element of the group of rotations, (17.47)
d^i	Representation of the subgroup, Sect. 3.1
d^{i_g}	Representation conjugate to d^i under g, Sect. 3.2
$d^{\mathbf{k}j}$	Small representation of the group $G_{\mathbf{k}}$, Sect. 5.6
E	Unit matrix, Sect. 2.2
E	Identity transformation (Einheit), Sect. 6.1
e	Unit element of a group, Sect. 2.1

$f_\kappa^{(p)}$	Basis function belonging to the κ-th row of the irreducible representation $\mathcal{D}_p$, Sect. 7.1
G	Group, Sect. 2.1
$\tilde{G}$	Covering group, Sect. 4.6
$G_\mathbf{k}$	Group of the wavevector $\mathbf{k}$ (little group), Sect. 5.5
G_0	Point group, Sect. 5.3
$G_\mathbf{k}^0$	Point group of the wavevector $\mathbf{k}$, Sect. 5.4
G_p^l	Group of the vector $\mathbf{Q}_p^l$, Sect. 13.1.9
G/H	Factor group, Sect. 2.1
g	Element of the group G, Sect. 2.1
$\tilde{g}$	Element of the covering group $\tilde{G}$, Sect. 4.7
g_i	Generator of the group G, Sect. 4.4
$H = T + V$	Hamiltonian, Sect. 7.2
H	Invariant subgroup (normal divisor), Sect. 2.1
H_{ij}	Matrix element of the Hamiltonian, Sect. 14.1
I	Inversion in the space group, Sect. 6.1
$I_{\mu\mu'}^{jj'}$	Energy integral, Sect. 13.1.2
i	Inversion in the abstract group ($i\mathbf{r} = -\mathbf{r}$), Sect. 5.1
K	Complex conjugation operator, (8.35)
$\mathbf{K}_\mu$	Reciprocal lattice vector, Sect. 5.2
$\mathbf{k}$	Reduced wavevector, Sect. 5.3
l_p	Dimension of the representation $\mathcal{D}_p$, Sect. 2.3
m_g	Order of the element g, Sect. 4.4
m_i	Order of the generator g_i, Sect. 6.1
N_i	Normalizer of the representation d^i, Sect. 3.2
N	Total number of random points in Ω_{BZ}, (14.19)
$N(\varepsilon_j)$	Number of random points with the energy ε_j, (14.19)
n	Index of the energy band, (8.3)
n, l, m	Atomic quantum numbers, Sect. 11.1.2
$\mathbf{n}$	Unit vector of the rotation axis, Sect. 6.1
$n, n(G)$	Order of the group G, Sect. 2.1
n_p	Multiplicity of the representation $\mathcal{D}_p$ in $\mathcal{D}$, Sect. 2.3
$n(\varepsilon)$	Density of states, (14.17)
$O_{\mu\mu}^p$	Operator that generates the basis function belonging to the μ-th row of the representation $\mathcal{D}_p$ (projection operator), Sect. 7.1
$O_{\mu\nu}^p$	Operator that generates a basis partner function of the representation $\mathcal{D}_p$ ($\mu \neq \nu$), Sect. 7.1
P_a, $P_{\{\alpha\|\mathbf{t}\}}$	Symmetry operator, (7.10)
$P_{[\alpha\|\mathbf{t}]}$	Symmetry operator that describes transformation rule of a two-component spinor, (6.31)
$P_l^{\|m\|}(\cos\vartheta)$	Associated Legendre polynomial, (17.43)
p	Number of generators in the group G, Sect. 4.4
$\mathbf{Q}_p^l$	Fixed vector of a coordination sphere, Sect. 13.1.8

$\mathbf{R}_{\{\alpha\beta\gamma\}}$	Transformation matrix associated with the rotation by the Euler angles α, β, γ, Sect. 17.1.3
$\mathbf{R}_n, \mathbf{R}_p$	Crystal lattice vector, (5.4)
$R^s_{nl}(r)$	Atomic radial function, (9.14)
r	Number of classes of conjugate elements, Sect. 2.1
r_i	Number of elements in the class C_i, Sect. 2.3
$S(\alpha)$	2×2 matrix that acts on two-component spinors, (6.32)
$\hat{S}^j_{\sigma\sigma}$	Projection operator that generates the function belonging to the σ-th row of the j-th irreducible representation, (17.4)
$\mathbf{S}^j_{\sigma\sigma}$	Projection matrix with the elements $(S^j_{\sigma\sigma})_{\mu\nu}$, (17.7)
T	Translation subgroup of the space group, Sect. 5.3
T	Kinetic energy, Sect. 7.2
$\mathbf{t}$	Translation vector, Sect. 5.1
$u_{n\mathbf{k}}(\mathbf{r})$	Periodic function, (8.3)
$V(\mathbf{r})$	Potential (potential energy), Sect. 7.2
$V_C(\mathbf{r})$	Coulomb part of the crystal potential, Sect. 9.2.3
$V_{ex}(\mathbf{r})$	Exchange part of the crystal potential, (9.29)
$Y_{lm}(\vartheta, \varphi)$	Spherical function, (9.14)
Z	Atomic number, Sect. 15.6
Z_s	Charge of the s-th nucleus, (9.31)
z_0	Crystallographic parameter for the $2H$ structure, Sect. 15.2
α	Point transformation (rotation, inversion or their combination), Sect. 5.1
$\{\alpha\|0\}$	Rotation (proper or improper), Sect. 5.1
$\{\alpha\|\,\mathbf{t}\}$	Spatial transformation, (5.1)
$\{\alpha_m\|\boldsymbol{\tau}_{\alpha_m}\}$	Representative in the (right) coset decomposition of $G_\mathbf{k}$ with respect to T, Sect. 5.5
$\{\alpha_M\|\boldsymbol{\tau}_{\alpha_M}\}$	Representative in the (right) coset decomposition of G with respect to $G_\mathbf{k}$, Sect. 5.5
$\{\alpha_M\mathbf{k}\}$	Star of the wavevector $\mathbf{k}$, Sect. 5.7
$\{\alpha, \beta, \gamma\}$	Proper rotation by the Euler angles α, β, γ, Sect. 17.1.3
$\Delta \equiv \nabla^2$	Laplace operator, (8.1)
$\Delta^{\mathbf{k}j}$	Irreducible projective representation of the factor group $G_\mathbf{k}/T$, Sect. 5.6
δ_i	Gauge factor, Sect. 6.1
δ_{ij}	Dirac delta-symbol, Sect. 2.3.4
$\delta(\mathbf{r} - \boldsymbol{\tau}_s)$	Dirac delta-function, (9.31)
$\varepsilon, \varepsilon_n(\mathbf{k})$	Eigenvalue, (8.1)
$\{\varepsilon\|0\}$	Identical transformation, Sect. 5.1
$\{\varepsilon\|\mathbf{t}\}$	Translation, Sect. 5.1
θ	Rotation angle, Sect. 6.1
θ	Time-reversal operator, (8.35)
ϑ	Spherical coordinate, (17.41)
ξ, η, ζ	Values ak_x, ak_y, ak_z, Table 13.2
ξ, η, ζ	Values ak_x, ak_y, ck_z, Sect. 15.4

$\rho^{e}(\mathbf{r})$	Crystal electron density, (9.30)
$\boldsymbol{\sigma}$	Vector of the Pauli matrices $\sigma_x, \sigma_y, \sigma_z$, (6.33)
$\boldsymbol{\tau}_{\alpha}$	Nonprimitive translation, Sect. 5.1
$\boldsymbol{\tau}_{s}$	Basis vector indicating the position of atom in a unit cell, Sect. 9.2
$\boldsymbol{\tau}_{t}^{s}$	Basis vector indicating the position of the t-th atom of the s-th type in the unit cell, Sect. 11.1
$\Phi_{nlm}^{\mathbf{k}s}(\mathbf{r})$	Bloch sum, (9.12)
$\Phi_{\mu}^{sj}(\mathbf{r})$	Bloch sum, (13.7)
φ	Spherical coordinate, (17.41)
$\varphi_{\mu}^{j}(\mathbf{r})$	Function belongs to the μ-th row of the representation Γ_j of the point subgroup G_0', Sect. 13.1.2
$\varphi_{nlm}^{s}(\mathbf{r})$	Atomic function, (9.12)
$\chi(a)$	Character of an element $D(a)$, Sect. 2.3
χ_i	Character of the ith class, Sect. 2.3
χ^{i}	Character of the representation d^i, Sect. 3.2
χ^{i_g}	Character of the representation d^{i_g}, Sect. 3.2
$\chi^{l\pm}(\alpha)$	Character of the representation $\mathcal{D}^{l\pm}$, Sect. 12.5.1
χ_{μ}	Basis function, (8.10)
ψ	Eigen function, (8.1)
$\psi_{n\mathbf{k}}(\mathbf{r})$	Bloch function, (5.1)
Ω_{BZ}	Volume of the Brillouin zone, Sect. 5.2
Ω_{WS}	Volume of the Wigner–Seitz cell, Sect. 5.2
$\omega_{\alpha,\beta}$	Spin factor taking values ± 1, Sect. 6.1
$\omega_{g,g'}$	Projective factor, Sect. 4.1
$\omega_{ij}, \omega_{ij}'$	Character factor, Sect. 4.5

Chapter 1
Scope and Overview

I have no patience with books so thick they can serve as doorstops; such excessive bulk, I feel, can only result from a lack of clear thinking. (A.D. Sakharov, Memoir, Knopf, New York, 1990).

Nowadays both calculations and experiment provide an enormous amount of data that cannot be systematically studied without taking symmetry into account. In particular, many problems in solid-state physics including studies of electron and phonon spectra, structural and magnetic phase transitions, etc. benefit a lot from using space group representations (for an introduction see, e.g. [1–7] and refs. therein).[1] Representation theory also helps in search for new materials with specific characteristics based on symmetry properties. Many of these materials have space groups with compound symmetry elements (nonsymmorphic groups).

In this book, we present the theory and tables that work equally well with symmorphic and nonsymmorphic space groups. The use of the theory and tables is illustrated in problems of solid-state physics and energy band calculations by concrete examples. In full accord with the title, the book consists of three parts: theory (Chaps. 2–5), tables (Chap. 6) and applications (Chaps. 7–17).

Part I consists of one introductory chapter and three core chapters. Chapter 2 gives a brief review of the group theory, linear algebra and representation theory. Construction of the irreducible representations of the space groups is based on induced and projective representations. In Chap. 3 we provide the necessary facts on induced representations. In Chap. 4 we present elements of the theory of projective representations, which is a key to constructing and understanding the tables.

We tried to make a trade-off between textbooks that mention induced and projective representations within a very limited space (see, e.g. [13–16]) and thick volumes like [17–20]. An important difference in our description of the induced representations is that only elementary matrix theory is used. Unlike other textbooks such as

[1] For applications of the space group in crystallography beyond the representation theory, see, e.g. [8–12].

N. B. Melnikov and B. I. Reser, *Space Group Representations*,
https://doi.org/10.1007/978-3-031-13991-8_1

[17–19], we do not work with basis functions. This, in our view, makes the theory more straightforward and easier to understand.

To make the presentation self-contained, in Chaps. 3 and 4 we prove nearly all statements and theorems. The proofs are constructive and concise. The theory applies not only to the space groups and can be useful in other areas beyond the solid-state physics. This comes at a price. The level of abstraction of Chaps. 3 and 4 is higher than in the rest of the book. An application-oriented reader can either skip them completely or come back to them later.

In Chap. 5 we summarize the results of two previous chapters and apply this theory to irreducible representations of the space groups. The main results are as follows.

- We give a complete description of irreducible representations of the space groups and provide an effective method for constructing them.
- Namely, we show that the problem reduces to constructing irreducible *projective* representations of the 32 crystallographic point groups. We present a convenient way of deriving these representations.
- The matrices of all irreducible (single- and double-valued) representations of a space group with an arbitrary crystallographic structure are obtained from the irreducible projective representations of the 32 crystallographic point groups by multiplying them by easily determined factors.
- We show that the structure of the matrices of irreducible representations of the space groups allows to express many properties of these representations in terms of small representations of the group of the wavevector.
- We also discuss the role of spin and effects of time-reversal symmetry.

Part II consists of Chap. 6. It presents complete tables of irreducible representations of the space groups. In its importance (and in volume), Chap. 6 is one of the main chapters of the book and can be used independently. Section 6.1 gives a short summary of the above theory and guide to tables. Sections 6.2–6.19 present compact tables of irreducible projective representations of 32 crystallographic point groups, which are used to obtain all irreducible representations, including double-valued, of all 230 space groups. Section 6.20 explains how to use these tables in practice. At this point, we finish the development of group theory and return to discussing physics.

Part III presents applications to solid-state physics and energy band calculations (for an introduction, see, e.g. [7, 21–23]). This part can be broken up into four subparts according to the application area. The *first subpart of Part III* consists of Chaps. 7–9. Our aim here is to build a bridge with the previous material.

In Chap. 7 we explain the role of symmetry in the classification of the eigenstates of quantum mechanical systems. The use of group-theoretical methods to quantum mechanics is demonstrated in a *general* form, without focusing on particular problems. At the end of the chapter, we consider three important applications: classification of states, splitting of states and selection rules.

In Chap. 8 we proceed to applications of symmetry in the solid state. First we give a mathematical formulation of one-electron problem in a periodic potential and apply crystal symmetry to energy band structure. Then we take spin into account and discuss double-valued representations and time-reversal symmetry.

In Chap. 9 we present the basics of the electronic energy band calculations and discuss in detail the use of symmetry in a simple but universal method, the orthogonal plane wave (OPW) method.

Since representation theory often seems too abstract to fully grasp, it can be difficult to start using it without reviewing a number of solid "real-life" examples. This is especially the case with nonsymmorphic space groups. The *second subpart of Part III*, consisting of Chaps. 10–12, provides the reader with a number of such examples. The simplest and most popular example of the nonsymmorphic group in the literature is the diamond structure, the space group O_h^7 (see, e.g. [17, 19, 24]). We consider the diamond structure and more complex examples with several atoms per unit cell, including FeS_2 (pyrite) and Cr_3Si (chromium silicide) structures, the space groups T_h^6 and O_h^3, respectively.

In Chap. 10 we present the example of the diamond structure as the reference point. In Chap. 11 we explain how to make a correspondence between electron states in free atoms and solids. The method can be used for any crystal structures and any set of atomic wave functions. As an example, we consider the FeS_2 crystal structure.

Chapter 12 gives a group-theoretical analysis of a crystal structure of the Cr_3Si type. By the example of a fairly complex and important crystal structure, we demonstrate how the group-theoretical analysis can be carried out in practice. We present compatibility relations, study the effect of time-reversal symmetry on band degeneracy and make a correspondence between the electronic states in the crystal and in free atoms.

The *third subpart of Part III*, consisting of Chaps. 13–15, deals with the tight-binding method, also called the linear combination of atomic orbitals (LCAO) method.

In Chap. 13 we present the tight-binding method. Our treatment applies to any crystal structure (including those having nonprimitive translations) and large number of atomic wave functions. Matrix elements of the Hamiltonian in the tight-binding approximation are determined group theoretically using space symmetry of the problem, time-reversal symmetry and Hermiticity of the Hamiltonian.

In Chap. 14 we use symmetry in the tight-binding method to interpolate the energy bands. The tight-binding interpolation method is illustrated in the example of the diamond structure.

In Chap. 15 the tight-binding method is applied to a more complex example with several atoms per unit cell. We obtain the analytical form of a low-dimensional effective Hamiltonian describing the conduction band of the quasi 2D dichalcogenides with the $2H$-$TaSe_2$ structure.

The *fourth subpart of Part III* is devoted to the so-called symmetrized functions. Since all initial functions in energy band calculations are certain combinations of either plane waves or Bloch sums, it suffices to construct only symmetrized plane waves (SPW) and symmetrized Bloch sums (SBS). In Chaps. 16 and 17, we describe the methods of constructing SPW and SBS, respectively, for crystals with arbitrarily complex symmetry. Finally, in Chap. 18 we compare the theory and tables presented in this book with exiting ones and refer to the literature on related topics.

In Appendices we collected proofs of useful mathematical theorems, additional material on projective representations and double-valued representations.

Our notation mainly follows Schönflies [25], Seitz [26], Bouckaert, Smoluchowski and Wigner [27], Koster [28], Faddeyev [29] and Hurley [30]. For brevity we omitted elementary material and proofs of standard theorems in group theory, which can be found in textbooks [24, 31–37].

References

1. P. Jacobs, *Group Theory with Applications in Chemical Physics* (Cambridge University Press, Cambridge, 2005)
2. M. Dresselhaus, G. Dresselhaus, A. Jorio, *Group Theory: Application to the Physics of Condensed Matter* (Springer, Berlin, 2008)
3. M. El-Batanouny, F. Wooten, *Symmetry and Condensed Matter Physics: A Computational Approach* (Cambridge University Press, Cambridge, 2008)
4. T. Wolfram, S. Ellialtioglu, *Applications of Group Theory to Atoms, Molecules, and Solids* (Cambridge University Press, Cambridge, 2014)
5. C. Malgrange, C. Ricolleau, M. Schlenker, *Symmetry and Physical Properties of Crystals* (Springer, Berlin, 2014)
6. M. Ladd, *Symmetry of Crystals and Molecules* (Oxford University Press, Oxford, 2014)
7. W. Hergert, R.M. Geilhufe, *Group Theory in Solid State Physics and Photonics: Problem Solving with Mathematica* (Wiley, New York, 2018)
8. R.W.G. Wyckoff, *Crystal Structures*, Vols. 1–6, 2nd edn. (Interscience, New York, 1963–1968)
9. N.V. Belov, Y.O. Zagal'skaya, G.P. Litvinskaya, Y.K. Egorov-Tismenko, *Atlas of Space Groups of the Cubic System* [in Russian] (Nauka, Moscow, 1980)
10. G. Burns, A.M. Glazer, *Space Groups for Solid State Scientists*, 3rd edn. (Elsevier, Amsterdam, 2013)
11. U. Müller, *Symmetry Relationships between Crystal Structures: Applications of Crystallographic Group Theory in Crystal Chemistry* (Oxford University Press, Oxford, 2013)
12. M.I. Aroyo (ed.), *International Tables for Crystallography, Volume A, Space-group symmetry*, 6th edn. (Wiley, New York, 2016)
13. T. Inui, Y. Tanabe, Y. Onodera, *Group Theory and Its Applications in Physics* (Springer, Berlin, 1990)
14. W. Ludwig, C. Falter, *Symmetries in Physics Group Theory Applied to Physical Problems*, 2nd edn. (Springer, Berlin, 1996)
15. R.A. Evarestov, V.P. Smirnov, *Site Symmetry in Crystals: Theory and Applications*, 2nd edn. (Springer, Berlin, 1997)
16. J.F. Cornwell, *Group Theory in Physics: An Introduction* (Academic, San Diego, 1997)
17. C.J. Bradley, A.P. Cracknell, *The Mathematical Theory of Symmetry in Solids. Representation Theory for Point Groups and Space Groups* (Clarendon, Oxford, 1972). Reprinted: Clarendon, Oxford, 2010
18. G.L. Bir, G.E. Pikus, *Symmetry and Strain-induced Effects in Semiconductors* [in Russian] (Nauka, Moscow, 1972) Engl. transl.: Wiley, New York, 1974
19. J.L. Birman, *Theory of Crystal Space Groups and Lattice Dynamics: Infra-Red and Raman Optical Processes of Insulating Crystals* (Springer, Berlin, 1974)
20. S.K. Kim, *Group Theoretical Methods and Applications to Molecules and Crystals*, 2nd edn. (Cambridge University Press, Cambridge, 2003)
21. H. Jones, *Theory of Brillouin Zones and Electronic States in Crystals* (North-Holland, Amsterdam, 1960). Reprinted: LLC, Whitefish, MT, 2013

22. V. Heine, *Group Theory in Quantum Mechanics* (Pergamon, Oxford, 1960). Reprinted: Dover, New York, 2009
23. J. Callaway, *Quantum Theory of the Solid State*, 2nd edn. (Academic, New York, 1991)
24. H.W. Streitwolf, *Group Theory in Solid State Physics* (MacDonald, London, 1971)
25. A. Schönflies, *Theorie der Kristallstruktur* (Gebr. Borntraeger, Berlin, 1923)
26. F. Seitz, Ann. Math. **37**, 17 (1936)
27. L.P. Bouckaert, R. Smoluchowski, E. Wigner, Phys. Rev. **50**, 58 (1936)
28. G.F. Koster, Solid State Phys. **5**, 173 (1957)
29. D.K. Faddeev, *Tables of the Principal Unitary Representations of Fedorov Groups* [in Russian] (Acad. Sci. USSR, Moscow, 1961). Engl. transl.: Pergamon, Oxford, 1964
30. A.C. Hurley, Phil. Trans. Roy. Soc. London **A260**, 1 (1966)
31. B.F. Bayman, *Some Lectures on Groups and their Applications to Spectroscopy* (Nordita, Copenhagen, 1957)
32. E.P. Wigner, *Group Theory and its Application to the Quantum Mechanics of Atomic Spectra*, 2nd edn. (Academic, New York, 1959)
33. M. Hamermesh, *Group Theory and Its Applications to Physical Problems* (Addison-Wesley, Reading, MA, 1962). Reprinted: Dover, New York, 1989
34. R.S. Knox, A. Gold, *Symmetry in the Solid State* (Benjamin, New York, 1964)
35. M. Tinkham, *Group Theory and Quantum Mechanics* (McGraw-Hill, New York, 1964). Reprinted: Dover, New York, 2003
36. M.I. Petrashen, E.D. Trifonov, *Applications of Group Theory in Quantum Mechanics* (MIT Press, Cambridge, MA, 1969). Reprinted: Dover, New York, 2009
37. A. Nussbaum, *Applied Group Theory* (Prentice-Hall, Englewood Cliffs, NJ, 1971)

Chapter 2
Mathematical Preliminaries

In mathematics you don't understand things. You just get used to them. (J. von Neumann; from G. Zukav, The Dancing Wu Li Masters: An Overview of the New Physics, Bantam, New York, 1979)

This chapter gives a brief introduction to the representation theory of finite groups. We give only the most important proofs (for the rest we refer the reader to [1, 2]).

2.1 Elements of Group Theory

2.1.1 Definition of Groups

A *group* G is a set of elements $a, b, c, \dots$ satisfying the following conditions.

1. The group multiplication is defined, which assigns to any ordered pair of elements a and b a third element $c = ab$, called the *product* of a and b. The multiplication rule can be presented conveniently in the form of a *group table*, which lists the product of every pair. The group is uniquely defined by its group table.
2. The group multiplication is associative: $(ab)c = a(bc)$.
3. There is a unique *right identity* in the group, i.e. an element e such that $ae = a$ for all elements a.
4. Every element a in the group has a *right inverse element* a^{-1} such that $aa^{-1} = e$.

By using properties 1–4 one can prove that the right identity e is also a left identity, i.e. $ea = a$, and is unique, and that the right inverse of a is also a left inverse, i.e. $a^{-1}a = e$, and is uniquely determined by a. We now need some more definitions.

A group G is called *finite* if it contains a finite number of elements $n(G)$, and the number $n(G)$ is called the *order* of the group G. A group G is called *commutative* (or *Abelian*) if the group multiplication is commutative: $ab = ba$. A group G is called

N. B. Melnikov and B. I. Reser, *Space Group Representations*,
https://doi.org/10.1007/978-3-031-13991-8_2

cyclic if there is a *generating element* $a \in G$ such that its powers a^ν run over all elements of the group when the power ν run over all integers. If there is a smallest positive integer n such that $a^n = e$, then n is called the *order of the element* a. In finite groups each element has a well-defined order. In a cyclic group, the order of a generating element n is equal to the order of the group itself $n(G)$. Any cyclic group is commutative.

For any given order $n(G)$ there is a finite number of groups and finite number of commutative groups. For instance, for $n(G) = 48$ there are 32 finite groups, 5 of them are commutative. For any order there is at least one (commutative) group, namely the cyclic group of order $n(G)$.

Groups G and G' are called *homomorphic* if there is a mapping f of G to G' that preserves the multiplication: $f(ab) = f(a)f(b)$. If the mapping is also one-to-one, the groups are called *isomorphic*. In this case, the group table of images $f(a) \in G'$ is the same as the group table of initial elements $a \in G$.

2.1.2 Subgroups

If a subset G' of a group G is a group itself, then G' is called a *subgroup* of the group G. The set of elements of the form ab, where b runs over all elements of the subgroup G', is called a *left coset* $L_a = aG'$. Any subgroup G' of a group G generates a decomposition of G into left cosets L_a. Indeed, each element of the group G occurs in one left coset and any two cosets are either identical or have no elements in common.

A coset $L_a = aG'$ consists of distinct elements. Hence each coset L_a has the same number of elements, namely the same as in the subgroup itself: $L_e = eG' = G'$. So the number of elements in L_a is equal to the order $n(G')$ of the subgroup G'. The number of distinct left cosets j is called the *index* of the subgroup G' in the group G. Thus, $j = n(G)/n(G')$ is an integer; i.e. the order of any subgroup is a divisor of the order of the group. Therefore, a group G whose order is a prime number has only two subgroups: G and $\{e\}$.

Similar results hold for *right cosets* $R_a = G'a$.

2.1.3 Invariant Subgroup

A subgroup H of a group G is called an *invariant subgroup*, or *normal divisor*, if the left coset decomposition of G with respect to H coincides with the right coset decomposition. Clearly any subgroup of a commutative group is an invariant subgroup.

If H is an invariant subgroup of G, then the left cosets L_a themselves form a group, which is called the *factor group* G/H. By the product of two cosets L_a and L_b we understand the set of products of the form $a'b'$, where $a' \in L_a$ and $b' \in L_b$.

It is not difficult to show that the coset product $L_a L_b$ is the coset L_{ab}. Other group properties are also easily checked. The identity element of the factor group is the coset $L_e = H$, and the inverse to L_a is the coset $L_{a^{-1}}$. Thus, elements of the factor group G/H are the entire cosets of the group G.

A subgroup H of G is invariant if and only if for all elements $a \in G$ and $b \in H$ the element $b' = a^{-1}ba$ also belongs to the subgroup H. This theorem can be formulated differently if we use the notion of the class of conjugate elements.

Two elements $a, b \in G$ are said to be *conjugate* if there exits a third element $c \in G$ such that $a = c^{-1}bc$. Conjugacy satisfies the requirements for an equivalence relation:

1. a is conjugate to itself;
2. if a is conjugate to b, then b is conjugate to a;
3. if a is conjugate to b and b is conjugate to c, then a is conjugate to c.

Therefore, the group G can be split up into *classes of conjugate elements* (or briefly *classes*); the number of classes is denoted by r.

Now the above-mentioned theorem can be formulated as follows.

Theorem 2.1 *A subgroup H is an invariant subgroup of a group G if and only if the subgroup H consists of entire classes of G, i.e. each element belongs to H together with its class.*

This theorem can be illustrated in the example of the group D_3, the symmetry group of an equilateral triangle. The group consists of 6 elements: the identity e, two rotations C_3 and C_3^2 by the angle of $2\pi/3$ and $4\pi/3$, respectively, about the threefold axis perpendicular to the plane of the triangle and three rotations $C_2^{(1)}, C_2^{(2)}, C_2^{(3)}$ by the angle π about the twofold axes 1, 2 and 3 in the plane of the triangle (Fig. 2.1). There are three classes of conjugate elements: $\{e\}$, $\{C_3, C_3^2\}$ and $\{C_2^{(1)}, C_2^{(2)}, C_2^{(3)}\}$. The group has four subgroups: the subgroup $\{e, C_3, C_3^2\}$ and three subgroups $\{e, C_2^{(1)}\}$, $\{e, C_2^{(2)}\}$, $\{e, C_2^{(3)}\}$. The first subgroup is invariant, as it consists of two entire classes, while the other three subgroups are not.

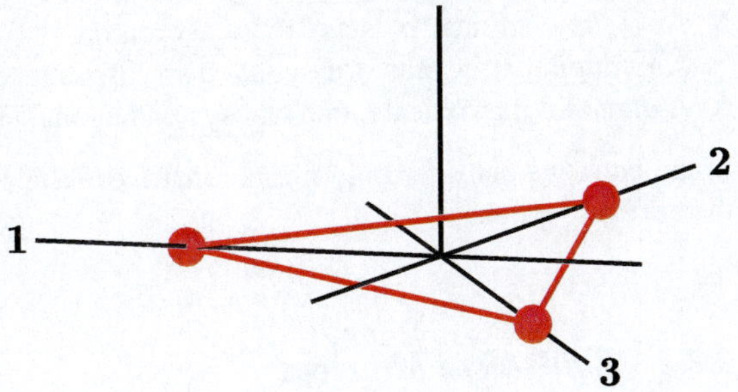

Fig. 2.1 Symmetry group of an equilateral triangle. The numbers 1, 2 and 3 denote the twofold axes in the plane of the triangle

2.1.4 *Classes of Conjugate Elements*

Let us mention some more properties of the classes of conjugate elements. In commutative groups, each element forms a class by itself. The number of classes in this case is equal to the order of the group, $r = n(G)$.

The identity element is always a class by itself. All elements in the same class have the same order. In particular, a class contains only rotations about axes of the same order (in the above example of the group D_3, it is the rotations $C_2^{(1)}, C_2^{(2)}, C_2^{(3)}$). The converse is not true: Two elements of the same order can belong to different classes. For example, consider the dihedral group D_2, which consists of the identity element e and three rotations C_{2x}, C_{2y} and C_{2z} through π about the coordinate axes. All rotations have the second order but, as elements of a commutative group, they belong to different classes.

2.2 Matrices: Definitions and Notation

2.2.1 *Special Types of Matrices*

The *dimension* of a square matrix is the number of its rows (or columns). Henceforth we consider square matrices of special form. A matrix $\tilde{D}$ obtained from a matrix D by switching its rows with its columns is called the *transpose* of D. The *adjoint* of a matrix D (or Hermitian conjugate), denoted by $D^\dagger$, is obtained by taking the transpose of D^*: $D^\dagger = \widetilde{D^*} = \tilde{D}^*$, where D^* is the complex conjugate of the matrix D.

A matrix D is *normal* if it commutes with its adjoint: $DD^\dagger = D^\dagger D$. Particular kinds of normal matrices are the following.

1. A *Hermitian* or *self-adjoint* matrix is such that $D = D^\dagger$.
2. A *symmetric* matrix is a real matrix D such that $D = \tilde{D}$.
3. A *unitary* matrix is the one whose inverse is the adjoint. That is $D^{-1} = D^\dagger$, so that $D^\dagger D = E$, where E is the unit matrix.
4. An *orthogonal* matrix is a real unitary matrix.
5. A *rotation* matrix is an orthogonal matrix with determinant equal to +1.
6. A *diagonal* matrix is the one whose nondiagonal elements are all zero.

In the book we consider only normal matrices. The reason for this is explained in the next subsection.

2.2.2 *Equivalent Matrices*

Two matrices D_1 and D_2 of the same dimension are called *equivalent* if there exists a *nonsingular* matrix S (i.e. $|S| \equiv \det S \neq 0$) such that $D_2 = S^{-1} D_1 S$. The transformation that maps D_1 to D_2 is called the *similarity transformation*. We write $D_1 \sim D_2$ if the two matrices are equivalent, and $D_1 \nsim D_2$ otherwise. The eigenvalues of a matrix (and hence the trace and determinant) are invariant under similarity transformations.

A matrix is called *diagonalizable* if it is equivalent to a diagonal matrix. A matrix D can be diagonalized by means of a unitary matrix S if and only if D is a normal matrix.

Of special importance for further considerations are *systems of matrices*. The number of matrices n in a *system of matrices* $\mathcal{D} = \{D_1, \ldots, D_n\}$ is called the *order of the system of matrices*. The notion of equivalence can be extended to systems of matrices. Two systems of matrices $\mathcal{D} = \{D_1, \ldots, D_n\}$ and $\mathcal{D}' = \{D'_1, \ldots, D'_n\}$ of the same order n are called *equivalent* if there is a nonsingular matrix S such that $D'_i = S^{-1} D_i S$ for all $i = 1, \ldots, n$. It is important here that the transformation matrix S is the *same* for all matrices D_i of the system $\mathcal{D}$.

2.2.3 Reducible Matrices

A matrix D is called the *direct sum of matrices* D_1 and D_2 if it has the form

$$D = \begin{pmatrix} D_1 & 0 \\ 0 & D_2 \end{pmatrix},$$

i.e. if D is a block-diagonal matrix with matrices D_1 and D_2 at the diagonal. In this case, we write $D = D_1 \oplus D_2$. Clearly, the dimension of D is equal to the sum of dimensions of D_1 and D_2, and

$$\mathrm{Tr}\, D = \mathrm{Tr}\, D_1 + \mathrm{Tr}\, D_2 .$$

Analogously, we can define the direct sum $\mathcal{D} \oplus \mathcal{D}'$ of two systems of matrices $\mathcal{D}$ and $\mathcal{D}'$. By $\mathcal{D} \oplus \mathcal{D}'$ we will understand the system of matrices $D_i \oplus D'_i$.

A matrix D is said to be *reducible* if it is equivalent to a direct sum of two matrices, and D is said to be *irreducible* otherwise. Analogously, a system of matrices is said to be reducible if it is equivalent to a direct sum of two systems of matrices, and *irreducible* otherwise.

Thus, a reducible system of matrices can be transformed by a single matrix S to a system consisting of block-diagonal matrices of the same form. If each system of blocks is itself an irreducible systems of matrices, then the initial system of matrices is said to be *completely reduced*. Any reducible system of matrices can be completely reduced. An important corollary of Schur's lemma is the following [1, 2].

Theorem 2.2 *If a nonzero matrix S commutes with all matrices in an irreducible system of matrices, then it is a multiple of the unit matrix: $S = \lambda E$.*

2.2.4 *Direct Matrix Product*

The *direct (Kronecker) product* $C = A \otimes B$ of a $m \times n$-matrix $A \equiv [a_{ij}]$ and $m' \times n'$-matrix $B \equiv [b_{i'j'}]$ is the $mm' \times nn'$-matrix whose elements are

$$c_{ii',jj'} = a_{ij} b_{i'j'}. \tag{2.1}$$

In particular, if A and B are both square matrices then $A \otimes B$ is also square, and

$$\mathrm{tr}(A \otimes B) = \mathrm{tr}\, A \cdot \mathrm{tr}\, B.$$

From the definition it immediately follows that the direct product of two diagonal matrices is again a diagonal matrix and direct product of two unit matrices is a unit matrix.

The matrix product of two direct products is the direct product of the two matrix products. More precisely, let $A \otimes B$ and $C \otimes D$ be two direct matrix products, and the number of rows in the matrix C is equal to the number of columns of A and number of rows in the matrix D is equal to the number of columns of B. Then

$$(A \otimes B)(C \otimes D) = AC \otimes BD. \tag{2.2}$$

2.3 Representation Theory

2.3.1 *Matrix Groups*

A *matrix group* is a system of matrices that forms a group, where the group multiplication is the (standard) matrix multiplication. In this case, the unit matrix is the identity element and inverse matrices are inverse elements in the group. Hence all matrices in a matrix group must be nondegenerate (nonsingular).

For matrix groups the following important theorem holds.

Theorem 2.3 *Every (finite) matrix group is equivalent to a unitary matrix group.*

Finally, we formulate the *uniqueness theorem* concerning the complete reduction of a reducible matrix group.

Theorem 2.4 *Irreducible parts of a reducible matrix group are uniquely determined up to order and equivalence (similarity transformation).*

2.3.2 *Group Representations*

A *representation of a group G* is a matrix group homomorphic to the group G. The dimension of the matrices is called the *dimension of the representation*.

In other words, each group representation of an abstract group consists of a system of matrices that has the same multiplication table as the group itself. The mapping of group elements to the matrix group is not necessarily one-to-one. Different elements in the group can be mapped to the same matrix. For example, all elements can be mapped to the d-dimensional unit matrix. The representation of a group by 1 (one-dimensional unit matrix) is called *identical* representation. If the mapping of the abstract group in a matrix group is an isomorphism, the representation is said to be *faithful* (true).

Now we can extend the notions of equivalence and reducibility to representations. Two representations $\mathcal{D}_1$ and $\mathcal{D}_2$ of a group G are said to be *equivalent*, if the representation matrices can be related by the same similarity transformation. We write $\mathcal{D}_1 \sim \mathcal{D}_2$ if the representation are equivalent, and $\mathcal{D}_1 \nsim \mathcal{D}_2$ otherwise. Since every representation is equivalent to a unitary one, we often consider only unitary representations if equivalence is unimportant.

A representation $\mathcal{D}$ of a group G is called *reducible* if it is equivalent to a direct sum of representations, and *irreducible* otherwise.

Again the following *uniqueness theorem* holds.

Theorem 2.5 *Every representation can be decomposed into a direct sum of irreducible representations and this decomposition is unique up to order and equivalence.*

Note some important properties. The identity element must be represented by a unit matrix. All irreducible representations of a commutative group are one-dimensional. All one-dimensional representations are unitary, i.e. consist of phase factors. Hence all irreducible representations of a commutative group are unitary. The following theorem is useful (but is rarely mentioned).

Theorem 2.6 *If $\mathcal{D}$ and $\mathcal{D}'$ are two equivalent unitary irreducible representations, then the matrix S, which is defined by $\mathcal{D}' = S^{-1}\mathcal{D}S$ only up to a constant factor, can be chosen unitary.*

It is convenient to have a characteristic of a representation which is invariant under similarity transformations. The character of a representation has exactly this property. The *character of a representation* is the set of traces of representation matrices. The function $\chi(a) = \mathrm{Tr}\, D(a)$ is called the *character of an element $D(a)$* in the group representation $\mathcal{D}$. Since similar matrices have the same trace, the character is a function of the class; i.e. conjugate elements in a representation have the same character. For this reason, it is sufficient to specify only the character of each class. A character of the identity element in a representation is equal to the dimension of the representation.

2.3.3 Regular Representation

We now consider a particular representation which can be constructed for any group, called the regular representation.

Let $a_1, \ldots, a_n$ denote the elements of a group G. If a is a fixed element of G and a_i runs over all group elements, then so does $a_i a$. Hence the right multiplication $a_i \mapsto a_i a$ is a one-to-one mapping of the group on itself. This mapping can be represented by a permutation matrix $D^r(a)$ as follows:

$$a_i a = \sum_{l=1}^{n} D^r_{il}(a) a_l \,. \tag{2.3}$$

This relation does not imply that elements of the group can be added as well as multiplied. Each row (column) of a permutation matrix has exactly one entry of 1 in each row (or column) and 0 elsewhere:

$$D^r_{ij}(a) = \delta_{a_i a a_j^{-1}, e} = \begin{cases} 1, & a_i a a_j^{-1} = e \,, \\ 0, & \text{otherwise} \,, \end{cases} \tag{2.4}$$

where e is the identity element. Therefore, the sum (2.3) has only one nonzero entry.

The matrices $D^r(a)$ form a representation. To show this, we consider the triple product $a_i a_j a_k$. Multiplying $a_i a_j$ by formula (2.3) and then multiplying by a_k on the right, we have

$$(a_i a_j) a_k = \left(\sum_{l=1}^{n} D^r_{il}(a_j) a_l \right) a_k = \sum_{l=1}^{n} D^r_{il}(a_j) \sum_{m=1}^{n} D^r_{lm}(a_k) a_m \,.$$

Changing the order of summation, we obtain

$$(a_i a_j) a_k = \sum_{m=1}^{n} \left[\sum_{l=1}^{n} D^r_{il}(a_j) D^r_{lm}(a_k) \right] a_m \,. \tag{2.5}$$

On the other hand, we can write

$$a_i (a_j a_k) = \sum_{m=1}^{n} D^r_{im}(a_j a_k) a_m \,. \tag{2.6}$$

The group multiplication is associative, $(a_i a_j) a_k = a_i (a_j a_k)$. Hence, equating the coefficient of a_m in (2.5) and (2.6), we obtain

$$\sum_{l=1}^{n} D^r_{il}(a_j)D^r_{lm}(a_k) = D^r_{im}(a_j a_k),$$

or in the matrix form,

$$D^r(a_j)D^r(a_k) = D^r(a_j a_k).$$

Thus, the system of $n \times n$ matrices $D^r(a)$ forms a faithful representation $\mathcal{D}^r$ of the group G, which is called the *regular representation* of G. Since permutation matrices are orthogonal, $\tilde{D}^r(a)D^r(a) = E$, the regular representation is unitary.

The character of an element a in the regular representation is equal to the number of elements a_j such that $aa_j = a_j$; i.e. the character of any element of the regular representation except for the identity is equal to zero. The character of the identity, as always, is equal to the dimension of the representation, which in this case is equal to n.

Theorem 2.7 *The regular representation is reducible and contains each irreducible representation of the group G a number of times equal to its dimension.*

2.3.4 Fundamental Theorems

We can now formulate two *fundamental theorems of the representation theory.*

Theorem 2.8 (The Number of Irreducible Representations) *A finite group G has as many irreducible representations as it has classes of conjugate elements.*

Theorem 2.9 ("The Great Orthogonality Theorem") *For any two irreducible representations $\mathcal{D}_q$ and $\mathcal{D}_p$ the following orthogonality relations hold*

$$\sum_{a\in G} D^{(q)}_{ij}(a)\, D^{(p)-1}_{kl}(a) = \frac{n}{l_p}\delta_{il}\delta_{jk}\delta'_{qp}\,, \tag{2.7}$$

where l_p is the dimension of the representation $\mathcal{D}_p$. The symbol δ'_{qp} means that $\delta'_{qp} = 0$ for $\mathcal{D}_q \nsim \mathcal{D}_p$ and $\delta'_{qp} = 1$ for $\mathcal{D}_q \equiv \mathcal{D}_p$. If $\mathcal{D}_q \sim \mathcal{D}_p$ but are not identical, then nothing can be said, just as in the case of a reducible representation. If the representations are unitary, then, clearly, the orthogonality relations (2.7) *take the form*

$$\sum_{a\in G} D^{(q)}_{ij}(a)\, D^{(p)*}_{lk}(a) = \frac{n}{l_p}\delta_{il}\delta_{jk}\delta'_{qp}\,. \tag{2.8}$$

2.3.5 Group Characters

In Sect. 2.3.2 we presented the definition and basic properties of characters. Here we discuss the properties of characters in more detail.

Most important are the characters of the *irreducible* representations. By the fundamental theorem of the representation theory the number of the irreducible representations $\mathcal{D}_p$ is equal to the number r of the classes C_i. Hence the characters $\chi_i^{(p)}$ of irreducible representations can be written down as a square array of numbers, the so-called *character table*. A character table can be determined for any group.

For characters of the irreducible representations the following orthogonality relations hold:

$$\sum_{i=1}^{r} r_i\, \chi_i^{(p)*} \chi_i^{(q)} = n\, \delta_{pq}\,, \tag{2.9}$$

$$\sum_{p=1}^{r} \chi_i^{(p)*} \chi_j^{(p)} = \frac{n}{r_i}\, \delta_{ij}\,, \tag{2.10}$$

where r_i is the number of elements in the class C_i. Since the characters of equivalent representations are equal, here we have $\delta_{pq} = 1$ even if $\mathcal{D}_p \sim \mathcal{D}_q$ but $\mathcal{D}_p \not\equiv \mathcal{D}_q$. Characters of *reducible* representations generally do not satisfy the orthogonality relations.

An arbitrary representation $\mathcal{D}$ of the group G can be expressed in terms of irreducible representations as

$$\mathcal{D} = n_1\, \mathcal{D}_1 \oplus \cdots \oplus n_r\, \mathcal{D}_r \equiv \bigoplus_{p=1}^{r} n_p\, \mathcal{D}_p, \tag{2.11}$$

where n_p is the multiplicity of the irreducible representation $\mathcal{D}_p$. Then the group characters can be expressed as

$$\chi_i = \sum_{q=1}^{r} n_q\, \chi_i^{(q)}\,. \tag{2.12}$$

Multiplying this equality by $r_i\, \chi_i^{(p)*}$, summing over i and using (2.9) we obtain

$$n_p = \frac{1}{n} \sum_{i=1}^{r} r_i\, \chi_i^{(p)*} \chi_i\,. \tag{2.13}$$

Thus, we use (2.13) to find the number of times a given irreducible representation occurs in the decomposition (2.11). This proves uniqueness of the decomposition up to order and equivalence.

Next, we multiply r_i by (2.12) times the complex conjugate equation and sum over i. By utilizing the orthogonality relation (2.9), we have

$$\sum_{i=1}^{r} r_i \, \chi_i^* \chi_i = \sum_{i=1}^{r} \sum_{p,q=1}^{r} r_i \, n_p \, n_q \, \chi_i^{(p)*} \chi_i^{(q)} = n \sum_{p=1}^{r} n_p^2 . \tag{2.14}$$

Once the characters χ_i and numbers r_i are known, this equality gives a simple criterion of reducibility. If the representation $\mathcal{D}$ is *irreducible*, it is equivalent to a irreducible representation $\mathcal{D}_{p_0}$, and the decomposition (2.11) will contain only the representation $\mathcal{D}_{p_0}$ with $n_{p_0} = 1$; all other n_p will be equal to 0, so that the right-hand side of (2.14) will be equal to n. If, on the other hand, the representation $\mathcal{D}$ is *reducible*, the sum on the right-hand side of (2.14) will be greater than 1. Hence the *irreducibility criterion* can be formulated as follows.

Theorem 2.10 (The Irreducibility Criterion) *A representation $\mathcal{D}$ of a group G is irreducible if and only if*

$$\sum_{i=1}^{r} r_i \, |\chi_i|^2 = n . \tag{2.15}$$

Otherwise, the sum (2.15) *is greater than n (but is always a multiple of n), and the representation $\mathcal{D}$ is reducible.*

Formulas (2.9) and (2.13) and criterion (2.15) can be used even if we do not know the classes of the group, but only know the representation matrices or even only their traces. In other words, it is sufficient to know the characters of all elements. Then formulas (2.9) and (2.13) become

$$\sum_{a \in G} \chi^{(p)*}(a) \chi^{(q)}(a) = n \, \delta_{pq} \tag{2.16}$$

and

$$n_p = \frac{1}{n} \sum_{a \in G} \chi^{(p)*}(a) \chi(a) . \tag{2.17}$$

Substituting the relation

$$\sum_{i=1}^{r} r_i \, |\chi_i|^2 = \sum_{a \in G} |\chi(a)|^2 ,$$

to the criterion (2.15), we can rewrite it as

$$\sum_{a \in G} |\chi(a)|^2 = n . \tag{2.18}$$

A special case of the orthogonally relation (2.10) is[1]

$$\sum_{p=1}^{r} l_p^2 = n. \tag{2.19}$$

This relation often allows us to determine dimensions of irreducible representations if we know the order of the group n and number of the classes r. For example, in a commutative group $r = n$, and we again obtain that commutative groups have only one-dimensional irreducible representations.

The relation (2.19) is also a completeness criterion for the system of irreducible representations.

Theorem 2.11 (Burnside's theorem) *The sum of the squares of the dimensions of* all nonequivalent *irreducible representations is equal to the order of the group.*

2.3.6 Reality Properties

In a number of applications, such as time-reversal symmetry (see Sect. 8.4), it is necessary to take the reality properties of the irreducible representations into account. As is known, the irreducible representations can be divided into three types: (potentially) real, complex and pseudoreal (see, e.g. [1, p. 287]). In terms of the representations $\mathcal{D}$, this can be stated more specifically as follows (see, e.g. [3, p. 138] and [4, p. 145]):

1. $\mathcal{D}$ and $\mathcal{D}^*$ are equivalent to the same real irreducible representation (real).
2. $\mathcal{D}$ and $\mathcal{D}^*$ are nonequivalent (complex).
3. $\mathcal{D}$ and $\mathcal{D}^*$ are equivalent but cannot be made real (pseudoreal).

The above cases are easier to distinguish by using the following criterion.

Theorem 2.12 (Frobenius–Schur) *Let $\mathcal{D}$ be an irreducible representation of a group G. Then*

$$\sum_{a} \chi(a^2) = \begin{cases} n(G) & \textit{real}, \\ 0 & \textit{complex}, \\ -n(G) & \textit{pseudoreal}. \end{cases} \tag{2.20}$$

A proof of the Frobenius–Schur theorem (in a more general case) is given in Appendix A.1.

[1] To see this, one needs to consider the first class, $i = j = 1$, consisting of only the identity element, $r_1 = 1$, whose character is $\chi_1^{(p)} = l_p$.

2.3.7 *Direct Product of Representations*

If $\mathcal{D}$ and $\mathcal{D}'$ are two systems of square matrices then $\mathcal{D} \otimes \mathcal{D}'$ consists of all direct products $D \otimes D'$. If $\mathcal{D}$ and $\mathcal{D}'$ are two representations of a group G, then $\mathcal{D} \otimes \mathcal{D}'$ is a new representation of G. The character of $\mathcal{D} \otimes \mathcal{D}'$ is the product of characters of $\mathcal{D}$ and $\mathcal{D}'$, i.e. $\chi(a)\chi'(a)$.

The main theorem about the direct product of representations $\mathcal{D} \otimes \mathcal{D}'$ is the following.

Theorem 2.13 *If $\mathcal{D}$ and $\mathcal{D}'$ are irreducible then so is $\mathcal{D} \otimes \mathcal{D}'$.*

References

1. E.P. Wigner, *Group Theory and its Application to the Quantum Mechanics of Atomic Spectra*, 2nd edn. (Academic, New York, 1959)
2. H.W. Streitwolf, *Group Theory in Solid State Physics* (MacDonald, London, 1971)
3. M. Hamermesh, *Group Theory and Its Applications to Physical Problems* (Addison-Wesley, Reading, MA, 1962). Reprinted: Dover, New York, 1989
4. M. Tinkham, *Group Theory and Quantum Mechanics* (McGraw-Hill, New York, 1964) Reprinted: Dover, New York, 2003

Chapter 3
Induced Representations

The construction of the [induced] modules for particular groups will probably remain more of an art than a science.
(C.W. Curtis, I. Reiner, Representation Theory of Finite Groups and Associative Algebras, Wiley, New York, 1962)

Construction of irreducible representations of finite groups remains a difficult problem. The problem can be substantially simplified if we use certain special group properties. In particular, as it was shown by [1, 2], for groups having an invariant subgroup (normal devisor) the problem of constructing irreducible representations of a group can be reduced to the problem of constructing irreducible representations of the invariant subgroup. In this chapter, we consider the necessary theory, formulate intermediate results and main theorem, which gives a recipe of constructing all irreducible representations. Our presentation closely follows the scheme of [3, 4], which we believe is the most straightforward. Alternative approaches to the induced representation theory can be found in the books [5–7].

3.1 Regular Continuation from a Subgroup to a Group

Let G be a finite group and H its subgroup. By $\{Hg_1, \ldots, Hg_r\}$ we denote the system of the (right) cosets, where $\{g_1, \ldots, g_r\}$ is a fixed system of the coset representatives, one from each coset, such that $g_1 = e$. Here $r = n(G)/n(H)$ is the number of cosets, where $n(G)$ and $n(H)$ are the orders of the group and subgroup, respectively. Then the group G has the coset decomposition

$$G = Hg_1 \cup Hg_2 \cup \ldots \equiv \bigcup_{l=1}^{r} Hg_l. \tag{3.1}$$

N. B. Melnikov and B. I. Reser, *Space Group Representations*,
https://doi.org/10.1007/978-3-031-13991-8_3

We can construct a representation of the entire group G from a representation d^i of the subgroup H analogously to constructing the regular representation (2.4). This time we consider a permutation of the cosets rather than of the group elements themselves and replace the nonzero matrix elements of the permutation matrix by the representation matrices d^i of the corresponding elements.

More precisely, for each $g \in G$ we construct the matrix $D^i(g)$ as follows:

$$D^i_{l\lambda,m\mu}(g) = \begin{cases} d^i_{\lambda\mu}(g_l g g_m^{-1}), & \text{if } \; g_l g g_m^{-1} \in H, \\ 0, & \text{if } \; g_l g g_m^{-1} \notin H. \end{cases}$$

or briefly,

$$D^i_{l\lambda,m\mu}(g) = d^i_{\lambda\mu}(g_l g g_m^{-1})\delta_{Hg_l g g_m^{-1},H} \,. \tag{3.2}$$

Here $1 \le l, m \le r$ and $1 \le \lambda, \mu \le l_i$, where l_i is the dimension of the representation d^i.

As we see from the above formulas, the matrix $D^i(g)$ is a $r \times r$ square block matrix, and for each position (l, m) there is a block $d^i(g_l g g_m^{-1})$ of the size $l_i \times l_i$. The matrix $d^i(g_l g g_m^{-1})$ is the only nonzero block in the lth row and the only nonzero block in the mth column. In particular, if we consider the subgroup H consisting of only the identity element e and consider the representation $d^i(e) = 1$ of H, then the regular continuation D^i becomes the regular representation (2.4) of the group G.

It is not difficult to prove that D^i is a representation of G:

$$\begin{aligned}
\left(D^i(g)\, D^i(g')\right)_{l\lambda,n\nu} &= \sum_{m\mu} D^i_{l\lambda,m\mu}(g)\, D^i_{m\mu,n\nu}(g') \\
&= \sum_m \sum_\mu d^i_{\lambda\mu}(g_l g g_m^{-1})\, d^i_{\mu\nu}(g_m g' g_n^{-1})\, \delta_{Hg_l g g_m^{-1},H}\, \delta_{Hg_m g' g_n^{-1},H} \\
&= \sum_m d^i_{\lambda\nu}(g_l g g' g_n^{-1})\, \delta_{Hg_l g,Hg_m}\, \delta_{Hg_m,Hg_n g'^{-1}} \\
&= d^i_{\lambda\nu}(g_l g g' g_n^{-1})\, \delta_{Hg_l g g' g_n^{-1},H} = D^i_{l\lambda,n\nu}(gg').
\end{aligned} \tag{3.3}$$

One can also prove that a different set of coset representatives gives an equivalent representation. We call the representation $\mathcal{D}^i$ the *regular continuation* of the representation d^i[1].

[1] We follow the terminology *regular continuation* coined by [3, 4], which indicates the similarity of this construction to the regular representation. Alternatively, one can say that $\mathcal{D}^i$ is induced from d^i, and that $\mathcal{D}^i$ is the *induced representation* (see, e.g. [7]).

If the representation d^i is unitary, so is the regular continuation $\mathcal{D}^i$[2]:

$$
\begin{aligned}
\left(D^i(g)\, D^{i\dagger}(g)\right)_{l\lambda,m\mu} &= \sum_{n\nu} D^i_{l\lambda,n\nu}(g)\, D^{i*}_{m\mu,n\nu}(g) \\
&= \sum_{n\nu} d^i_{\lambda\nu}(g_l g g_n^{-1})\, \delta_{Hg_lgg_n^{-1},H}\, d^{i*}_{\mu\nu}(g_m g g_n^{-1})\, \delta_{Hg_mgg_n^{-1},H} \\
&= \sum_{n} d^i_{\lambda\mu}(g_l g g_n^{-1} g_n g^{-1} g_m^{-1})\, \delta_{Hg_lg,Hg_n}\, \delta_{Hg_mg,Hg_n} \\
&= d^i_{\lambda\mu}(E)\, \delta_{Hg_l,Hg_m} = \delta_{\lambda\mu}\, \delta_{Hg_l,Hg_m} = \delta_{l\lambda,m\mu}.
\end{aligned}
\tag{3.4}
$$

Henceforth we consider only unitary representations d^i.

3.2 Normalizer of a Representation of an Invariant Subgroup

Let H be an invariant subgroup of G, i.e. for each element g in the group we have

$$gHg^{-1} = H. \tag{3.5}$$

Clearly, for a fixed element g the correspondence

$$h \mapsto ghg^{-1} \tag{3.6}$$

is an isomorphism of the group into itself. This isomorphism also acts on a representation d^i of the subgroup. Namely, if the mapping

$$h \mapsto d^i(h) \tag{3.7}$$

is a representation of H, the mapping generated by the isomorphism (3.6),

$$h \mapsto d^i(ghg^{-1}), \tag{3.8}$$

is also a representation of H, which is directly verified. The representation

$$d^{i_g}(h) = d^i(ghg^{-1}) \tag{3.9}$$

---

[2] Recall that the inverse to the product of elements is equal to the product of the inverses taken in the *reverse* order, i.e. $(c^{-1}b^{-1}a^{-1}) \cdot (abc) = e$.

is called *conjugate* to d^i under g. Obviously, conjugate representations may be equivalent or nonequivalent.

First of all, we characterize the set of elements in the group for which the representations d^{i_g} and d^i are equivalent,

$$N_i = \{g \in G : \; d^{i_g}(h) \sim d^i(h) \text{ for every } h \in H\}.$$

A necessary and sufficient criterion for equivalence of two representations is the equality of their characters (see Sect. 2.3.2). Hence, if d^{i_g} and d^i are equivalent, then for every element h in the subgroup the following relation must hold:

$$\chi^{i_g}(h) \equiv \chi^i(ghg^{-1}) = \chi^i(h). \tag{3.10}$$

It is easy to check that the set of the elements g satisfying the above condition forms a subgroup of G, which is called the *normalizer* (or the *little group* by some authors) of the representation d^i and denoted by N_i. Indeed, if $g, g' \in N_i$, so are $gg', g^{-1} \in N_i$, as can be proved directly from the definition:

$$\chi^i(gg'h(gg')^{-1}) = \chi^i(g(g'hg'^{-1})g^{-1}) = \chi^i(g'hg'^{-1}) = \chi^i(h), \tag{3.11}$$

and similarly,

$$\begin{aligned}\chi^i(g^{-1}hg) &= \chi^i\left((g^{-1}h^{-1}g)^{-1}\right) = \chi^{i*}(g^{-1}h^{-1}g) = \chi^{i*}(gg^{-1}h^{-1}gg^{-1}) \\ &= \chi^{i*}(h^{-1}) = \chi^i(h).\end{aligned}$$

Here we utilized relation (3.10) and the well-known fact that the character of the inverse element is complex conjugate with the character of the element itself.

Every element of H belongs to the normalizer N_i of a representation d^i. Indeed, if $g \in H$, then

$$d^i(ghg^{-1}) = d^i(g)d^i(h)d^i(g^{-1}) = d^i(g)d^i(h)d^i(g)^{-1}.$$

By the cyclic property of trace, from (3.10) we have

$$\chi^i(ghg^{-1}) = \mathrm{tr}(d^i(g)d^i(h)d^i(g)^{-1}) = \mathrm{tr}(d^i(h)) = \chi^i(h).$$

Since

$$H \subset N_i \subset G$$

and H is an invariant subgroup of G, we see that H is an invariant subgroup of the normalizer N_i. Thus, the normalizer has the coset decomposition

$$N_i = \bigcup_m Hg^i_m, \qquad m = 1, 2, \ldots, n(N_i)/n(H). \tag{3.12}$$

If N_i coincides with H, we call the representation d^i noncritical, otherwise we call it critical. If the normalizer coincides with the entire group G, the representation d^i is called completely critical.

Since the normalizer N_i is a subgroup of G, we have the decomposition of G into cosets with respect to N_i,

$$G = \bigcup_M N_i g^i_M, \qquad M = 1, 2, \ldots, n(G)/n(N_i), \tag{3.13}$$

where $\{g^i_M\}$ is the set of elements such that $d^{i_{g^i_M}}$ and d^i are nonequivalent. Let $g = g^i_M$ be one of those elements. We consider the conjugate representation d^{i_g} and its normalizer N_{i_g} (conjugate normalizer for short). The conjugate normalizers are related by

$$N_{i_g} = g^{-1} N_i g. \tag{3.14}$$

Indeed, if $n_i \in N_i$, by (3.10) we have

$$\begin{aligned}\chi^{i_g}\left((g^{-1}n_i g)h(g^{-1}n_i g)^{-1}\right) &= \chi^{i_g}(g^{-1}n_i ghg^{-1}n_i^{-1}g)\\ &= \chi^i(n_i ghg^{-1}n_i^{-1}) = \chi^i(ghg^{-1}) = \chi^{i_g}(h),\end{aligned} \tag{3.15}$$

i.e. $g^{-1}n_i g \in N_{i_g}$. Analogously, we can prove the inverse inclusion.

Relation (3.14) establishes an isomorphism between the elements of the conjugate normalizers N_{i_g} and N_i. From the decomposition (3.12) we see that every element in the normalizer N_i has the form hg_m. Therefore, the isomorphism between N_i and N_{i_g} can be written as

$$hg^i_m \mapsto g^{-1}(hg^i_m)g = h' g^{-1} g^i_m g = h' g^{i_g}_m. \tag{3.16}$$

Similarly to decompositions (3.12) and (3.13), we can now write the conjugate decompositions

$$N_{i_g} = \bigcup_m H g_m^{i_g}, \tag{3.17}$$

$$G = \bigcup_M N_{i_g} g_M^{i_g}, \tag{3.18}$$

where the representatives are conjugate to each other

$$g_m^{i_g} = g^{-1} g_m^i g, \tag{3.19}$$

$$g_M^{i_g} = g^{-1} g_M^i g. \tag{3.20}$$

Indeed, conjugating both sides of the decomposition (3.12) by g, we obtain

$$g^{-1} N_i g = \bigcup_m g^{-1} H g_m^i g = \bigcup_m g^{-1} H g (g^{-1} g_m^i g).$$

Recalling formula (3.14), invariance of the subgroup H and formula (3.19), we obtain relation (3.17). Relation (3.18) is proved analogously.

The set $\{d^{i_g}\}$ of nonequivalent conjugate representations (3.9) is called the *orbit*, and set of the indices $\{i_g\}$ is called the *star* of the representation d^i.

3.3 Unitary Transformation for Elements of a Normalizer

Next we study the relationship between *equivalent* conjugate representations in more detail.

By definition, for every $g \in N_i$ the representation d^{i_g} is equivalent d^i. Hence there exits a unitary matrix $U^i(g)$ such that[3]

$$U(g)\, d^i(h)\, U(g)^{-1} = d^{i_g}(h) = d^i(ghg^{-1}) \tag{3.21}$$

for all $h \in H$. Clearly, for $g \in H$ we have

$$U(g)\, d^i(h)\, U(g)^{-1} = d^i(g)\, d^i(h)\, d^i(g)^{-1}. \tag{3.22}$$

Hence we can take

$$U(h) = d^i(h). \tag{3.23}$$

Next, $U(g_m)$ is obtained as a solution of a matrix equation

[3] Index i is sometimes omitted for brevity.

$$U(g_m)\, d^i(h) = d^i(g_m h g_m^{-1})\, U(g_m). \tag{3.24}$$

For an *arbitrary* element hg_m of the normalizer N_i the matrix $U(hg_m)$ can be chosen as[4]

$$U(hg_m) = U(h)\, U(g_m) = d^i(h)\, U(g_m). \tag{3.25}$$

If the representation d^i of the invariant subgroup is *irreducible*, we can determine the multiplication rule for the unitary transformation matrices U. It is straightforward to verify that $U(g_m g_{m'})^{-1} U(g_m) U(g_{m'})$ commutes with all matrices $d^i(h)$. Indeed, taking (3.24) and (3.21) into account, we obtain

$$\begin{aligned} &U(g_m g_{m'})^{-1}\, U(g_m)\, U(g_{m'})\, d^i(h) \\ &\overset{(3.24)}{=} U(g_m g_{m'})^{-1}\, U(g_m)\, d^i(g_{m'} h g_{m'}^{-1}) U(g_{m'}) \\ &\overset{(3.21)}{=} U(g_m g_{m'})^{-1}\, d^i(g_m g_{m'} h g_{m'}^{-1} g_m^{-1})\, U(g_m)\, U(g_{m'}) \\ &\overset{(3.21)}{=} d^i(h)\, U(g_m g_{m'})^{-1}\, U(g_m)\, U(g_{m'}). \end{aligned} \tag{3.26}$$

By corollary of Schur's lemma (Theorem 2.2.), $U(g_m g_{m'})^{-1} U(g_m) U(g_{m'})$ is a scalar matrix. Thus, the matrices U satisfy the relation

$$U(g_m g_{m'}) = U(g_m)\, U(g_{m'})\, \omega_{g_m, g_{m'}}, \tag{3.27}$$

where $\{\omega_{g_m, g_{m'}}\}$ is a factor system.

Multiplying relation (3.27) by its Hermitian conjugate and recalling that U is unitary, we obtain

$$|\omega_{g_m, g_{m'}}| = 1. \tag{3.28}$$

Similarly, we check that the factors $\omega_{g_m, g_{m'}}$ satisfy the relation

$$\omega_{g_m, g_{m'}}\, \omega_{g_m g_{m'}, g_{m''}} = \omega_{g_m, g_{m'} g_{m''}}\, \omega_{g_{m'}, g_{m''}} \tag{3.29}$$

for all $g_m, g_{m'}, g_{m''} \in N_i$ (compare with (4.5)).

Furthermore, we can extend the factor system to arbitrary elements of the normalizer, $\omega_{hg_m, h' g_{m'}}$. Indeed, we can write

[4] Indeed, this choice of $U(hg_m)$ agrees with (3.21):

$$\begin{aligned} U(hg_m)\, d^i(h')\, U(hg_m)^{-1} &\overset{(3.25)}{=} d^i(h)\, U(g_m)\, d^i(h')\, U(g_m)^{-1}\, d^i(h)^{-1} \\ &\overset{(3.24)}{=} d^i(h)\, d^i(g_m h' g_m^{-1})\, d^i(h)^{-1} = d^i(h g_m h' g_m^{-1} h^{-1}) = d^i((hg_m) h' (hg_m)^{-1}) = d^{i_{hg_m}}(h'). \end{aligned}$$

$$
\begin{aligned}
U(hg_m h' g_{m'}) &= U(hg_m)\, U(h' g_{m'})\, \omega_{hg_m, h' g_{m'}} \\
&\overset{(3.25)}{=} d(h)\, U(g_m)\, d(h')\, U(g_{m'})\, \omega_{hg_m, h' g_{m'}} \\
&\overset{(3.24)}{=} d(h)\, d(g_m h' g_m^{-1})\, U(g_m)\, U(g_{m'})\, \omega_{hg_m, h' g_{m'}} .
\end{aligned}
\tag{3.30}
$$

On the other hand,

$$
\begin{aligned}
U(hg_m h' g_{m'}) &\overset{(3.25)}{=} d(h)\, U(g_m h' g_m^{-1} g_m g_{m'}) \\
&\overset{(3.25)}{=} d(h)\, d(g_m h' g_m^{-1})\, U(g_m g_{m'}) \\
&= d(h)\, d(g_m h' g_m^{-1})\, U(g_m)\, U(g_{m'})\, \omega_{g_m, g_{m'}} .
\end{aligned}
\tag{3.31}
$$

Here we used the fact that H is an invariant subgroup and hence $g_m h' g_m^{-1} \in H$. Comparing relations (3.30) and (3.31), we have

$$
\omega_{hg_m, h' g_{m'}} = \omega_{g_m, g_{m'}} \tag{3.32}
$$

for all $h, h' \in H$. In other words, $\omega_{hg_m, h' g_{m'}}$ is a function of only the cosets represented by g_m and $g_{m'}$.

3.4 Regular Continuation to the Normalizer

Let d^i be an irreducible representation of the invariant subgroup H. Utilizing the decomposition (3.12), we construct the regular continuation of the representation d^i from H to the normalizer N_i. By formula (3.2), for any element of the normalizer $hg_{m'}$, we have[5]

$$
D^i_{l\lambda, m\mu}(hg_{m'}) = d^i_{\lambda\mu}(g_l h g_{m'} g_m^{-1})\, \delta_{Hg_l h g_{m'} g_m^{-1}, H}, \qquad \begin{array}{l} 1 \le l, m \le n(N_i)/n(H), \\ 1 \le \lambda, \mu \le l_i . \end{array} \tag{3.33}
$$

We transform the matrix of the regular continuation by a block-diagonal unitary matrix S, with the unitary blocks U:

$$
S_{l\lambda, m\mu} = U_{\lambda\mu}(g_l)\, \delta_{Hg_l, Hg_m} . \tag{3.34}
$$

[5] Do not confuse with the regular continuation to the group G in Sect. 3.1.

Using the relations (3.25) and (3.27), we obtain

$$
\begin{aligned}
(S^{-1}\, D(hg_{m'})\, S)_{l\lambda,m\mu} &= \sum_{n\nu,t\tau} S^{-1}_{l\lambda,n\nu}\, D_{n\nu,t\tau}(hg_{m'})\, S_{t\tau,m\mu} \\
&= \sum_{n\nu,t\tau} U_{\lambda\nu}(g_l)^{-1}\, \delta_{Hg_l,Hg_n}\, d^i_{\nu\tau}(g_n h g_{m'} g_t^{-1})\, \delta_{Hg_n hg_{m'} g_t^{-1},H}\, U_{\tau\mu}(g_t)\, \delta_{Hg_t,Hg_m} \\
&\overset{(3.25)}{=} \sum_{n\nu,t} U_{\lambda\nu}(g_l)^{-1}\, U_{\nu\mu}(g_n h g_{m'})\, \delta_{Hg_l,Hg_n}\, \delta_{Hg_n hg_{m'} g_t^{-1},H}\, \delta_{Hg_t,Hg_m} \\
&= \sum_{\nu} U_{\lambda\nu}(g_l)^{-1}\, U_{\nu\mu}(g_l h g_{m'})\, \delta_{Hg_l hg_{m'} g_m^{-1},H} \\
&\overset{(3.27)}{=} U_{\lambda\mu}(hg_{m'})\, \omega_{g_l,hg_{m'}}\, \delta_{Hg_l hg_{m'} g_m^{-1},H} \\
&\equiv U_{\lambda\mu}(hg_{m'})\, V_{lm}(hg_{m'}),
\end{aligned}
\tag{3.35}
$$

where

$$
V_{lm}(hg_{m'}) = \omega_{g_l,hg_{m'}}\, \delta_{Hg_l hg_{m'} g_m^{-1},H}\, . \tag{3.36}
$$

Thus, we presented the matrices of the regular continuation D^i as the direct products of the matrices U^i and V^i,

$$
(S^i)^{-1}\, D^i(hg_{m'})\, S^i = U^i(hg_{m'}) \otimes V^i(hg_{m'})\, .
$$

Next, we study the properties of the matrices V^i omitting the index i where it does not lead to confusion.

3.5 Projective Representations of the Factor Group of a Normalizer

The matrices V defined in the previous section are equal for elements of the normalizer that belong to the same coset in the decomposition (3.12). This is easily verified, using (3.32) and (3.5),

$$
\begin{aligned}
V_{lm}(hg_{m'}) &= \omega_{g_l,hg_{m'}}\, \delta_{Hg_l hg_{m'} g_m^{-1},H} \overset{(3.32)}{=} \omega_{g_l,g_{m'}}\, \delta_{Hg_l hg_l^{-1} g_l g_{m'} g_m^{-1},H} \\
&\overset{(3.5)}{=} \omega_{g_l,g_{m'}}\, \delta_{Hg_l g_{m'} g_m^{-1},H} = V_{lm}(g_{m'}).
\end{aligned}
\tag{3.37}
$$

Therefore, it is convenient to define the matrices V only on the cosets of H in the normalizer N_i, i.e. on the elements of the factor group N_i/H of the normalizer N_i over the invariant subgroup H.

The matrices $V(g_m)$ form a unitary *projective representation* of the factor group N_i/H (see (4.1)),

$$V(g_{m'})\,V(g_{m''}) = \omega_{g_{m'},g_{m''}}\,V(g_{m'}g_{m''})\,.$$

Indeed, taking the relation (3.29) into account, we have

$$\begin{aligned}(V(g_{m'})\,V(g_{m''}))_{lm} &\overset{(3.36)}{=} \sum_n \omega_{g_l,g_{m'}}\,\delta_{Hg_lg_{m'}g_n^{-1},H}\,\omega_{g_n,g_{m''}}\,\delta_{Hg_ng_{m''}g_m^{-1},H}\\ &= \omega_{g_l,g_{m'}}\,\omega_{g_lg_{m'},g_{m''}}\,\delta_{Hg_lg_{m'},Hg_mg_{m''}^{-1}}\\ &\overset{(3.29)}{=} \omega_{g_l,g_{m'}g_{m''}}\,\omega_{g_{m'},g_{m''}}\,\delta_{Hg_lg_{m'}g_{m''}g_m^{-1},H}\\ &\overset{(3.36)}{=} \omega_{g_{m'},g_{m''}}\,V_{lm}(g_{m'}g_{m''}).\end{aligned} \tag{3.38}$$

The projective representations are described in detail in Chap. 4. According to the definition (4.18), the representation V is regular and hence the number of times it contains each irreducible projective representation Δ^j (with the same factor system ω as V) is equal to the dimension l_j of Δ^j. Thus, by applying a unitary transformation, the representation V can be decomposed into a direct sum of irreducible representations:

$$V(g_{m'}) \sim l_1\Delta^1(g_{m'}) \oplus l_2\Delta^2(g_{m'}) \oplus \ldots \tag{3.39}$$

Taking into account (3.35), we see that by applying two successive unitary transformations the regular continuation of the irreducible representation d^i of the invariant subgroup H to the normalizer of this representation N_i can be represented as a direct sum of representations of a special form:

$$D^i(hg_{m'}) \sim [l_1U(hg_{m'}) \otimes \Delta^1(g_{m'})] \oplus [l_2U(hg_{m'}) \otimes \Delta^2(g_{m'})] \oplus \ldots \tag{3.40}$$

3.6 Small Representations of a Normalizer

The building blocks of the decomposition (3.40),

$$d^{ij}(hg_{m'}) = U^i(hg_{m'}) \otimes \Delta^{ij}(g_{m'}), \tag{3.41}$$

are said to be *small* representations of the normalizer N_i. The fact that the matrices d^{ij} do form a representation of the normalizer, can also be checked directly. Using (3.30) and (3.32), we have

$$U^i(hg_m h' g_{m'}) = U^i(hg_m)\, U^i(h' g_{m'}) \omega_{hg_m, h' g_{m'}} = U^i(hg_m)\, U^i(h' g_{m'}) \omega_{g_m, g_{m'}},$$

and Δ^j corresponds to the same factor system ω as V, i.e. satisfy the relation

$$\Delta^j(g_m)\, \Delta^j(g_{m'}) = \omega_{g_m, g_{m'}}\, \Delta^j(g_m g_{m'}).$$

Hence

$$\begin{aligned} d^{ij}(hg_m)\, d^{ij}(h' g_{m'}) &= U^i(hg_m)\, U^i(h' g_{m'}) \otimes \Delta^j(g_m)\, \Delta^j(g_{m'}) \\ &= U^i(hg_m h' g_{m'}) \otimes \Delta^j(g_m g_{m'}) = d^{ij}(hg_m h' g_{m'}). \end{aligned} \tag{3.42}$$

A characteristic property of the small representations (3.41) is the quasidiagonality of the matrices that correspond to elements of the invariant subgroup,

$$d^{ij}(h) = U^i(h) \otimes \Delta^j(e) = d^i(h) \otimes E,$$

where E is the $l_j \times l_j$ unit matrix.

We can prove the irreducibility of small representations and nonequivalence of small representations constructed from distinct irreducible projective representations of the factor group N_i/H. Indeed,

$$\begin{aligned} &\sum_{g_{m''}} \sum_{h} \chi_d^{ij}(hg_{m''})\, \chi_d^{ij'*}(hg_{m''}) \\ &\overset{(3.41)}{=} \sum_{g_{m''}} \sum_{h} \sum_{ll'} U_{ll}(hg_{m''})\, U^*_{l'l'}(hg_{m''}) \sum_{\lambda\lambda'} \Delta^j_{\lambda\lambda}(g_{m''})\, \Delta^{j'*}_{\lambda'\lambda'}(g_{m''}) \\ &\overset{(3.25)}{=} \sum_{g_{m''}} \sum_{ll'mm'} U_{ml}(g_{m''})\, U^*_{m'l'}(g_{m''}) \chi^j(g_{m''})\, \chi^{j'*}(g_{m''}) \sum_{h} d^i_{lm}(h)\, d^{i*}_{l'm'}(h), \end{aligned} \tag{3.43}$$

where $\chi^j(g_{m''})$ is the character of $\Delta^j(g_{m''})$. Using the orthogonality relations (2.8) for the matrices of irreducible representation $d^i(h)$, we have

$$\begin{aligned} &\sum_{g_{m''}} \sum_{h} \chi_d^{ij}(hg_{m''})\, \chi_d^{ij'*}(hg_{m''}) \\ &\overset{2.8}{=} \sum_{g_{m''}} \sum_{ll'mm'} U_{ml}(g_{m''})\, U^*_{m'l'}(g_{m''}) \chi^j(g_{m''})\, \chi^{j'*}(g_{m''}) \frac{n(H)}{l_i} \delta_{ll'}\, \delta_{mm'} \\ &= \frac{n(H)}{l_i} \sum_{g_{m''}} \sum_{lm} U_{ml}(g_{m''})\, U_{lm}(g_{m''})^{-1} \chi^j(g_{m''})\, \chi^{j'*}(g_{m''}) \\ &= \frac{n(H)}{l_i} \sum_{g_{m''}} l_i \chi^j(g_{m''})\, \chi^{j'*}(g_{m''}) = n(H) \sum_{g_{m''}} \chi^j(g_{m''})\, \chi^{j'*}(g_{m''}), \end{aligned} \tag{3.44}$$

where l_i is the dimension of d^i. Finally, using the orthogonality relations (4.16) for the characters χ^j of irreducible projective representations Δ^j of the factor group N_i/H, we obtain

$$\sum_{g_{m''}}\sum_{h} \chi_d^{ij}(hg_{m''})\,\chi_d^{ij'*}(hg_{m''}) = n(H)\,\frac{n(N_i)}{n(H)}\,\delta_{jj'} = n(N_i)\,\delta_{jj'}\,. \tag{3.45}$$

The irreducibility criterion (2.18),

$$\sum_{g_{m''}}\sum_{h} |\chi_d^{ij}(hg_{m''})|^2 = n(N_i),$$

gives the result.

Note that the system of small representations of the normalizer of a fixed ith representation is not complete, i.e. d^{ij} are not all irreducible representations of N_i. Indeed, let l_{ij} be the dimension of the representation d^{ij}. Since d^{ij} is the direct product (3.41), we have $l_{ij} = l_i l_j$, where l_i is the dimension of the representation d^i and l_j is the dimension of Δ^j. Hence, using the completeness criterion (4.23) of the system of projective representations Δ^j of the factor group N_i/H, we obtain

$$\sum_j l_{ij}^2 = \sum_j l_i^2 l_j^2 = l_i^2 \sum_j l_j^2 = l_i^2\,\frac{n(N_i)}{n(H)} = \frac{l_i^2}{\sum_i l_i^2}\,n(N_i) \neq n(N_i)\,. \tag{3.46}$$

This violates the completeness criterion (2.19) for the system of representations d^{ij}.

Next, we consider the small representations of the conjugate normalizers. The isomorphism of the conjugate normalizers (3.16) allows to establish a one-to-one correspondence between their small representations. Indeed, assuming

$$U^{i_g}(g^{-1}hg_m^i g) = U^i(hg_m^i) \tag{3.47}$$

and

$$\Delta^{i_g j}(g^{-1}g_m^i g) = \Delta^{ij}(g_m^i)\,, \tag{3.48}$$

it is easy to check that the matrices U^{i_g} and $\Delta^{i_g j}$ satisfy all the necessary properties. Utilizing (3.21), we have

$$\begin{aligned} &U^{i_g}(g^{-1}hg_m^i g)\,d^{i_g}(h)\,U^{i_g}(g^{-1}hg_m^i g)^{-1} = U^i(hg_m^i)\,d^i(ghg^{-1})\,U^i(hg_m^i)^{-1} \\ &= d^i\left(hg_m^i ghg^{-1}(hg_m^i)^{-1}\right) = d^{i_g}\left((g^{-1}hg_m^i g)h(g^{-1}hg_m^i g)^{-1}\right); \end{aligned} \tag{3.49}$$

for the matrices of the projective representation, we obtain

$$\begin{aligned}\Delta^{i_g j}(g^{-1}g^i_m g)\,\Delta^{i_g j}(g^{-1}g^i_{m'}g) &= \Delta^{ij}(g^i_n)\,\Delta^{ij}(g^i_{m'}) = \omega^i_{g^i_m,g^i_{m'}}\,\Delta^{ij}(g^i_m g^i_{m'})\\ &= \omega^{i_g}_{g^{-1}g^i_m g,g^{-1}g^i_{m'}g}\,\Delta^{i_g j}(g^{-1}g^i_m g g^{-1}g^i_{m'}g)\,, \qquad (3.50)\end{aligned}$$

where the equality of the factors ω follows from relation (3.27):

$$\begin{aligned}\omega^i_{g_m,g_{m'}} &= U^i(g^i_{m'})^{-1}U^i(g^i_m)^{-1}U^i(g^i_m g^i_{m'})\\ &= U^{i_g}(g^{-1}g^i_{m'}g)^{-1}U^{i_g}(g^{-1}g^i_m g)^{-1}U^{i_g}(g^{-1}g^i_m g g^{-1}g^i_{m'}g) = \omega^{i_g}_{g^{-1}g^i_m g,g^{-1}g^i_{m'}g}\,.\end{aligned}$$

In equality (3.48), the indices i and i_g reflect the fact that Δ^j are the representations of the factor groups N_i/H and N_{i_g}/H of conjugate normalizers.

Relations (3.47) and (3.48) allow to index consistently the small representations of all conjugate normalizers. To do this, it suffices to index the small representations for one of the normalizers, fix an element that establishes the isomorphism and utilize relations (3.47) and (3.48). In practice, instead of relation (3.48) it is more convenient to use

$$\chi^{ij}(g^i_m) = \chi^{i_g j}(g^{-1}g^i_m g)\,. \qquad (3.51)$$

The small representations of conjugate normalizers having the same index j are clearly equivalent.

The above procedure also allows to obtain the small representations of conjugate normalizers if we know the small representations of only one of them.

3.7 Special Case of an Abelian Invariant Subgroup

In the special case of an Abelian invariant subgroup H the situation simplifies.

As we know, all irreducible representations d^i of an Abelian group are one-dimensional and coincide with the characters χ^i. Hence the matrices $U(hg_m)$ are just (complex) numbers. According to (3.24) and (3.25), for a chosen system of the coset representatives g_m we can take

$$U(g_m) = 1\,, \qquad (3.52)$$

and for an arbitrary element hg_m of the normalizer N_i we have

$$U(hg_m) = d^i(h)\,U(g_m) = \chi^i(h). \qquad (3.53)$$

Then

$$\begin{aligned}\omega_{g_m,g_{m'}} &= U(g_m g_{m'})\,U(g_{m'})^{-1}\,U(g_m)^{-1}\\ &= U(g_m g_{m'}) = U(h_{mm'}g_{mm'}) = \chi^i(h_{mm'})\,, \qquad (3.54)\end{aligned}$$

where $g_{mm'}$ is the representative of the coset $Hg_m g_{m'}$ and $h_{mm'}$ is the element of H defined by

$$g_m g_{m'} = h_{mm'}\, g_{mm'}\,. \tag{3.55}$$

Relation (3.38) now takes the form

$$V(g_m)\, V(g_{m'}) = \chi^i(h_{mm'})\, V(g_m g_{m'}), \tag{3.56}$$

and small representations d^{ij} of the normalizer (3.41) are defined as

$$d^{ij}(hg_m) = \chi^i(h)\, \Delta^{ij}(g_m)\,. \tag{3.57}$$

Finally, if the representatives g_m can be chosen in such a way that they themselves form a subgroup of N_i, then clearly $\chi^i(h_{mm'}) = 1$ for all $g_m, g_{m'} \in N_i/H$, and the projective representation Δ^{ij} turns into an ordinary (vector) representation of the factor group N_i/H.

3.8 Main Theorem

We now turn to our main task of constructing all irreducible representations of a finite group. Using the decomposition of a group with respect to the normalizer (3.13), we construct the regular continuation of the small representation d^{ij} of the normalizer N_i from N_i to the group G:

$$D^{ij}_{Ll\lambda,Mm\mu}(g) = d^{ij}_{l\lambda,m\mu}(g_L g g_M^{-1})\, \delta_{N_i g_L g g_M^{-1}, N_i}\,. \tag{3.58}$$

It turns out that the representations $D^{ij}_{Ll\lambda,Mm\mu}$ completely solve our problem. We formulate the main result as a theorem.

Theorem 3.1 *Let*

(i) G be a finite group with an invariant subgroup H,
(ii) d^i be an irreducible representation of H,
(iii) N_i be the normalizer of d^i,
(iv) d^{ij} be a small representation of N_i .

Then

(i) the regular continuation of d^{ij} from N_i to G is irreducible,

(ii) the regular continuations of equivalent small representations from the conjugate normalizers to G are equivalent,

(iii) the regular continuations of nonequivalent small representations from the normalizer N_i to G are nonequivalent,

(iv) the system of representations constructed by regular continuation of all small representations from all nonconjugated normalizers is complete.

Proof For the characters χ_D^{ij} and $\chi_D^{i'j'}$ of the representations D^{ij} and $D^{i'j'}$, using the definition (3.58), we write

$$\begin{aligned}
\sum_g \chi_D^{ij}(g)\, \chi_D^{i'j'*}(g) &= \sum_g \sum_{Ll\lambda} d^{ij}_{l\lambda,l\lambda}(g_L g g_L^{-1})\, \delta_{N_i g_L g g_L^{-1}, N_i}\, \chi_D^{i'j'*}(g) \\
&= \sum_{g'} \sum_{Ll\lambda} d^{ij}_{l\lambda,l\lambda}(g') \delta_{N_i g', N_i}\, \chi_D^{i'j'*}(g_L^{-1} g' g_L) \\
&= \sum_{g'} \sum_{Ll\lambda} d^{ij}_{l\lambda,l\lambda}(g') \delta_{N_i g', N_i}\, \chi_D^{i'j'*}(g') \\
&= \frac{n(G)}{n(N_i)} \sum_{g' \in N_i} \sum_{l\lambda} d^{ij}_{l\lambda,l\lambda}(g') \chi_D^{i'j'*}(g'). \qquad (3.59)
\end{aligned}$$

To deduce the above formula we used the following. If g runs over all elements of the group then so does $g' = g_L g g_L^{-1}$ for any g_L. The characters χ_D^{ij} of the conjugate elements g and $g' = g_L^{-1} g g_L$ of the group G are equal (see Sect. 2.3.2). The number of elements in the sum over L is the number of the coset representatives g_L in the group decomposition (3.18), which is equal to $n(G)/n(N_i)$.

Each element of the normalizer N_i can be represented as $h g_m^i$ using the coset decomposition (3.12). Thus, we rewrite (3.59) as

$$\begin{aligned}
\sum_g \chi_D^{ij}(g)\, \chi_D^{i'j'*}(g) &\overset{(3.58)}{=} \frac{n(G)}{n(N_i)} \sum_{g \in N_i} \sum_{\substack{l\lambda \\ Mm\mu}} d^{ij}_{l\lambda,l\lambda}(g)\, d^{i'j'*}_{m\mu,m\mu}(g_M g g_M^{-1})\, \delta_{N_{i'} g_M g g_M^{-1}, N_{i'}} \\
&= \frac{n(G)}{n(N_i)} \sum_{g^i_{m'}} \sum_h \sum_{\substack{l\lambda \\ Mm\mu}} d^{ij}_{l\lambda,l\lambda}(h g^i_{m'})\, d^{i'j'*}_{m\mu,m\mu}(g_M h g_M^{-1} g_M g^i_{m'} g_M^{-1})\, \delta_{N_{i'} g_M h g_M^{-1} g_M g^i_{m'} g_M^{-1}, N_{i'}},
\end{aligned}$$

where $g_M h g_M^{-1}$ is an element of the invariant subgroup H (see (3.5)). Utilizing expression (3.41) for the matrices of small representations of a normalizer and taking (3.25) into account, we have

$$
\begin{aligned}
&\sum_{g} \chi_D^{ij}(g)\, \chi_D^{i'j'*}(g) \\
&= \frac{n(G)}{n(N_i)} \sum_{g^i_{m'}} \sum_{\substack{l \\ Mm}} \sum_{nk} U^i_{nl}(g^i_{m'})\, U^{i'*}_{km}(g_M g^i_{m'} g_M^{-1}) \sum_{\lambda} \Delta^{ij}_{\lambda\lambda}(g^i_{m'}) \sum_{\mu} \Delta^{i'j'*}_{\mu\mu}(g_M g^i_{m'} g_M^{-1}) \\
&\times \delta_{N_{i'} g_M g^i_{m'} g_M^{-1}, N_{i'}} \sum_{h} d^i_{ln}(h)\, d^{i'*}_{mk}(g_M h g_M^{-1}).
\end{aligned} \tag{3.60}
$$

The last sum in the above expression can be calculated explicitly. Using the definition of the conjugate representation (3.9) and orthogonality relations for the matrices of irreducible representations of the invariant group H, we obtain

$$
\sum_{h} d^i_{ln}(h)\, d^{i'*}_{mk}(g_M h g_M^{-1}) = \sum_{h} d^i_{ln}(h)\, d^{i'_{g_M}*}_{mk}(h) = \frac{n(H)}{l_i}\, \delta_{lm}\, \delta_{nk}\, \delta_{i,i'_{g_M}}\,. \tag{3.61}
$$

Hence expression (3.60) can be nonzero if and only if $i = i'_{g_M}$; i.e. when the irreducible representations d^i and $d^{i'}$ are conjugate. Because of the factor $\delta_{N_{i'} g_M g^i_{m'} g_M^{-1}, N_{i'}}$ in expression (3.60), the nonzero terms must be such that $g_M g^i_{m'} g_M^{-1} \in N_{i'}$. Therefore, taking (3.47) and (3.48) into account, we obtain

$$
U^{i'}_{km}(g_M g^i_{m'} g_M^{-1}) = U^{i'_{g_M}}_{km}(g_M^{-1} g_M g^i_{m'} g_M^{-1} g_M) = U^i_{km}(g^i_{m'}), \tag{3.62}
$$

$$
\Delta^{i'j'}_{\mu\mu}(g_M g^i_{m'} g_M^{-1}) = \Delta^{i'_{g_M} j'}_{\mu\mu}(g_M^{-1} g_M g^i_{m'} g_M^{-1} g_M) = \Delta^{ij'}_{\mu\mu}(g^i_{m'})\,. \tag{3.63}
$$

Now relation (3.60) takes the form

$$
\begin{aligned}
&\sum_{g} \chi_D^{ij}(g)\, \chi_D^{i'j'*}(g) \\
&= \frac{n(G)n(H)}{n(N_i)l_i} \sum_{g^i_{m'}} \sum_{km} U^i_{km}(g^i_{m'})\, U^{i*}_{km}(g^i_{m'})\, \chi^j(g^i_{m'})\, \chi^{j'*}(g^i_{m'}) \sum_{M} \delta_{i,i'_{g_M}} \\
&= \frac{n(G)n(H)}{n(N_i)l_i}\, \delta'_{ii'} \sum_{g^i_{m'}} \sum_{km} U^i_{km}(g^i_{m'})\, U^i_{mk}(g^i_{m'})^{-1}\, \chi^j(g^i_{m'})\, \chi^{j'*}(g^i_{m'}) \\
&= \frac{n(G)n(H)}{n(N_i)}\, \delta'_{ii'} \sum_{g^i_{m'}} \chi^j(g^i_{m'})\, \chi^{j'*}(g^i_{m'}) \overset{(4.16)}{=} n(G)\, \delta'_{ii'}\, \delta_{jj'},
\end{aligned} \tag{3.64}
$$

where the delta-symbol $\delta'_{ii'}$ means that $\delta'_{ii'} = 1$ if d^i and $d^{i'}$ are conjugate and $\delta'_{ii'} = 0$ otherwise (see the comment to (3.61)). Application of the irreducibility criterion (2.18) completes the proof of (i), (ii) and (iii).

The dimension L_{ij} of the representation D^{ij}, obtained by regular continuation of the small representation d^{ij} from N_i to G, is equal to the product of the number of cosets $n(G)/n(N_i)$ and dimension of d^{ij},

$$L_{ij} = \frac{n(G)}{n(N_i)} l_i(H) l_j(N_i/H). \tag{3.65}$$

To prove completeness of the system of representations D^{ij} (statement (iv)), we calculate the sum of squares of L_{ij} over all representations,

$$\sum_i{}' \sum_j L_{ij}^2 = \sum_i{}' \frac{n^2(G)}{n^2(N_i)} l_i^2(H) \sum_j l_j^2(N_i/H). \tag{3.66}$$

The prime means that we take only one representation D^{ij} out of each orbit (the number of representations in an orbit is equal to $n(G)/n(N_i)$). Using statement (ii) and completeness of the system of representations d^i and Δ^{ij},

$$\sum_i l_i^2(H) = n(H), \qquad \sum_j l_j^2(N_i/H) = \frac{n(N_i)}{n(H)},$$

we have

$$\sum_i{}' \sum_j L_{ij}^2 = \frac{n^2(G)}{n(H)} \sum_i{}' \frac{l_i^2(H)}{n(N_i)} = \frac{n^2(G)}{n(H)} \sum_i \frac{n(N_i)}{n(G)} \frac{l_i^2(H)}{n(N_i)}$$
$$= \frac{n(G)}{n(H)} \sum_i l_i^2(H) = n(G). \tag{3.67}$$

According to the completeness criterion (2.19) the sum of the squares of the dimensions of all irreducible representations is equal to the order of the group. Since the irreducibility and nonequivalence of the representations D^{ij} is proved, equality (3.67) proves completeness of the system of these representations.

3.9 Reality Properties of the Induced Representations

In some applications it is necessary to take into account the reality properties of the irreducible representations. The irreducible representations can be divided into three types: (potentially) real, complex and pseudoreal (for details, see Sect. 2.3.6). The real and pseudoreal representations have real characters and complex have the complex ones. Matrices of a real representation can be made real by a unitary transformation; for the two other types this cannot be done. To determine the type, it is sufficient to calculate the sum of the characters of the squares of all group elements:

$$\sigma_D \equiv \sum_g \chi(g^2) = \begin{cases} n(G) & \text{real}, \\ 0 & \text{complex}, \\ -n(G) & \text{pseudoreal}. \end{cases} \tag{3.68}$$

For the induced representations this criterion can be simplified and reduced to a criterion for the small representations. Using the definition (3.58), we transform the left-hand side of expression (3.68) as

$$\begin{aligned}\sigma_D &= \sum_{g} \sum_{Ll\lambda} d^{ij}_{l\lambda,l\lambda}(g_L g^2 g_L^{-1})\, \delta_{N_i g_L g^2 g_L^{-1}, N_i} \\ &= \sum_{g} \sum_{Ll\lambda} d^{ij}_{l\lambda,l\lambda}(g^2)\, \delta_{N_i g^2, N_i} = \frac{n(G)}{n(N_i)} \sum_{g} \sum_{l\lambda} d^{ij}_{l\lambda,l\lambda}(g^2)\, \delta_{N_i g^2, N_i}. \end{aligned} \tag{3.69}$$

By utilizing the group decomposition (3.1), we represent each element in the group as $g = hg_n$ and use the explicit expression for matrices of the small representation (3.41). Taking (3.25) into account, we have

$$\begin{aligned}\sigma_D &= \frac{n(G)}{n(N_i)} \sum_{g_n} \delta_{N_i g_n^2, N_i} \sum_{lmk} \sum_{h} d^i_{lm}(h)\, d^i_{mk}(g_n h g_n^{-1})\, U^i_{kl}(g_n^2) \sum_{\lambda} \Delta^j_{\lambda\lambda}(g_n^2) \\ &= \frac{n(H)}{l_i} \frac{n(G)}{n(N_i)} \sum_{g_n} \delta_{N_i g_n^2, N_i}\, \delta_{i, i_{g_n}*} \sum_{lmk} \delta_{lm}\, \delta_{mk}\, U^i_{kl}(g_n^2) \sum_{\lambda} \Delta^j_{\lambda\lambda}(g_n^2). \end{aligned} \tag{3.70}$$

Here we used the orthogonality relations for the irreducible representations of an invariant subgroup. The representation d^{i*} is the complex conjugate of the representation d^i. The delta-symbol $\delta_{i,i_{g_n}*}$ means that the result of the summation over g_n is nonzero if and only if

$$\chi^i(h) = \chi^{i_{g_n}*}(h). \tag{3.71}$$

Passing in this equality to representations conjugate by g_n, we obtain

$$\chi^{i_{g_n}}(h) = \chi^{i_{g_n^2}*}(h). \tag{3.72}$$

Comparing relation (3.71) with the complex conjugate of (3.72), we have

$$\chi^i(h) = \chi^{i_{g_n^2}}(h). \tag{3.73}$$

According to the definition (3.10), that means g_n^2 is an element of the normalizer N_i, and hence

$$\delta_{N_i g_n^2, N_i} = 1. \tag{3.74}$$

Then

$$\begin{aligned}\sigma_D &= \frac{n(G)}{l_i n(N_i/H)} \sum_{g_n} \delta_{i,i_{g_n}*} \sum_{l\lambda} U^i_{ll}(g_n^2)\, \Delta^j_{\lambda\lambda}(g_n^2) \\ &= \frac{n(G)}{l_i n(N_i/H)} \sum_{g_n} \chi^{ij}_d(g_n^2)\, \delta_{i,i_{g_n}*}. \end{aligned} \tag{3.75}$$

Now the criterion (3.68) can be written as

$$\sigma_d \equiv \sum_{g_n} \chi_d^{ij}(g_n^2) = \begin{cases} l_i n(N_i/H) & \text{real}, \\ 0 & \text{complex}, \\ -l_i n(N_i/H) & \text{pseudoreal}, \end{cases} \tag{3.76}$$

where the summation is carried out only over the representatives in the coset decomposition of the group with respect to an invariant subgroup such that

$$\chi^{i_{g_n}*}(h) = \chi^i(h) \tag{3.77}$$

for all h in the subgroup.

Thus, expressions (3.76) and (3.77) allow us to use only the character of the *small* representation χ_d^{ij} to determine the reality property of the corresponding representation of the entire group.

3.10 Algorithm for Constructing Irreducible Representations

So the problem of constructing irreducible representations of a finite group G is reduced to construction of irreducible representations d^i of its invariant subgroup H and irreducible projective representations Δ^{ij} of the factor groups of the normalizers N_i with respect to H.

It is convenient to use the following algorithm:

(a) Choose an invariant subgroup using relation (3.5).
(b) Construct all its irreducible representations. If the invariant subgroup is commutative, this step is trivial; otherwise this problem can be solved using the same method.
(c) Classify the irreducible representations by orbits (see Sect. 3.2).
(d) Determine the normalizer of one representation from each orbit (see Sect. 3.2).
(e) Construct the factor group of the normalizer over the invariant subgroup using the subgroup decomposition (3.12).
(f) Obtain the unitary transformation matrices for the elements of the factor group and calculate the factor system using relations (3.24) and (3.27).
(g) Construct all irreducible projective representations of the factor group that correspond to the obtained factor system using Chap. 4.
(h) Obtain matrices of the small representations of the normalizer using expression (3.41).
(i) Construct the regular continuation from the normalizer to the group using expression (3.58).

For special types of irreducible representations of an invariant subgroup the algorithm can be simplified. If d^i is a noncritical representation, i.e. $N_i = H$, then from step (d) one should go directly to step (i). In this case, the representation d^i is already the small representation of the normalizer. If d^i is completely critical, i.e. $N_i = G$, then step (i) is redundant because the small representations of the normalizer are already the irreducible representations of the group.

References

1. F.G. Frobenius, Sitzber. Preuss. Akad. Wiss. pp. 501–515 (1898)
2. A.H. Clifford, Ann. Math. **38**, 533 (1937)
3. D.K. Faddeev, *Tables of the Principal Unitary Representations of Fedorov Groups* [in Russian] (Acad. Sci. USSR, Moscow, 1961). Engl. transl.: Pergamon, Oxford, 1964
4. R.F. Egorov, V.K. Zvezdin, B.I. Reser, *Representations of space groups* [in Russian] (1971). Dep. VINITI, No. 2865–71
5. C.W. Curtis, I. Reiner, *Representation Theory of Finite Groups and Associative Algebras* (Wiley, New York, 1962)
6. H.W. Streitwolf, *Group Theory in Solid State Physics* (MacDonald, London, 1971)
7. C.J. Bradley, A.P. Cracknell, *The Mathematical Theory of Symmetry in Solids. Representation Theory for Point Groups and Space Groups* (Clarendon, Oxford, 1972). Reprinted: Clarendon, Oxford, 2010

Chapter 4
Projective Representations

Just go on, and faith will soon return. (J. d'Alembert; from P. J. Davis, R. Hersh, The Mathematical Experience, Birkhäuser, Boston, 1981)

In some problems of quantum mechanics it is necessary to consider, along with ordinary (vector) representations, the so-called projective representations of finite groups (also known as ray representations). For example, the double-valued representations are one of them.

The theory of projective representations was developed by Schur [1–3]. He showed that irreducible projective representations can be obtained by constructing ordinary representations of an *extended* group, which contains the original group.

In quantum mechanics the projective (ray) representations were introduced by Weyl [4]. They appear quite naturally since a quantum state is determined by a ray rather than a vector. Therefore, two unitary transformations U and εU, which differ only by a numerical factor ε of modulus 1, are considered as equivalent, $U \sim \varepsilon U$, because they determine the same *ray field* rotation. If $U(s)$ form a group of rotations, then the condition for a representation is now only $U(s)U(t) \sim U(st)$, i.e. $U(s)U(t) = \delta(s,t)U(st)$, where $\delta(s,t)$ is a numerical factor of modulus 1 depending on elements s and t.

In the study of space groups the projective representations were introduced by Lyubarskii [5] (as weighted representations) and Döring [6] (as ray representations). The appearance of the projective representations is motivated by the method of induced representations, which we described in the previous chapter.

Main definitions and properties of projective representations can be found in [5, 7–9]. Construction of the projective representations is greatly simplified if the group is described by means of its generators [6, 10]. This approach proved to be very efficient and was used by many authors [11–14]. Our presentation is close to [11].

N. B. Melnikov and B. I. Reser, *Space Group Representations*,
https://doi.org/10.1007/978-3-031-13991-8_4

4.1 The Factor System

Let G be a finite group consisting of n elements. The matrices of a projective representation satisfy the following relation:

$$D(g)\,D(g') = \omega_{g,g'}\,D(gg')\,, \tag{4.1}$$

for all elements g and g' in the group. The complex numbers $\omega_{g,g'}$, called *projective factors*, must be nonzero for all g and g'. The set of all n^2 such factors $\omega_{g,g'}$ is called the *factor system*. The set of matrices D satisfying (4.1) is called the *projective representation* belonging to the factor system $\omega_{g,g'}$. The projective factors can be chosen such that [7]

$$|\omega_{g,g'}| = 1. \tag{4.2}$$

In particular, if $\omega_{g,g'} = 1$ for all g and g', we have a vector representation.

From relation (4.1) we have

$$D(g)\,D(g')\,D(g'') = \omega_{g,g'}\,D(gg')\,D(g'') = \omega_{g,g'}\,\omega_{gg',g''}\,D(gg'g''). \tag{4.3}$$

On the other hand, from the associativity of the group multiplication we obtain

$$D(g)\,D(g')\,D(g'') = D(g)\,\omega_{g',g''}\,D(g'g'') = \omega_{g,g'g''}\,\omega_{g',g''}\,D(gg'g''). \tag{4.4}$$

Comparing the two results, we come to the relation

$$\omega_{g,g'}\,\omega_{gg',g''} = \omega_{g,g'g''}\,\omega_{g',g''}. \tag{4.5}$$

From the fact that the matrix $D(e)$ can be chosen to be a unit matrix,

$$D(e) = E\,, \tag{4.6}$$

it follows that

$$\begin{aligned} D(e)\,D(g) &= \omega_{e,g}\,D(g)\,,\\ D(g)\,D(e) &= \omega_{g,e}\,D(g). \end{aligned} \tag{4.7}$$

Hence

$$\omega_{g,e} = \omega_{e,g} = 1. \tag{4.8}$$

By the definition of the inverse element,

$$D(g)\,D(g^{-1}) = \omega_{g,g^{-1}}\,D(e)\,,$$
$$D(g^{-1})\,D(g) = \omega_{g^{-1},g}\,D(e). \tag{4.9}$$

Multiplying the first relation by $D(g)^{-1}$ from the left and the second by $D(g)^{-1}$ on the right and comparing the results, we obtain

$$\omega_{g,g^{-1}} = \omega_{g^{-1},g}. \tag{4.10}$$

Equations (4.5) have an infinite number of solutions. If $\omega_{g,g'}$ is a solution and ω_g is a set of nonzero complex numbers, then

$$\omega'_{g,g'} = \frac{\omega_g\,\omega_{g'}}{\omega_{gg'}}\,\omega_{g,g'} \tag{4.11}$$

is also a solution. We say that the two solutions $\omega_{g,g'}$ and $\omega'_{g,g'}$ belong to the same class (they are also called associated or equivalent factor systems). If D is a representation belonging to the factor system $\omega_{g,g'}$, the matrices

$$D'(g) = \frac{1}{\omega_g}\,D(g) \tag{4.12}$$

form a representation belonging to the factor system $\omega'_{g,g'}$. Thus, if two factor systems are equivalent, the representations belonging to the one are immediately determined from the representations belonging to the other by means of the *gauge transformation* (4.12). Representations belonging to nonequivalent factor systems are said to be of different type.

In particular, if the factor system $\omega_{g,g'}$ is such that

$$\omega_{g,g'} = \frac{\omega_g\,\omega_{g'}}{\omega_{gg'}}$$

for all g and g', then the new matrices D' defined by (4.12) form a vector representation according to the relation (4.1). In this case, we see that the matrices of the projective representation D can be obtained from the matrices of a vector representation D' by simple multiplication.

4.2 Main Properties of Projective Representations

The concepts of equivalence and reducibility are the same as for the vector representations. For a particular factor system, there is a finite number of nonequivalent irreducible representations. Here we are concerned with the properties of the nonequivalent irreducible representations that correspond to a given factor system.

Many properties of the vector representations have their analogues for the projective representations.

(a) The matrices of the projective representations can be chosen *unitary*,

$$D(g)^{-1} = D(g)^{\dagger}. \tag{4.13}$$

The proof of this fact can be carried out in complete analogy with the proof of the corresponding statement for vector representations. Condition (4.2) is essential. Henceforth we consider only unitary projective representations.

(b) Projective representations can be reducible and irreducible.
Maschke's theorem [15]. Any projective representation can be expressed as a direct sum of irreducible projective representations:

$$D(g) \sim \sum_j n_j \, D^j(g)\,, \tag{4.14}$$

where n_j is the multiplicity of D^j in D.

(c) *Schur's lemma.* A matrix commuting with all matrices of an irreducible projective representation is a scalar matrix.
The proof is literally the same as for vector representations.

(d) Using Schur's lemma, it is easy to prove that irreducible projective representations satisfy the *orthogonality relations*,

$$\sum_g D^j_{\mu\nu}(g)\, D^{j'*}_{\mu'\nu'}(g) = \frac{n}{l_j}\, \delta_{jj'}\, \delta_{\mu\mu'}\, \delta_{\nu\nu'}\,, \tag{4.15}$$

where n is the order of the group and l_j is the dimension of the jth irreducible projective representation.
From relation (4.15) we obtain the *orthogonality relations* for the characters of irreducible projective representations,

$$\sum_g \chi^j(g)\, \chi^{j'*}(g) = n\, \delta_{jj'}. \tag{4.16}$$

Hence for $j = j'$ we have the equality

$$\sum_g |\chi^j(g)|^2 = n\,, \tag{4.17}$$

which can be used as an irreducibility criterion of a projective representation.

However, not all statements valid for vector representations can be carried over without change to projective representations. For example, the characters of elements

$\chi(g)$ in general are not functions of the classes though, as can be shown, their absolute values have this property.[1]

4.3 Regular Representation and Completeness Criterion

We consider the mapping that sends a group element g into a matrix of the form

$$D_{lm}(g) = \omega_{g_l,g}\, \delta_{g_l g, g_m}\,, \qquad 1 \le l, m \le n. \tag{4.18}$$

These matrices form a projective representation. Indeed,

$$\begin{aligned}\big(D(g)\,D(g')\big)_{lm} &= \sum_k D_{lk}(g)\, D_{km}(g') \\ &= \sum_k \omega_{g_l,g}\, \delta_{g_l g, g_k}\, \omega_{g_k,g'}\, \delta_{g_k g', g_m} = \omega_{g_l,g}\, \omega_{g_l g, g'}\, \delta_{g_l g g', g_m} \\ &= \omega_{g,g'}\, \omega_{g_l, gg'}\, \delta_{g_l gg', g_m} = \omega_{g,g'}\, D_{lm}(gg').\end{aligned} \tag{4.19}$$

Here we used the relation (4.5).

The projective representation (4.18) is called *regular* projective representation. Using relation (4.8), we calculate the characters of elements of this representation,

$$\chi(g) = \sum_l \omega_{g_l,g}\, \delta_{g_l g, g_l} = \delta_{g,e} \sum_l \omega_{g,e} = \delta_{g,e}\, n. \tag{4.20}$$

The representation (4.18) is generally reducible. The multiplicity of the jth irreducible projective representation n_j in the direct sum (4.14) is calculated by the formula

$$n_j = \frac{1}{n} \sum_g \chi(g)\, \chi^{j*}(g). \tag{4.21}$$

[1] One more example. The orthogonality relation for vector representations,

$$\sum_g \chi^j(g)\, \chi^{j'}(g^{-1}) = n\, \delta_{jj'}\,,$$

for projective representations will take the form

$$\sum_g \omega^{-1}_{g,g^{-1}}\, \chi^j(g)\, \chi^{j'}(g^{-1}) = n\, \delta_{jj'}.$$

The latter is obtained from (4.16) by taking (4.13) and (4.9) into account,

$$\sum_g \chi^j(g)\, \chi^{j'*}(g) = \sum_g \mathrm{Tr}(D^j(g))\ \mathrm{Tr}(D(g)^{-1}) = \sum_g \omega^{-1}_{g,g^{-1}}\, \chi^j(g)\, \chi^{j'}(g^{-1}).$$

Substituting the relation (4.20) in (4.21), for the regular projective representation, we have

$$n_j = \frac{1}{n} \sum_g n\, \delta_{g,e}\, \chi^{j*}(g) = \chi^{j*}(e) = l_j. \tag{4.22}$$

Thus, the number of times an irreducible projective representation enters the regular projective representation is equal to the dimension of the irreducible projective representation. So, on the one hand, the dimension of the regular projective representation is equal to n by the definition, and on the other hand, it is equal to the sum of squares of the dimensions of all irreducible representations that are contained in the regular projective representation,

$$n = \sum_j l_j^2. \tag{4.23}$$

This relation is a generalization of Burnside's theorem (2.19) to projective representations.

Note that the possibility of constructing the regular projective representation for a given factor system ensures the very existence of projective representations with this factor system.

4.4 Group Generators

Our next step is to describe a method for constructing irreducible projective representations of a finite group for all possible factor systems compatible with the structure of this group.

The structure of a group can be described in two ways. The usual way is to obtain the group multiplication table. Then it is necessary to know the multiplication rule for all elements of the group. For our purpose, it is more convenient to describe the structure of the group by a set of relations involving certain elements (generators) from which the whole multiplication table can be constructed. It is then sufficient to know the multiplication rule only for the generators. This in turn ensures that the projective representation can be described by a smaller number of factors.

The choice of the generators is generally nonunique. In practice, one can do as follows. Choose an arbitrary element g and take all successive powers of g up to the m_gth power, where m_g is the order of g. We have thus obtained m_g elements. From what is left we choose another element g' and take all its powers up to the $m_{g'}$th power. Next we take all the products of the form $g^{\mu} g'^{\mu'}$, where $1 \le \mu \le m_g$ and $1 \le \mu' \le m_{g'}$. If we have not exhausted the group, we continue the procedure. Since the group is finite, sooner or later, we obtain all the elements.

If we now order the generators, say in the order of their appearance, then every element of the group can be presented as a product of the generators,

$$g = g_1^{\mu_1} g_2^{\mu_2} \dots g_p^{\mu_p} \equiv \prod_{i=1}^{p} g_i^{\mu_i}, \tag{4.24}$$

where $\{\mu_i\}$ is the set of exponents that characterize the element g and p is the number of the generators. To present a product of two elements (or an inverse element) as an *ordered* product of the generators (4.24), it is necessary to use the relation[2]

$$g_i g_j = \prod_{k=1}^{p} g_k^{\mu_k(i,j)}, \qquad i > j. \tag{4.25}$$

and

$$g_i^{m_i} = e. \tag{4.26}$$

These relations are called the *generating relations* (or defining relations). Specification of generators and generating relations completely determines the group.

4.5 Character Factors

We rewrite the relations (4.1) that determine the projective representation in the case when the group is defined by its generators. First of all, since each element has the form (4.24), we write the matrix $D(g)$ as

$$D(g) \equiv D(\prod_{i=1}^{p} g_i^{\mu_i}) = \omega_g^* \prod_{i=1}^{p} D(g_i)^{\mu_i}, \tag{4.27}$$

where

$$\omega_g \equiv \omega_{\prod_{i=1}^{p} g_i^{\mu_i}} = \prod_{i=1}^{p} \prod_{j=1}^{\mu_i - 1} \omega_{g_i^j, g_i} \prod_{k=1}^{p-1} \omega_{\prod_{l=1}^{k} g_l^{\mu_l}, g_{k+1}^{\mu_{k+1}}}. \tag{4.28}$$

The expression for the factors ω_g can be obtained in a number of different ways depending on the order in which we multiply the matrices of the generators. Due to associativity of the group multiplication and equality (4.5), which follows from it, the resulting expressions will be identical. We use the following algorithm. First we obtain the matrices $D(g_i^{\mu_i})$ of every generator by combining them as

$$\begin{aligned} D(g_i)^2 = D(g_i)D(g_i) &= \omega_{g_i, g_i} D(g_i^2), \\ D(g_i)^3 = D(g_i)^2 D(g_i) &= \omega_{g_i, g_i} \omega_{g_i^2, g_i} D(g_i^3), \\ &\dots \\ D(g_i)^{\mu_i} = D(g_i)^{\mu_i - 1} D(g_i) &= \textstyle\prod_{j=1}^{\mu_i - 1} \omega_{g_i^j, g_i} D(g_i^{\mu_i}). \end{aligned} \tag{4.29}$$

[2] The notation $\mu_k(i, j)$ means that the set of exponents μ_k is different for each pair (i, j).

Then we multiply the matrices $D(g_i^{\mu_i})$ in the order of their appearance on the right-hand side of (4.27). Using the property $\omega^{-1}_{g,g'} = \omega^*_{g,g'}$, we come to expression (4.28).

Next we obtain the defining relations for the matrices of the generators. Using the generating relations (4.26) and (4.25) and taking formulas (4.1) and (4.27) into account, we obtain

$$D(g_i)^{m_i} = \omega_i\, E\,, \tag{4.30}$$

$$D(g_i)D(g_j) = \omega_{ij} \prod_{k=1}^{p} D(g_k)^{\mu_k}\,, \qquad i > j\,, \tag{4.31}$$

where E is the unit matrix and the factors ω_i and ω_{ij} are given by

$$\omega_i = \prod_{j=1}^{m_i-1} \omega_{g_i^j, g_i}\,, \tag{4.32}$$

$$\omega_{ij} = \omega_{g_i, g_j}\, \omega^*_{\prod_{k=1}^{p} g_k^{\mu_k}}\,. \tag{4.33}$$

We call ω_{ij} the *character factors.*

The defining relations (4.30) and (4.31) can be simplified further if we replace the matrices $D(g_i)$ using the gauge transformation[3]

$$D'(g_i) = \frac{1}{\sqrt[m_i]{\omega_i}}\, D(g_i). \tag{4.34}$$

Then the defining relations for the matrices of the generators are written in the following final form:

$$D'(g_i)^{m_i} = E\,, \tag{4.35}$$

$$D'(g_i)D'(g_j) = \omega'_{ij} \prod_{k=1}^{p} D'(g_k)^{\mu_k}\,, \qquad i > j\,, \tag{4.36}$$

where

$$\omega'_{ij} = \omega_{ij} \frac{\prod_{l=1}^{p} \left(\sqrt[m_l]{\omega_l}\right)^{\mu_l}}{\sqrt[m_i]{\omega_i}\, \sqrt[m_j]{\omega_j}}\,. \tag{4.37}$$

[3] By the mth root $\sqrt[m]{\omega}$ we mean a complex number z such that $z^m = \omega$. The complex number z can be chosen arbitrarily out of m solutions of this equation but then we assume it to be fixed once and for all.

From expressions (4.35) and (4.36) we see that the projective representations of a group are completely determined by a relatively small number of character factors ω'_{ij}, which are products of original projective factors $\omega_{g,g'}$. It is easy to calculate that, for a group with p generators, the number of character factors is $p(p-1)/2$.[4]

In the general case, not all character factors ω'_{ij} are independent. Using the relations (4.35) and different combinations of the relations (4.36), we obtain

$$\prod_{i,j} \omega'^{\;\mu_{ij}}_{ij} = 1. \tag{4.38}$$

In each specific case, the formulas (4.38) allow to choose a system of independent character factors that satisfy the relations

$$\omega'^{\;m_{ij}}_{ij} = 1 \tag{4.39}$$

and express all other character factors in terms of the independent ones. Note that some character factors can simply be equal to 1.

The set of all solutions to equations (4.39) gives the set of all character factors admissible for this group, and they completely determine the structure of the irreducible projective representations. Clearly, among these solutions there is one such that all ω'_{ij} are equal to 1. It corresponds to the vector representation.

For any fixed factor system $\omega_{g,g'}$, once we defined the matrices of the irreducible projective representation for the generators, we can then calculate the matrices of all other elements of the group using relations (4.34) and (4.27).

4.6 Covering Group and Its Representations

Following Schur [1–3], we consider the matrices $D(g_i)$ and matrices of independent character factors $\omega_{ij}E$ as generators of a new extended group $\tilde{G}$, which is called the *covering group*. (For brevity, here and later in this chapter we omit the prime in the notation of the independent character factors (4.39).)

We need to make an assumption about the order of appearance of the generators in the presentation of an element of the group $\tilde{G}$ (analogous to relation (4.24) in the original group G). Since the matrices $\omega_{ij}E$ are scalar, the order of their appearance can be chosen arbitrarily. We assume that the matrices $\omega_{ij}E$ appear before the generators $D(g_i)$ for which we retain the same order of appearance as above. Then an element of the covering group is written as

---

[4] In the simplest example of the cyclic group of order m, we have only one generator g and one generating relation $g^m = e$. Then the defining relation (4.35) becomes $D'(g)^m = E$, and $D'(g^k) = D'(g)^k$ is a vector representation. Since an irreducible vector representation of a cyclic group is one-dimensional, so is an irreducible projective representation. In the case of a commutative group with already two generators, the irreducible projective representations can be two-dimensional [4, 8]. Clearly, this irreducible projective representation is nonequivalent to an irreducible vector representation of a commutative group, which is always one-dimensional.

$$\prod_{i,j} (\omega_{ij} E)^{\mu_{ij}} \prod_k D(g_k)^{\mu_k}. \tag{4.40}$$

To define the structure of the covering group completely, we need relations for the generators $\omega_{ij} E$, analogous to relations (4.35) and (4.36). Since the matrices $\omega_{ij} E$ are scalar, we have

$$\omega_{ij} E \; D(g_k) = D(g_k) \; \omega_{ij} E \,, \tag{4.41}$$

$$\omega_{ij} E \; \omega_{i'j'} E = \omega_{i'j'} E \; \omega_{ij} E \,, \tag{4.42}$$

and by utilizing (4.39), we obtain

$$(\omega_{ij} E)^{m_{ij}} = E. \tag{4.43}$$

Relations (4.35) and (4.36), together with (4.41), (4.42) and (4.43), form a complete system of defining relations of the covering group $\tilde{G}$.

To construct irreducible vector representations of the covering group $\tilde{G}$ we can use the theory of induced representations introduced in Chap. 3. Indeed, the covering group always has an invariant subgroup that consists of all elements $\omega_{ij} E$ and their products. In addition, we can include elements of an invariant subgroup of the initial group G if such a subgroup exists. To simplify the calculations it is practical to choose the invariant subgroup of the covering group to be commutative and of maximal order.

Note that, in any irreducible representation of the covering group, the matrices corresponding to $\omega_{ij} E$ will be scalar due to Schur's lemma. From relation (4.43) we see that each of those matrices will be a unit matrix times a coefficient, which is a solution to (4.39).

4.7 Irreducible Projective Representations

Once the vector representations of the covering group are obtained, we need to return to the problem of constructing the irreducible projective representations of the original group.

Let us consider an arbitrary irreducible representation of the covering group. The matrices of this representation satisfy the relations analogous to (4.35) and (4.36) and relations (4.41)–(4.43). If we confine ourselves to considering only the matrices that correspond to the generators of the original group, then from (4.35) and (4.36) we see that these matrices form a projective representation. This projective representation belongs to the factor system consisting of the coefficients in the matrices of the generators $\omega_{ij} E$ in the chosen irreducible representation of the covering group. Thus, for each vector representation of the covering group there is a projective representation of the original group.

The converse is also true, i.e. for each projective representation of the original group belonging to a fixed factor system $\omega_{g,g'}$ there is a vector representation of the covering group. To set up this correspondence, we map $D(g_i)$ to themselves and $\omega_{ij}E$ to the unit matrices multiplied by the ω_{ij} which are calculated based on $\omega_{g,g'}$. Clearly, the obtained matrices satisfy relations (4.35), (4.36) and (4.41)–(4.43) and hence form a representation of the covering group.

Since there exists a projective representation belonging to any factor system (Sect. 4.3), according to the described mapping, there exists a vector representation of the covering group that corresponds to the same factor system. Hence, among the representations of the covering group there are representations corresponding to all solutions of equations (4.39).

Let us show that the projective representations obtained from the irreducible representations of the covering group are also irreducible. Indeed, applying the irreducibility criterion (2.18) to an irreducible representation of the covering group, we have

$$\sum_{\tilde{g}\in\tilde{G}} |\chi(\tilde{g})|^2 = n(\tilde{G}). \tag{4.44}$$

Using relation (4.40), we have

$$\chi(\tilde{g}) = \chi\left(\prod_{ij}(\omega_{ij}E)^{\mu_{ij}}\, D(g)\right) = \prod_{ij}\omega_{ij}^{\mu_{ij}}\,\chi\,(D(g)) \equiv \prod_{ij}\omega_{ij}^{\mu_{ij}}\,\chi(g). \tag{4.45}$$

Decomposing the covering group with respect to the original group, we have

$$\tilde{G} = \sum_k \prod_{ij}(\omega_{ij}E)^{\mu_k}\, G\,, \qquad 1 \le k \le n(\tilde{G})/n(G). \tag{4.46}$$

Then, taking relations (4.45) and (4.2) into account, we transform the left-hand side of equality (4.44) as

$$\sum_{\tilde{g}\in\tilde{G}} |\chi(\tilde{g})|^2 = \sum_k \left|\prod_{ij}\omega_{ij}^{\mu_k}\right|^2 \sum_{g\in G} |\chi(g)|^2 = \frac{n(\tilde{G})}{n(G)} \sum_{g\in G} |\chi(g)|^2. \tag{4.47}$$

By comparing the equalities (4.44) and (4.47), we obtain

$$\sum_{g\in G} |\chi(g)|^2 = n(G). \tag{4.48}$$

According to (4.17) this proves the irreducibility of the projective representation.

The converse is also true, i.e. the representation of the covering group obtained from an irreducible projective representation of the original group is irreducible.

Let us prove completeness of the system of irreducible projective representations obtained from the irreducible representations of the covering group for a fixed factor system. We prove this by contradiction. Assume that there is an irreducible projective representation that cannot be obtained from any irreducible representation of the covering group. Then, extending it to the covering group, we obtain an irreducible (vector) representation of the covering group. Due to completeness of the irreducible representations of the covering group, the latter must coincide with one of its representations. The projective representation obtained from it coincides with the original one, which contradicts the assumption.

4.8 Outline of the Algorithm

The construction of all irreducible projective representations of a finite group can be carried out as follows.

(a) Choose a set of generators of the group (see Sect. 4.4).
(b) Obtain the defining relations (4.25) and (4.26).
(c) For the matrices of the projective representation write the defining relations (4.35) and (4.36) with arbitrary character factors.
(d) Analyze properties of the system of character factors. Obtain equations (4.38) and (4.39).
(e) Construct the covering group and pick an invariant subgroup (see Sect. 4.6).
(f) Determine all irreducible vector representations of the covering group (see Chap. 3).
(g) From each irreducible vector representation of the covering group obtain an irreducible projective representation of the original group and classify them according to the values of the character factors.

Once the representations are obtained, the problem of constructing the irreducible projective representations belonging to a fixed factor system $\omega_{g,g'}$ consists in the following. Calculate all ω_g by formula (4.28) and character factors by formula (4.37). According to the values of the character factors, choose a system of matrices of projective representations for the generators of the group. Using (4.27) and (4.34), calculate the matrices of all group elements.

Checking the irreducibility and completeness criterions and orthogonality relations is an effective way to control the results.

References

1. J. Schur, J. Reine Angew. Math. **127**, 20 (1904)
2. J. Schur, J. Reine Angew. Math. **132**, 85 (1907)
3. J. Schur, J. Reine Angew. Math. **139**, 155 (1911)

4. H. Weyl, *The Theory of Groups and Quantum Mechanics* (Dover, New York, 1931). Reprinted: Dover, New York, 1950
5. G.Y. Lyubarskii, *Application of Group Theory in Physics* [in Russian] (Fizmatgiz, Moscow, 1958). Engl. transl.: Pergamon, New York, 1960
6. W. Döring, Z. Naturforschg. **14 a**, 343 (1959)
7. C.W. Curtis, I. Reiner, *Representation Theory of Finite Groups and Associative Algebras* (Wiley, New York, 1962)
8. M. Hamermesh, *Group Theory and Its Applications to Physical Problems* (Addison-Wesley, Reading, MA, 1962). Reprinted: Dover, New York, 1989
9. P. Rudra, J. Math. Phys. **6**, 1278 (1965)
10. A.C. Hurley, Phil. Trans. Roy. Soc. London **A260**, 1 (1966)
11. R.F. Egorov, V.K. Zvezdin, B.I. Reser, *Representations of space groups* [in Russian] (1971). Dep. VINITI, No. 2865–71
12. C.J. Bradley, A.P. Cracknell, *The Mathematical Theory of Symmetry in Solids. Representation Theory for Point Groups and Space Groups* (Clarendon, Oxford, 1972). Reprinted: Clarendon, Oxford, 2010
13. G.L. Bir, G.E. Pikus, *Symmetry and Strain-induced Effects in Semiconductors* [in Russian] (Nauka, Moscow, 1972). Engl. transl.: Wiley, New York, 1974
14. J.L. Birman, *Theory of Crystal Space Groups and Lattice Dynamics: Infra-Red and Raman Optical Processes of Insulating Crystals* (Springer, Berlin, 1974)
15. G.M. Mackey, *The Theory of Unitary Group Representations* (The University of Chicago, Chicago, 1976)

Chapter 5
Representations of the Space Groups

In group theory, despite the extraordinary penetrating theorems which have been proved, one cannot say that we have a general method of attack... one always has the impression that one is working like a craftsman, by accumulating a series of stratagems. (J. A. Dieudonne, The American Mathematical Monthly, 77, 34, (1970))

Practical importance of symmetry analysis in quantum systems is demonstrated by the fundamental Wigner's theorem (see Chap. 7). According to this theorem, we can associate an irreducible representation of the symmetry group with each energy level of the system. The knowledge of irreducible representations allows to classify the energy levels and make a number of conclusions about properties of the system, which do not depend on the model. Therefore, our primary goal is to describe all irreducible representations of the space groups.

The theory presented in the two preceding chapters has a general character and can be applied to constructing irreducible representations of any finite group. Here we apply this technique to describe the structure of irreducible representations of the space groups.

5.1 Space Groups and Their Properties

A crystal symmetry is a mapping of the three-dimensional Euclidian space that leaves the crystal unchanged. Therefore, a spatial symmetry transformation must preserve the distance between any two points. We consider only this kind of spatial transformations without further notice.

In the Seitz notation [1], any spatial transformation is written as $\{\alpha\,|\,\mathbf{t}\}$ and its action on a vector $\mathbf{r}$ is given by

$$\{\alpha\,|\,\mathbf{t}\}\mathbf{r} = \alpha\mathbf{r} + \mathbf{t}. \tag{5.1}$$

N. B. Melnikov and B. I. Reser, *Space Group Representations*,
https://doi.org/10.1007/978-3-031-13991-8_5

Here α is a *point transformation* corresponding to a rotation, inversion,[1] or their combination (called an *improper rotation*), and $\mathbf{t}$ is a *translation vector*. In this notation $\{\varepsilon|0\}$ is the identical transformation, $\{\varepsilon|\mathbf{t}\}$ represents a translation and $\{\alpha|0\}$ is a rotation (proper or improper). It is easy to check that

$$\{\alpha|\mathbf{t}\}\{\beta|\mathbf{t}'\} = \{\alpha\beta|\alpha\mathbf{t}' + \mathbf{t}\}\,, \tag{5.2}$$

$$\{\alpha|\mathbf{t}\}^{-1} = \{\alpha^{-1}| - \alpha^{-1}\mathbf{t}\}. \tag{5.3}$$

The set of all space transformations that leave the crystal unchanged forms a *space group G*. Note that the entire crystal must be left unchanged, not just the crystal lattice. The stationary positions of the atoms in the crystal do not necessarily coincide with the lattice sites. In general there are more atomic positions than the lattice sites.

An important property of a space group is that it has the *translation subgroup T* consisting of the transformations

$$\{\varepsilon|\mathbf{R}_n\}, \qquad \mathbf{R}_n = n_1\mathbf{a}_1 + n_2\mathbf{a}_2 + n_3\mathbf{a}_3, \tag{5.4}$$

where $\mathbf{a}_i$ are three linearly independent vectors, called the *basic vectors (primitive translations)* of the lattice, and n_i are integers. Space is divided by a lattice into (primitive) *unit cells* , i.e. the regions containing no two points $\mathbf{r}$ and $\mathbf{r}'$ that differ by a lattice vector, $\mathbf{r}' = \mathbf{r} + \mathbf{R}_n$. The shape of the unit cell is not uniquely determined by the lattice. A symmetrical unit cell is called *Wigner–Seitz cell*. This is a polyhedron around a given lattice point; each bounding face is perpendicular to the line segment joining the given lattice point with another lattice point and crosses this line segment through the midpoint.

The factor group of the space group over the translation subgroup G/T is isomorphic to the point group G_0 of the lattice. Symmetry transformations of the form (5.4) lead to restrictions on the position, order and number of the symmetry axes. It follows that there are only 32 crystallographic point groups, which define the so-called 32 crystal classes. This leads to restrictions on the possible values of basic vectors, which gives only 14 essentially different translation subgroups, the so-called 14 Bravais lattices (for details, see, e.g. [2, 3]).

It turns out that a space group is not yet fully determined by specifying the Bravais lattice and crystal class. In addition, for all rotations of α, it is necessary to indicate some characteristic vectors $\boldsymbol{\tau}_\alpha$, the so-called *nonprimitive translations*, which are associated with possible presence of the helical axes and slip planes in the space group. Thus, any transformation of a space group can be represented as

$$\{\alpha|\mathbf{R}_n + \boldsymbol{\tau}_\alpha\} = \{\varepsilon|\mathbf{R}_n\}\{\alpha|\boldsymbol{\tau}_\alpha\}\,, \tag{5.5}$$

---

[1] Under the inversion i, each vector $\mathbf{r}$ reverses its direction, $i\mathbf{r} = -\mathbf{r}$.

and the coset decomposition of the group G over the subgroup T is written as

$$G = \bigcup_i T\{\alpha_i|\boldsymbol{\tau}_{\alpha_i}\}, \qquad i = 1, \ldots, n(G)/n(T), \tag{5.6}$$

where $\{\alpha_i|\boldsymbol{\tau}_{\alpha_i}\}$ is the ith coset representative .

A space group for which all $\boldsymbol{\tau}_\alpha = 0$ is called a *symmorphic* group. A space groups for which there exist $\boldsymbol{\tau}_\alpha \neq 0$ is called a group with nonprimitive translations (*nonsymmorphic*). Nonsymmorphic space groups correspond to crystals that have more stationary positions of the atoms than the lattice sites. Such crystals can be monatomic, like diamond (Chap. 10), or complex, like Cr_3Si (Chap. 12). A detailed analysis shows that overall there are 230 space groups, of which 73 are symmorphic and 157 are nonsymmorphic [4–7].

5.2 Reciprocal Lattice and Brillouin Zone

For any crystal lattice with the lattice sites $\mathbf{R}_n$ it is useful to introduce the concept of the *reciprocal lattice*, which was developed by Gibbs [8] and Ewald [9].[2] The sites of the reciprocal lattice $\mathbf{K}_\mu$ are defined by the relation

$$e^{i\mathbf{K}_\mu \mathbf{R}_n} = 1, \qquad \text{or equivalently,} \qquad \mathbf{K}_\mu \mathbf{R}_n = 2\pi \times \text{integer}, \tag{5.7}$$

which must hold for all $\mathbf{R}_n$. This relation means that $e^{i\mathbf{K}_\mu \mathbf{r}}$ is a periodic function with the periodicity of the lattice,

$$e^{i\mathbf{K}_\mu (\mathbf{r}+\mathbf{R}_n)} = e^{i\mathbf{K}_\mu \mathbf{r}}.$$

Hence the exponentials $e^{i\mathbf{K}_\mu \mathbf{r}}$ can be used to expand any periodic function $f(\mathbf{r} + \mathbf{R}_n) = f(\mathbf{r})$ into the Fourier series,

$$f(\mathbf{r}) = \sum_\mu c_\mu \, e^{i\mathbf{K}_\mu \mathbf{r}}.$$

The reciprocal lattice is completely determined by linear combinations of the basic (primitive) vectors of the reciprocal space $\mathbf{b}_j$, which are defined as

$$\mathbf{a}_i \cdot \mathbf{b}_j = 2\pi \delta_{ij} \qquad (i, j = 1, 2, 3). \tag{5.8}$$

The vectors $\mathbf{b}_j$ are uniquely determined by equalities (5.8) and are linearly independent. An arbitrary reciprocal lattice vector $\mathbf{K}_\mu$ can be written as

$$\mathbf{K}_\mu = \mu_1 \mathbf{b}_1 + \mu_2 \mathbf{b}_2 + \mu_3 \mathbf{b}_3,$$

[2] History of the development of the reciprocal lattice has been reviewed by [10].

where μ_i are integers and

$$\mathbf{b}_1 = \frac{2\pi(\mathbf{a}_2 \times \mathbf{a}_3)}{\mathbf{a}_1 \cdot (\mathbf{a}_2 \times \mathbf{a}_3)}, \qquad \mathbf{b}_2 = \frac{2\pi(\mathbf{a}_3 \times \mathbf{a}_1)}{\mathbf{a}_1 \cdot (\mathbf{a}_2 \times \mathbf{a}_3)}, \qquad \mathbf{b}_3 = \frac{2\pi(\mathbf{a}_1 \times \mathbf{a}_2)}{\mathbf{a}_1 \cdot (\mathbf{a}_2 \times \mathbf{a}_3)}. \tag{5.9}$$

It is convenient to choose the unit cell of the reciprocal lattice to be symmetric, such a unit cell is called the *Brillouin zone*. (The Brillouin zone in the reciprocal lattice is the same as the Wigner–Seitz cell in the direct lattice.) The volume of the Brillouin zone is obtained by the formula

$$\Omega_{\mathrm{BZ}} = \mathbf{b}_1 \cdot (\mathbf{b}_2 \times \mathbf{b}_3).$$

Using the well-known vector triple product identity (also known as the BAC-CAB identity), $\mathbf{A} \times (\mathbf{B} \times \mathbf{C}) = \mathbf{B} \cdot (\mathbf{A} \cdot \mathbf{C}) - \mathbf{C} \cdot (\mathbf{A} \cdot \mathbf{B})$, with $\mathbf{A} = \mathbf{a}_3 \times \mathbf{a}_1$, $\mathbf{B} = \mathbf{a}_1$ and $\mathbf{C} = \mathbf{a}_2$, we have

$$\begin{aligned}\mathbf{b}_1 \cdot (\mathbf{b}_2 \times \mathbf{b}_3) &= \frac{(2\pi)^3}{(\Omega_{\mathrm{WS}})^3}\left((\mathbf{a}_2 \times \mathbf{a}_3) \cdot ((\mathbf{a}_3 \times \mathbf{a}_1) \times (\mathbf{a}_1 \times \mathbf{a}_2))\right) \\ &= \frac{(2\pi)^3}{(\Omega_{\mathrm{WS}})^3}(\mathbf{a}_2 \times \mathbf{a}_3) \cdot (\mathbf{a}_1 \cdot ((\mathbf{a}_3 \times \mathbf{a}_1) \cdot \mathbf{a}_2) - \mathbf{a}_2 \cdot ((\mathbf{a}_3 \times \mathbf{a}_1) \cdot \mathbf{a}_1))\,.\end{aligned}$$

Taking into account

$$(\mathbf{a}_1 \times \mathbf{a}_2) \cdot \mathbf{a}_3 = (\mathbf{a}_2 \times \mathbf{a}_3) \cdot \mathbf{a}_1 = (\mathbf{a}_3 \times \mathbf{a}_1) \cdot \mathbf{a}_2 = \Omega_{\mathrm{WS}},$$

and $(\mathbf{a}_i \times \mathbf{a}_j) \cdot \mathbf{a}_j = 0$, we obtain

$$\Omega_{\mathrm{BZ}} = \frac{(2\pi)^3}{\Omega_{\mathrm{WS}}}. \tag{5.10}$$

Note that the reciprocal lattice has the same point group as the direct lattice. Indeed, orthogonality of a point transformation α implies $\alpha\dagger = \alpha^{-1}$. Then from (5.8) it follows that

$$(\mathbf{a}_i, \alpha\mathbf{b}_j) = (\alpha^{-1}\mathbf{a}_i, \mathbf{b}_j) = 2\pi \times \text{integer}.$$

By (5.7) the latter means $\alpha\mathbf{b}_j$ is a vector of the reciprocal lattice. Similarly, the opposite is also true: If α is a point symmetry transformation of the reciprocal lattice, then $\alpha\mathbf{a}_i$ is a vector of the direct lattice.

5.3 Translation Symmetry and Its Consequences

The translation group T is an *invariant* subgroup of a space group G.[3] Indeed, for every $\{\alpha|\mathbf{t}\}$ in the space group, according to (5.2) and (5.3), we have

$$\{\alpha|\mathbf{t}\}\{\varepsilon|\mathbf{R}_n\}\{\alpha|\mathbf{t}\}^{-1} = \{\alpha|\mathbf{t}\}\{\varepsilon|\mathbf{R}_n\}\{\alpha^{-1}| - \alpha^{-1}\mathbf{t}\} = \{\varepsilon|\alpha\mathbf{R}_n\}. \tag{5.11}$$

In order to use the representation theory of finite groups, we have to make the translation group (and the space group!) finite. For this, we assume that a crystal is of finite extent and apply the (Born–Karman) cyclic boundary conditions [11]

$$\{\varepsilon|\mathbf{t}_j\}^N = \{\varepsilon|0\}, \qquad j = 1, 2, 3, \tag{5.12}$$

where N is a sufficiently large integer (such that boundary conditions do not affect bulk properties).

As a result of the above construction, the translation group becomes a finite group of order N^3. Since all translations commute with each other, the translation subgroup T is commutative (Abelian). Hence all representations of T are one-dimensional, i.e. coincide with their characters

$$\chi^{\mathbf{k}}(\{\varepsilon|\mathbf{R}_n\}) = \mathrm{e}^{-\mathrm{i}\mathbf{k}\mathbf{R}_n}. \tag{5.13}$$

Here different representations are labeled by the *reduced wavevector* $\mathbf{k}$, i.e. the vector $\mathbf{k}$ takes N^3 discrete values in the Brillouin zone [2].

Let us see what translation symmetry gives for the representations of space groups. Every space group representation is a representation of the translation subgroup (in general, reducible). A representation of the translation subgroup can be presented as a direct sum of irreducible representations with the vectors $\mathbf{k}_1, \ldots, \mathbf{k}_s$ (each vector can appear more than once). Therefore, in each space group representation the matrices of the translation elements are diagonal:

$$D(\{\varepsilon|\mathbf{R}_n\}) = \begin{pmatrix} \mathrm{e}^{-\mathrm{i}\mathbf{k}_1\mathbf{R}_n} & \ldots & 0 \\ \vdots & \ddots & \vdots \\ 0 & \ldots & \mathrm{e}^{-\mathrm{i}\mathbf{k}_s\mathbf{R}_n} \end{pmatrix}. \tag{5.14}$$

From the invariance of the translation group it is straightforward to prove the following result (see, e.g. [2]).

Theorem 5.1 *A necessary condition for a space group representation to be irreducible is that*

$$\{\alpha_i\mathbf{k} : \alpha_i \in G_0\} = \{\mathbf{k}_1, \ldots, \mathbf{k}_s\} \tag{5.15}$$

[3] This property is often taken as a formal definition of a space group.

for every $\mathbf{k}$ *from* $\{\mathbf{k}_1, \ldots, \mathbf{k}_s\}$. *That means, given an admissible wavevector* $\mathbf{k}$, *the set of vectors obtained by applying all rotations of the point group* $\{\alpha_i \mathbf{k} : \alpha_i \in G_0\}$ *must be equal to the set of admissible wavevectors* $\{\mathbf{k}_1, \ldots, \mathbf{k}_s\}$.

The set of different wavevectors $\{\alpha_i \mathbf{k}\}$ is called the *star of the wavevector* $\mathbf{k}$, and the set $\chi^{\alpha_i \mathbf{k}}$ is called the *orbit*. It is convenient to label the irreducible representations of a space group by the vector $\mathbf{k}$. In order to make this labeling unambiguous, the vector $\mathbf{k}$ must be chosen in an irreducible Brillouin zone (IBZ) [12] which we introduce below (Sect. 5.4).

The number of different arms of a star is called the *order of the star*. The order of a star is less or equal to $n(G_0)$, where $n(G_0)$ is the order of the point group G_0. If the order of a star is equal to $n(G_0)$, we call it a *general star*. Symmetry will "cost an arm and a leg": The more symmetric is the wavevector, the smaller is the number of arms (smaller the order of the star). For instance, the star of the most symmetric vector, $\mathbf{k} = 0$, has only one arm.

5.4 Types of the k Vectors in the Brillouin Zone

To construct the orbits of irreducible representations of the translation subgroup T with respect to the group G, we need to consider all stars of vectors $\mathbf{k}$ in the Brillouin zone. Let us analyze possible types of stars and their arrangement in the Brillouin zone.

Let $\mathbf{k}$ be a wavevector in the Brillouin zone, and $G^0_{\mathbf{k}}$ be the point group of this wavevector, i.e. $G^0_{\mathbf{k}} = \{\alpha \in G_0 : \alpha\mathbf{k} = \mathbf{k} + \mathbf{K}_\mu\}$. A vector $\mathbf{k}$ such that the space group of the wavevector $G_{\mathbf{k}}$ coincides with the translation subgroup T is called a *point in a general position* (or a general point). A point in a general position can be only an interior point of the Brillouin zone. Hence the points in a symmetrical position (i.e. not in a general position) can occur only as the boundary points of regions of the points in a general position. Since the elements of a point group are rotations or rotations with inversions, the regions of the points in a general position must be bounded by planes. Hence the $\mathbf{k}$ points in a symmetrical position having the same point groups $G^0_{\mathbf{k}}$ can form *symmetry planes*, *lines* and *points*.

If we choose one arm in each star[4] in such a way that the arms corresponding to infinitesimally close stars are infinitesimally close, then the arms that are $\mathbf{k}$ points in a general position will fill a closed region that is equal to one $n(G_0)$-th of the Brillouin zone. We call each of those regions an *irreducible Brillouin zone*, as is usual in energy band calculations (also called an *irreducible wedge* or *elementary region* of the Brillouin zone). If we fix one irreducible Brillouin zone, then all other ones will be transformations by means of the point group elements. The boundaries of irreducible Brillouin zones will partially overlap because they consist of points in a symmetrical position.

---

[4] To obtain each irreducible representation of a space group only once, it is necessary to choose one representation from each orbit.

5.5 Group of the Wavevector

In terms of Chap. 3, given a space group G the translation subgroup T is the (commutative) invariant subgroup H having a finite number of elements $h = \mathbf{R}_n$, and the characters of the translation subgroup $\chi^{\mathbf{k}}$ are the representations d^i of H. Then, according to relation (3.9), the conjugate representation d^{i_g} takes the form

$$\begin{aligned}\chi^{\mathbf{k}_{\{\alpha|\mathbf{R}_n+\tau_\alpha\}^{-1}}}(\{\varepsilon|\mathbf{R}_n\}) &= \chi^{\mathbf{k}}(\{\alpha|\mathbf{R}_n+\tau_\alpha\}^{-1}\{\varepsilon|\mathbf{R}_n\}\{\alpha|\mathbf{R}_n+\tau_\alpha\})\\ &= \chi^{\mathbf{k}}(\{\varepsilon|\alpha^{-1}\mathbf{R}_n\}) = \mathrm{e}^{-\mathrm{i}\mathbf{k}\alpha^{-1}\mathbf{R}_n} = \mathrm{e}^{-\mathrm{i}\alpha\mathbf{k}\mathbf{R}_n} = \chi^{\alpha\mathbf{k}}(\{\varepsilon|\mathbf{R}_n\}),\end{aligned} \tag{5.16}$$

where $g = \{\alpha|\mathbf{R}_n + \tau_\alpha\}^{-1}$ is an arbitrary element of the space group.

Next, we obtain the normalizer $G_{\mathbf{k}}$ of the **k**th representation (the normalizer N_i of the representation d^i in Chap. 3). Following (3.10), we equate (5.13) and (5.16). Then, the element $g = \{\alpha|\mathbf{R}_n + \tau_\alpha\}^{-1}$ belongs to the normalizer if $\mathrm{e}^{-\mathrm{i}\mathbf{k}\mathbf{R}_n} = \mathrm{e}^{-\mathrm{i}\alpha\mathbf{k}\mathbf{R}_n}$ for every $\mathbf{R}_n$, or equivalently,

$$\alpha\mathbf{k} = \mathbf{k} + \mathbf{K}_\mu, \tag{5.17}$$

where $\mathbf{K}_\mu$ is the reciprocal lattice vector. Hence the normalizer $G_{\mathbf{k}}$ consists of all elements of the space group with rotational parts α_m leaving the wavevector **k** invariant up to a reciprocal lattice vector. Clearly, $G_{\mathbf{k}}$ contains all translations, $T \subset G_{\mathbf{k}} \subset G$. The normalizer $G_{\mathbf{k}}$ is also called the group of the wavevector.[5]

For elements of the space group $\{\alpha_M|\tau_{\alpha_M}\}$ that do *not* belong to $G_{\mathbf{k}}$ we have

$$\alpha_M\mathbf{k} \neq \mathbf{k} + \mathbf{K}_\mu. \tag{5.18}$$

The set of different representations $\chi^{\alpha_M\mathbf{k}}$ with $\alpha_M\mathbf{k}$ satisfying the above condition is the *orbit of the wavevector* **k** (or any other vector $\alpha_M\mathbf{k}$). The corresponding set of wavevectors $\alpha_M\mathbf{k}$, including the vector **k** itself, is the *star* of the representation, or *the star of the wavevector* **k** (or any other vector $\alpha_M\mathbf{k}$).

According to (3.12) and (3.13), we have the following decompositions[6]:

$$G_{\mathbf{k}} = \bigcup_m T\{\alpha_m|\tau_{\alpha_m}\}, \qquad m = 1, 2, \ldots, n(G_{\mathbf{k}})/n(T), \tag{5.19}$$

$$G = \bigcup_M G_{\mathbf{k}}\{\alpha_M|\tau_{\alpha_M}\}, \qquad M = 1, 2, \ldots, n(G)/n(G_{\mathbf{k}}). \tag{5.20}$$

[5] Sometimes called the *little group* (see, e.g. [13]).

[6] Strictly speaking, the representatives α_m in decomposition (5.19) should have been denoted by $\alpha_m^{\mathbf{k}}$ to avoid confusion with the representatives in decomposition (5.6) (just as we did in Chap. 3). But, for brevity, we omit the superscript **k**. In fact, the representatives in decomposition (5.6) will never appear in the following text.

Here all representatives α_m in the coset decomposition of $G_{\mathbf{k}}$ satisfy relation (5.17), and all representatives α_M in the coset decomposition of G with respect to $G_{\mathbf{k}}$ satisfy the opposite relation (5.18). We always reserve the index m to refer to one of the coset representatives in (5.19) and index M to refer to one of the coset representatives in (5.20). Clearly $\{\alpha_M|\boldsymbol{\tau}_{\alpha_M}\}$ is an element of G but *not* of $G_{\mathbf{k}}$. This point is obvious, but will be important in the next section.

5.6 Small Representations of the Group of the Wavevector

As it follows from the induced representation theory, to construct all irreducible representations of a space group, it is necessary to choose one vector from each star. It is convenient to choose the wavevectors $\mathbf{k}$ from the irreducible part of the Brillouin zone.

According to decomposition (5.19), every element of the group $G_{\mathbf{k}}$ has the form $\{\varepsilon|\mathbf{R}_n\}\{\alpha_m|\boldsymbol{\tau}_{\alpha_m}\}$. Using relations (3.57) and (5.13), we write the matrices of the small representations of the group $G_{\mathbf{k}}$ as

$$d^{\mathbf{k}j}(\{\varepsilon|\mathbf{R}_n\}\{\alpha_m|\boldsymbol{\tau}_{\alpha_m}\}) = \mathrm{e}^{-\mathrm{i}\mathbf{k}\mathbf{R}_n}\,\Delta^{\mathbf{k}j}(\{\alpha_m|\boldsymbol{\tau}_{\alpha_m}\}). \tag{5.21}$$

Clearly, for translation elements the matrices of the small representations are scalar,

$$d^{\mathbf{k}j}(\{\varepsilon|\mathbf{R}_n\}) = \mathrm{e}^{-\mathrm{i}\mathbf{k}\mathbf{R}_n}\,E\,, \tag{5.22}$$

which is a characteristic property of the small representations. The representations $d^{\mathbf{k}j}$, of course, do not exhaust all irreducible representations of the group $G_{\mathbf{k}}$ (see (3.46)).

The problem of obtaining small representations of the group $G_{\mathbf{k}}$ is thus reduced to the problem of obtaining irreducible projective representations $\Delta^{\mathbf{k}j}$ of the factor group $G_{\mathbf{k}}/T$. It is easy to check that $G_{\mathbf{k}}/T$ is isomorphic to a point group $G^0_{\mathbf{k}}$, which consists of rotational parts α_m of the representatives in the decomposition (5.19) of the group $G_{\mathbf{k}}$ and is a subgroup of the crystal class. Generally speaking, to construct the small representations of the groups $G_{\mathbf{k}}$ of all wavevectors $\mathbf{k}$ and all space groups it is necessary to construct irreducible projective representations of all 32 crystallographic point groups. Note that all these (point) groups possess the property that a consecutive application of the induced representation theory and projective representation theory reduces the problem of constructing their projective representations to an analogous problem for commutative groups, for which it is solved elementary [1].

The irreducible projective representations $\Delta^{\mathbf{k}j}$ of the group $G^0_{\mathbf{k}}$, according to (3.54) and (3.55), correspond to the factor system

$$\omega^{\mathbf{k}}_{\{\alpha_m|\boldsymbol{\tau}_{\alpha_m}\},\{\alpha_{m'}|\boldsymbol{\tau}_{\alpha_{m'}}\}} = \chi^{\mathbf{k}}(\{\varepsilon|\mathbf{R}_{mm'}\}) = \mathrm{e}^{-\mathrm{i}\mathbf{k}\mathbf{R}_{mm'}}\,, \tag{5.23}$$

where $\mathbf{R}_{mm'}$ is determined from the relation

$$\{\alpha_m|\boldsymbol{\tau}_{\alpha_m}\}\{\alpha_{m'}|\boldsymbol{\tau}_{\alpha_{m'}}\} = \{\varepsilon|\mathbf{R}_{mm'}\}\{\alpha_m\alpha_{m'}|\boldsymbol{\tau}_{\alpha_m\alpha_{m'}}\}, \tag{5.24}$$

and is equal to

$$\mathbf{R}_{mm'} = \alpha_m\boldsymbol{\tau}_{\alpha_{m'}} + \boldsymbol{\tau}_{\alpha_m} - \boldsymbol{\tau}_{\alpha_m\alpha_{m'}}. \tag{5.25}$$

If the space group is symmorphic, it is necessary to construct ordinary (vector) representations, because in this case there is no nonprimitive translations, $\boldsymbol{\tau}_\alpha = 0$, the representatives $\{\alpha_m|0\}$ form a subgroup of the group $G_{\mathbf{k}}$, and all projective factors (5.23) are equal to 1.

If the space group is nonsymmorphic but wavevector $\mathbf{k}$ is *inside* the Brillouin zone, the problem can be simplified (see Sect. 4.1). Indeed, by the definition of the Brillouin zone, two internal points cannot differ by a reciprocal lattice vector, hence the relation (5.17) becomes

$$\alpha_m\mathbf{k} = \mathbf{k}. \tag{5.26}$$

Substituting (5.25) in (5.23), we have

$$\omega^{\mathbf{k}}_{\{\alpha_m|\boldsymbol{\tau}_{\alpha_m}\},\{\alpha_{m'}|\boldsymbol{\tau}_{\alpha_{m'}}\}} = \mathrm{e}^{-\mathrm{i}\mathbf{k}(\alpha_m\boldsymbol{\tau}_{\alpha_{m'}}+\boldsymbol{\tau}_{\alpha_m}-\boldsymbol{\tau}_{\alpha_m\alpha_{m'}})} = \frac{\mathrm{e}^{-\mathrm{i}\mathbf{k}\boldsymbol{\tau}_{\alpha_m}}\mathrm{e}^{-\mathrm{i}\mathbf{k}\boldsymbol{\tau}_{\alpha_{m'}}}}{\mathrm{e}^{-\mathrm{i}\mathbf{k}\boldsymbol{\tau}_{\alpha_m\alpha_{m'}}}}. \tag{5.27}$$

Then the new matrices

$$\Delta'^{\mathbf{k}j}(\{\alpha_m|\boldsymbol{\tau}_{\alpha_m}\}) = \frac{1}{\mathrm{e}^{-\mathrm{i}\mathbf{k}\boldsymbol{\tau}_{\alpha_m}}}\Delta^{\mathbf{k}j}(\{\alpha_m|\boldsymbol{\tau}_{\alpha_m}\}) \tag{5.28}$$

form a vector representation of the point group $G^0_{\mathbf{k}}$. From this it immediately follows that the matrices of the projective representation $\Delta^{\mathbf{k}j}$ can be obtained from the matrices of the vector representation $\Delta'^{\mathbf{k}j}$ by means of the gauge transformation with the gauges $\mathrm{e}^{-\mathrm{i}\mathbf{k}\boldsymbol{\tau}_{\alpha_m}}$.

If the space group is nonsymmorphic and vector $\mathbf{k}$ is at the boundary of the Brillouin zone, the problem cannot be simplified and, as we mentioned above, it reduces to constructing projective representations of the corresponding point group $G^0_{\mathbf{k}}$.

Sometimes it is necessary to construct small representations of different vectors of a star, based on the known representations of one of them, or to establish a correspondence between small representations of different vectors of a star, which are obtained independently. Using relations (3.47), (3.48) and (3.57), we have

$$\begin{aligned} d^{\mathbf{k}j}(\{\alpha_m|\boldsymbol{\tau}_{\alpha_m}\}) &\overset{(3.57)}{=} \Delta^{\mathbf{k}j}(\{\alpha_m|\boldsymbol{\tau}_{\alpha_m}\}) \\ &\overset{(3.48)}{=} \Delta^{\alpha_M\mathbf{k},j}(\{\alpha_M|\boldsymbol{\tau}_{\alpha_M}\}\{\alpha_m|\boldsymbol{\tau}_{\alpha_m}\}\{\alpha_M|\boldsymbol{\tau}_{\alpha_M}\}^{-1}) \\ &\overset{(3.47),\,(3.48)}{=} d^{\alpha_M\mathbf{k},j}(\{\alpha_M|\boldsymbol{\tau}_{\alpha_M}\}\{\alpha_m|\boldsymbol{\tau}_{\alpha_m}\}\{\alpha_M|\boldsymbol{\tau}_{\alpha_M}\}^{-1}). \end{aligned}$$

From relations (5.2), (5.3) and (5.21) it follows that

$$d^{\alpha_M \mathbf{k},j}(\{\alpha_M|\boldsymbol{\tau}_{\alpha_M}\}\{\alpha_m|\boldsymbol{\tau}_{\alpha_m}\}\{\alpha_M|\boldsymbol{\tau}_{\alpha_M}\}^{-1})$$
$$= \mathrm{e}^{-\mathrm{i}\alpha_M\mathbf{k}(-\boldsymbol{\tau}_{\alpha_M\alpha_m\alpha_M^{-1}} - \alpha_M\alpha_m\alpha_M^{-1}\boldsymbol{\tau}_{\alpha_M} + \alpha_M\boldsymbol{\tau}_{\alpha_m} + \boldsymbol{\tau}_{\alpha_M})}\, d^{\alpha_M\mathbf{k},j}(\{\alpha_M\alpha_m\alpha_M^{-1}|\boldsymbol{\tau}_{\alpha_M\alpha_m\alpha_M^{-1}}\})$$
$$\equiv \mathrm{e}^{-\mathrm{i}\alpha_M\mathbf{k}\mathbf{R}_{mM}}\, d^{\alpha_M\mathbf{k},j}(\{\alpha_M\alpha_m\alpha_M^{-1}|\boldsymbol{\tau}_{\alpha_M\alpha_m\alpha_M^{-1}}\}).$$

Combining the results, we obtain

$$d^{\mathbf{k}j}(\{\alpha_m|\boldsymbol{\tau}_{\alpha_m}\}) = \mathrm{e}^{-\mathrm{i}\alpha_M\mathbf{k}\mathbf{R}_{mM}}\, d^{\alpha_M\mathbf{k},j}(\{\alpha_M\alpha_m\alpha_M^{-1}|\boldsymbol{\tau}_{\alpha_M\alpha_m\alpha_M^{-1}}\}), \tag{5.29}$$

where

$$\mathbf{R}_{mM} = -\boldsymbol{\tau}_{\alpha_M\alpha_m\alpha_M^{-1}} - \alpha_M\alpha_m\alpha_M^{-1}\boldsymbol{\tau}_{\alpha_M} + \alpha_M\boldsymbol{\tau}_{\alpha_m} + \boldsymbol{\tau}_{\alpha_M}. \tag{5.30}$$

5.7 Irreducible Representations of Space Groups

We utilize the coset decomposition (5.20) of the space group G with respect to a group of the wavevector $G_{\mathbf{k}}$. Using relation (3.58), we write the matrix of an irreducible representation of the space group as

$$D^{\mathbf{k}j}_{L\lambda,M\mu}(\{\varepsilon|\mathbf{R}_n\}\{\alpha|\boldsymbol{\tau}_\alpha\}) = d^{\mathbf{k}j}_{\lambda\mu}(\{\alpha_L|\boldsymbol{\tau}_{\alpha_L}\}\{\varepsilon|\mathbf{R}_n\}\{\alpha|\boldsymbol{\tau}_\alpha\}\{\alpha_M|\boldsymbol{\tau}_{\alpha_M}\}^{-1})$$
$$\times\, \delta_{G_{\mathbf{k}}\{\alpha_L|\boldsymbol{\tau}_{\alpha_L}\}\{\varepsilon|\mathbf{R}_n\}\{\alpha|\boldsymbol{\tau}_\alpha\}\{\alpha_M|\boldsymbol{\tau}_{\alpha_M}\}^{-1},G_{\mathbf{k}}}, \tag{5.31}$$

where $d^{\mathbf{k}j}_{\lambda\mu}$ is the small representation (5.21) of $G_{\mathbf{k}}$. From formula (5.31) we see that $D^{\mathbf{k}j}$ can be viewed as a block-matrix of dimension $n(G)/n(G_{\mathbf{k}})$, where each block is a matrix of the small representation $d^{\mathbf{k}j}$ of dimension l_j; thus the total dimension $l_{\mathbf{k}j}$ of the matrix $D^{\mathbf{k}j}$ is equal to

$$l_{\mathbf{k}j} = \frac{n(G)\,l_j}{n(G_{\mathbf{k}})}. \tag{5.32}$$

Each row (column) of the block-matrix has only one nonzero block, i.e. a matrix of the small representation $d^{\mathbf{k}j}$. Since we agreed that the first representative in the coset decomposition (5.20) is always the identity element, from (5.31) we obtain

$$D^{\mathbf{k}j}_{1\lambda,1\mu}(\{\varepsilon|\mathbf{R}_n\}\{\alpha|\boldsymbol{\tau}_\alpha\}) = d^{\mathbf{k}j}_{\lambda\mu}(\{\varepsilon|\mathbf{R}_n\}\{\alpha|\boldsymbol{\tau}_\alpha\})\,\delta_{G_{\mathbf{k}}\{\varepsilon|\mathbf{R}_n\}\{\alpha|\boldsymbol{\tau}_\alpha\},G_{\mathbf{k}}}, \tag{5.33}$$

and this element is nonzero only for elements of the group $G_{\mathbf{k}}$.

Let us consider the representation matrices (5.31) for translations. Analogously to the above, we obtain that only the diagonal elements are nonzero. Indeed,

$$D^{\mathbf{k}j}_{L\lambda,M\mu}(\{\varepsilon|\mathbf{R}_n\}) = d^{\mathbf{k}j}_{\lambda\mu}(\{\alpha_L|\boldsymbol{\tau}_{\alpha_L}\}\{\varepsilon|\mathbf{R}_n\}\{\alpha_M|\boldsymbol{\tau}_{\alpha_M}\}^{-1})\delta_{G_{\mathbf{k}}\{\alpha_L|\boldsymbol{\tau}_{\alpha_L}\}\{\varepsilon|\mathbf{R}_n\}\{\alpha_M|\boldsymbol{\tau}_{\alpha_M}\}^{-1},G_{\mathbf{k}}}.$$

Using (5.2) and (5.3), we have

$$\{\alpha_L|\boldsymbol{\tau}_{\alpha_L}\}\{\varepsilon|\mathbf{R}_n\}\{\alpha_M|\boldsymbol{\tau}_{\alpha_M}\}^{-1} = \{\alpha_L\alpha_M^{-1}| - \alpha_L\alpha_M^{-1}\boldsymbol{\tau}_{\alpha_M} + \alpha_L\mathbf{R}_n + \boldsymbol{\tau}_{\alpha_L}\}.$$

The condition $\{\alpha_L|\boldsymbol{\tau}_{\alpha_L}\}\{\varepsilon|\mathbf{R}_n\}\{\alpha_M|\boldsymbol{\tau}_{\alpha_M}\}^{-1} \in G_{\mathbf{k}}$ implies $\alpha_L\alpha_M^{-1}\mathbf{k} = \mathbf{k} + \mathbf{K}_\mu$. Hence $\alpha_L\alpha_M^{-1} = \varepsilon$. Then $\alpha_L = \alpha_M$ and so $\boldsymbol{\tau}_{\alpha_L} = \boldsymbol{\tau}_{\alpha_M}$. In this case, we have

$$\{\alpha_L|\boldsymbol{\tau}_{\alpha_L}\}\{\varepsilon|\mathbf{R}_n\}\{\alpha_M|\boldsymbol{\tau}_{\alpha_M}\}^{-1} = \{\varepsilon|\alpha_L\mathbf{R}_n\},$$

and hence

$$\begin{aligned} D^{\mathbf{k}j}_{L\lambda,M\mu}(\{\varepsilon|\mathbf{R}_n\}) &= d^{\mathbf{k}j}_{\lambda\mu}(\{\varepsilon|\alpha_L\mathbf{R}_n\})\delta_{LM} \\ &= \mathrm{e}^{-\mathrm{i}\mathbf{k}\alpha_L\mathbf{R}_n}\,\delta_{LM}\,\delta_{\lambda\mu} = \mathrm{e}^{-\mathrm{i}\alpha_L^{-1}\mathbf{k}\mathbf{R}_n}\,\delta_{LM}\,\delta_{\lambda\mu}. \end{aligned} \tag{5.34}$$

Due to (5.18), the vector $\alpha_L^{-1}\mathbf{k}$ for $L \neq 1$ is one of the vectors in the star of **k**, different from **k**. Thus, the matrices of the irreducible representations for translations are diagonal matrices; the diagonal includes all vectors of the star, and the number of appearance of each vector is equal to the dimension of the small representation. The above facts are characteristic to irreducible representations of the space groups and usually form the basis of any discussion of the space group representations.

Relation between the general representation theory presented in Chap. 3 and its application to space groups is summarized in Table 5.1.

Table 5.1 Relation between the general representation theory and its application to space groups

General representation theory		Application to space groups	
Finite group	G	Space group	G
Invariant subgroup	H	Translation subgroup	T
Representation of H	d^i	Representation of T	$\chi^{\mathbf{k}}$
Normalizer of d^i	N_i	Group of the wavevector **k**	$G_{\mathbf{k}}$
Elements of the factor group N_i/H	Hg^i_m	Elements of the factor group $G_{\mathbf{k}}/T$	$T\{\alpha_m|\boldsymbol{\tau}_{\alpha_m}\}$
Conjugate representation of d^i	d^{i_g}	Conjugate representation of $\chi^{\mathbf{k}}$	$\chi^{\alpha_m\mathbf{k}}$
Elements of the factor group G/N_i	$N_i g^i_M$	Elements of the factor group $G/G_{\mathbf{k}}$	$G_{\mathbf{k}}\{\alpha_M|\boldsymbol{\tau}_{\alpha_M}\}$
Star of the small representation d^i	$\{i_g\}$	Star of the wavevector **k**	$\{\alpha_M\mathbf{k}\}$
Small representation of N^i	d^{ij}	Small representation of $G_{\mathbf{k}}$	$d^{\mathbf{k}j}$
Projective representation of the factor group N_i/H	Δ^{ij}	Projective representation of the factor group $G_{\mathbf{k}}/T$	$\Delta^{\mathbf{k}j}$
Regular continuation of d^{ij}	$\mathcal{D}^{ij}$	Regular continuation of $d^{\mathbf{k}j}$	$\mathcal{D}^{\mathbf{k}j}$

According to Sect. 3.8, irreducible representations of the space groups have the following properties:

1. the regular continuation (5.31) of a small representation (5.21) from a group of the wavevector to the space group is irreducible;
2. the regular continuations (5.31) of equivalent small representations from groups of the wavevectors in the same star to the space group are equivalent;
3. the regular continuations (5.31) of nonequivalent small representations from a group of the wavevector to the space group are nonequivalent;
4. the system of representations obtained by regular continuation of all small representations of groups of the wavevector taken one from each stars is complete.

Note that, once we know the structure of matrices of irreducible representations of space groups (5.31), we can express all the properties of these representations in terms of certain properties of the small representations of the groups of the wavevectors. Therefore, in a majority of problems related to space groups, it is sufficient to obtain only small representations of the groups of the wavevectors.

5.8 Compatibility Relations

It is often of interest to explore the relation between small representations $d^{\mathbf{k}i}$ and $d^{\mathbf{k}+\boldsymbol{\kappa},j}$ corresponding to close vectors $\mathbf{k}$ and $\mathbf{k}+\boldsymbol{\kappa}$ in the Brillouin zone. In particular, these *compatibility relations* can be used to analyze the nature (but not the value) of the energy bands splitting solely from the symmetry considerations, as it was first demonstrated by Bouckaert et al. [14].

If $\mathbf{k}$ is a general point, then all neighboring points are also general, the groups of the wavevectors $G_{\mathbf{k}}$ and $G_{\mathbf{k}+\boldsymbol{\kappa}}$ are equal to the translation subgroup T, and the small representations are one-dimensional

$$d^{\mathbf{k}+\boldsymbol{\kappa},j}(\{\varepsilon|\mathbf{R}_n\}) = \mathrm{e}^{-\mathrm{i}(\mathbf{k}+\boldsymbol{\kappa})\mathbf{R}_n}.$$

As $\boldsymbol{\kappa} \to 0$, we obtain

$$\lim_{\boldsymbol{\kappa}\to 0} d^{\mathbf{k}+\boldsymbol{\kappa},j}(\{\varepsilon|\mathbf{R}_n\}) = d^{\mathbf{k}j}(\{\varepsilon|\mathbf{R}_n\}) = \mathrm{e}^{-\mathrm{i}\mathbf{k}\mathbf{R}_n}$$

for any direction $\boldsymbol{\kappa}$. No compatibility relations appear in this case.

Next, let $\mathbf{k}$ be a high symmetry point (belongs to a symmetry axis or a symmetry plane). Then the neighboring point $\mathbf{k}+\boldsymbol{\kappa}$ either has the same symmetry or lower symmetry; i.e. it is sufficient to consider two cases:

(a) $G_{\mathbf{k}+\boldsymbol{\kappa}}$ is isomorphic to $G_{\mathbf{k}}$;
(b) $G_{\mathbf{k}+\boldsymbol{\kappa}}$ is a subgroup of $G_{\mathbf{k}}$.

In case (a) all small representations of $G_{\mathbf{k}}$ and $G_{\mathbf{k}+\boldsymbol{\kappa}}$ can be put into one-to-one correspondence, and each of $d^{\mathbf{k}+\boldsymbol{\kappa},j}$ tends to a certain $d^{\mathbf{k},i}$ as $\boldsymbol{\kappa} \to 0$. The situation is similar to the one for a general point: the degeneracy at $\mathbf{k}$ is not removed by moving along $\boldsymbol{\kappa}$. Case (a) occurs when $\mathbf{k}+\boldsymbol{\kappa}$ belongs to the same symmetry line (or plane) as $\mathbf{k}$.

Case (b) occurs when $\boldsymbol{\kappa}$ shifts the vector $\mathbf{k}$ off the symmetry line (or plane). It is this case that leads to the band splitting in solids.[7] In case (b), the restriction of the representation $d^{\mathbf{k},i}$ to the subgroup $G_{\mathbf{k}+\boldsymbol{\kappa}}$ of the group $G_{\mathbf{k}}$ (sometimes called subduction) will be a reducible representation of $G_{\mathbf{k}+\boldsymbol{\kappa}}$. The problem then is to determine which representations $d^{\mathbf{k}+\boldsymbol{\kappa},j}$ of $G_{\mathbf{k}+\boldsymbol{\kappa}}$ occur in the restriction of $d^{\mathbf{k},i}$ to $G_{\mathbf{k}+\boldsymbol{\kappa}}$.

In order to derive compatibility relations, we write the decomposition of each $d^{\mathbf{k},i}$ in terms of the irreducible representations $d^{\mathbf{k}+\boldsymbol{\kappa},j}$ of $G_{\mathbf{k}+\boldsymbol{\kappa}}$ for each element g of $G_{\mathbf{k}+\boldsymbol{\kappa}}$:

$$d^{\mathbf{k},i}(g)=\sum_j a_{i,j}^{\mathbf{k},\mathbf{k}+\boldsymbol{\kappa}}\,d^{\mathbf{k}+\boldsymbol{\kappa},j}(g).$$

Calculating trace of both sides, we express the character as

$$\chi^{\mathbf{k},i}(g)=\sum_j a_{i,j}^{\mathbf{k},\mathbf{k}+\boldsymbol{\kappa}}\,\chi^{\mathbf{k}+\boldsymbol{\kappa},j}(g).$$

Using the orthogonality relations for characters of irreducible representations (Sect. 2.3.5), we obtain the formula for the coefficients

$$a_{i,j}^{\mathbf{k},\mathbf{k}+\boldsymbol{\kappa}}=\frac{1}{n(G_{\mathbf{k}+\boldsymbol{\kappa}})}\sum_{g\in G_{\mathbf{k}+\boldsymbol{\kappa}}}\chi^{\mathbf{k},i}(g)\,\chi^{\mathbf{k}+\boldsymbol{\kappa},j*}(g) \tag{5.35}$$

(compare with formula (2.13)). To simplify this expression, we recall that an arbitrary element of $G_{\mathbf{k}+\boldsymbol{\kappa}}$ can be written as $g=\{\varepsilon|\mathbf{R}_n\}\{\alpha|\boldsymbol{\tau}_\alpha\}$. Taking (5.13) into account, we have

$$\chi^{\mathbf{k},i}(\{\varepsilon|\mathbf{R}_n\}\{\alpha|\boldsymbol{\tau}_\alpha\})=\mathrm{e}^{-\mathrm{i}\mathbf{k}\mathbf{R}_n}\,\chi^{\mathbf{k},i}(\{\alpha|\boldsymbol{\tau}_\alpha\}),$$
$$\chi^{\mathbf{k}+\boldsymbol{\kappa},j}(\{\varepsilon|\mathbf{R}_n\}\{\alpha|\boldsymbol{\tau}_\alpha\})=\mathrm{e}^{-\mathrm{i}(\mathbf{k}+\boldsymbol{\kappa})\mathbf{R}_n}\,\chi^{\mathbf{k}+\boldsymbol{\kappa},j}(\{\alpha|\boldsymbol{\tau}_\alpha\}).$$

Substituting these expressions into (5.35), we obtain

$$a_{i,j}^{\mathbf{k},\mathbf{k}+\boldsymbol{\kappa}}=\frac{1}{n(G_{\mathbf{k}+\boldsymbol{\kappa}})}\sum_{\alpha\in G^0_{\mathbf{k}+\boldsymbol{\kappa}}}\chi^{\mathbf{k},i}(\{\alpha|\boldsymbol{\tau}_\alpha\})\chi^{\mathbf{k}+\boldsymbol{\kappa},j*}(\{\alpha|\boldsymbol{\tau}_\alpha\})\sum_{\mathbf{R}_n}\mathrm{e}^{\mathrm{i}\boldsymbol{\kappa}\mathbf{R}_n}.$$

Using the identity

$$\sum_{\mathbf{R}_n}\mathrm{e}^{\mathrm{i}\boldsymbol{\kappa}\mathbf{R}_n}=n(T)\,\delta_{\boldsymbol{\kappa}0}$$

[7] This situation is similar to the splitting of eigenvalues caused by a perturbation lowering symmetry from a group G to its subgroup G' (see Sect. 7.3).

and formula $n(G^0_{\mathbf{k}+\boldsymbol{\kappa}}) = n(G_{\mathbf{k}+\boldsymbol{\kappa}})/n(T)$ for the order of the factor group $G_{\mathbf{k}+\boldsymbol{\kappa}}/T$ isomorphic to $G^0_{\mathbf{k}+\boldsymbol{\kappa}}$, we have

$$\lim_{\boldsymbol{\kappa}\to 0} a^{\mathbf{k},\mathbf{k}+\boldsymbol{\kappa}}_{i,j} = \lim_{\boldsymbol{\kappa}\to 0} \frac{1}{n(G^0_{\mathbf{k}+\boldsymbol{\kappa}})} \sum_{\alpha\in G^0_{\mathbf{k}+\boldsymbol{\kappa}}} \chi^{\mathbf{k},i}(\{\alpha|\boldsymbol{\tau}_\alpha\})\chi^{\mathbf{k}+\boldsymbol{\kappa},j*}(\{\alpha|\boldsymbol{\tau}_\alpha\}). \tag{5.36}$$

Since $G_{\mathbf{k}+\boldsymbol{\kappa}}$ and $\chi^{\mathbf{k}+\boldsymbol{\kappa},j}$ do not change along the direction of $\boldsymbol{\kappa}$, we can write the final form of the multiplicity coefficient as

$$a^{\mathbf{k},\mathbf{k}+\boldsymbol{\kappa}}_{i,j} = \frac{1}{n(G^0_{\mathbf{k}+\boldsymbol{\kappa}})} \sum_{\alpha\in G^0_{\mathbf{k}+\boldsymbol{\kappa}}} \chi^{\mathbf{k},i}(\{\alpha|\boldsymbol{\tau}_\alpha\})\chi^{\mathbf{k}+\boldsymbol{\kappa},j*}(\{\alpha|\boldsymbol{\tau}_\alpha\}). \tag{5.37}$$

The compatibility tables for symmorphic space groups are presented in Bouckaert et al. [14]. We will illustrate the use of the compatibility relations by a concrete example of the nonsymmorphic group O^3_h in Sect. 12.3.

References

1. F. Seitz, Ann. Math. **37**, 17 (1936)
2. G.F. Koster, Solid State Phys. **5**, 173 (1957)
3. H.W. Streitwolf, *Group Theory in Solid State Physics* (MacDonald, London, 1971)
4. F. Seitz, Z. Kristallogr. **88**, 433 (1934)
5. F. Seitz, Z. Kristallogr. **90**, 289 (1935)
6. F. Seitz, Z. Kristallogr. **91**, 336 (1935)
7. F. Seitz, Z. Kristallogr. **94**, 100 (1936)
8. J.W. Gibbs, E.B. Wilson, *Vector Analysis: A Textbook for the Use of Students of Mathematics and Physics, Founded upon the Lectures of JWG* (Yale University Press, New Haven, CT, 1901)
9. P.P. Ewald, Phys. Z. **14**, 465 (1913)
10. M. Fathi, Powder Diffr. **34**, 260 (2019)
11. M. Born, K. Huang, *Dynamical Theory of Crystal Lattices* (Oxford University Press, Oxford, 1954)
12. B.I. Reser, phys. stat. sol. (b) **116**, 31 (1983)
13. C.J. Bradley, A.P. Cracknell, *The Mathematical Theory of Symmetry in Solids. Representation Theory for Point Groups and Space Groups* (Clarendon, Oxford, 1972). Reprinted: Clarendon, Oxford, 2010
14. L.P. Bouckaert, R. Smoluchowski, E. Wigner, Phys. Rev. **50**, 58 (1936)

Chapter 6
Tables

...here 99% of the work is in implementation of the idea.
(A.D. Sakharov, Memoir, Knopf, New York, 1990)

This chapter summarizes the theoretical material of the previous chapters, presents compact tables of irreducible projective representations of 32 crystallographic point groups, which are used to obtain all irreducible representations, including double-valued, of all 230 space groups, and explains how to use this tables in practice [1]. This chapter is self-contained and can be used quite independently.

6.1 Summary of Theory and Guide to Tables

6.1.1 Summary of Theory

6.1.1.1 Representations of the Space Groups

An irreducible representation[1] $\mathcal{D}^{\mathbf{k}j}$ of a space group G is expressed in terms of the small representation $d^{\mathbf{k}j}$ of the group of the wavevector $G_{\mathbf{k}}$:

$$D^{\mathbf{k}j}_{L\lambda,M\mu}(\{\varepsilon|\mathbf{R}_n\}\{\alpha|\boldsymbol{\tau}_\alpha\}) = d^{\mathbf{k}j}_{\lambda\mu}(\{\alpha_L|\boldsymbol{\tau}_{\alpha_L}\}\{\varepsilon|\mathbf{R}_n\}\{\alpha|\boldsymbol{\tau}_\alpha\}\{\alpha_M|\boldsymbol{\tau}_{\alpha_M}\}^{-1}) \times \delta_{G_{\mathbf{k}}\{\alpha_L|\boldsymbol{\tau}_{\alpha_L}\}\{\varepsilon|\mathbf{R}_n\}\{\alpha|\boldsymbol{\tau}_\alpha\}\{\alpha_M|\boldsymbol{\tau}_{\alpha_M}\}^{-1},G_{\mathbf{k}}}\,, \qquad (6.1)$$

where $\{\alpha_L|\boldsymbol{\tau}_{\alpha_L}\}$ and $\{\alpha_M|\boldsymbol{\tau}_{\alpha_M}\}$ are the representatives in the coset decomposition of G with respect to $G_{\mathbf{k}}$:

[1] In this chapter we always consider *irreducible* representations, therefore the word “irreducible” will be omitted for brevity.

N. B. Melnikov and B. I. Reser, *Space Group Representations*,
https://doi.org/10.1007/978-3-031-13991-8_6

$$G = \sum_M G_{\mathbf{k}}\{\alpha_M|\boldsymbol{\tau}_{\alpha_M}\}, \qquad M = 1, 2, \ldots, n(G)/n(G_{\mathbf{k}}). \tag{6.2}$$

The small representation $d^{\mathbf{k}j}$ itself is expressed in terms of the projective representation $\Delta^{\mathbf{k}j}$ of the factor group $G_{\mathbf{k}}/T$ of the group $G_{\mathbf{k}}$ by the translation group T:

$$d^{\mathbf{k}j}(\{\varepsilon|\mathbf{R}_n\}\{\alpha_m|\boldsymbol{\tau}_{\alpha_m}\}) = e^{-i\mathbf{k}\mathbf{R}_n}\Delta^{\mathbf{k}j}(\{\alpha_m|\boldsymbol{\tau}_{\alpha_m}\}), \tag{6.3}$$

where $\{\alpha_m|\boldsymbol{\tau}_{\alpha_m}\}$ is the representative in the decomposition of $G_{\mathbf{k}}$ with respect to the translation group T:

$$G_{\mathbf{k}} = \sum_m T\{\alpha_m|\boldsymbol{\tau}_{\alpha_m}\}, \qquad m = 1, 2, \ldots, n(G_{\mathbf{k}})/n(T). \tag{6.4}$$

Thus, the problem of constructing the representation $\mathcal{D}^{\mathbf{k}j}$ of the space group G is reduced to constructing the projective representations $\Delta^{\mathbf{k}j}$ of the factor group $G_{\mathbf{k}}/T$ or the isomorphic point group $G^0_{\mathbf{k}}$, which consists of the rotational parts α_m.

The projective representation $\Delta^{\mathbf{k}j}$ corresponds to the factor system

$$\omega^{\mathbf{k}}_{\{\alpha_m|\boldsymbol{\tau}_{\alpha_m}\},\{\alpha_{m'}|\boldsymbol{\tau}_{\alpha_{m'}}\}} = e^{-i\mathbf{k}\mathbf{R}_{mm'}}, \tag{6.5}$$

where

$$\mathbf{R}_{mm'} = \alpha_m\boldsymbol{\tau}_{\alpha_{m'}} + \boldsymbol{\tau}_{\alpha_m} - \boldsymbol{\tau}_{\alpha_m\alpha_{m'}}. \tag{6.6}$$

6.1.1.2 Projective Representations of the Point Groups

Here we describe the structure of the projective representations in more detail and in a more convenient notation than in Chap. 4.

We use the description of a group by its generators $g_1, g_2, \ldots, g_p$. In this case an element of the group is written as

$$g_1^{\mu_1} g_2^{\mu_2} \cdots g_p^{\mu_p} = \prod_{i=1}^{p} g_i^{\mu_i}, \qquad 1 \leq \mu_i \leq m_i, \tag{6.7}$$

(m_i is the order of the generator g_i), and the matrix of an element is expressed in terms of the matrices of the generators:

$$D(\prod_{i=1}^{p} g_i^{\mu_i}) = \omega^*_{\prod_{i=1}^{p} g_i^{\mu_i}} \prod_{i=1}^{p} D(g_i)^{\mu_i}, \tag{6.8}$$

where

$$\omega_{\prod_{i=1}^{p} g_i^{\mu_i}} = \prod_{i=1}^{p}\prod_{j=1}^{\mu_i-1} \omega_{g_i^j, g_i} \prod_{k=1}^{p-1} \omega_{\prod_{l=1}^{k} g_l^{\mu_l}, g_{k+1}^{\mu_{k+1}}} \tag{6.9}$$

is the product of all projective factors that appear in the product of the matrices of the generators.

The structure of the group is determined by the defining relations

$$g_i^{m_i} = e, \qquad 1 \le i \le p, \tag{6.10}$$

$$g_i g_j = \prod_{k=1}^{p} g_k^{\mu_k(i,j)}, \qquad 1 \le j < i \le p. \tag{6.11}$$

Hence, taking into account (6.8) and the fact that $D(e) = E$ (E is the unit matrix), the defining relations for the matrices of the generators of the projective representation will take the form

$$D(g_i)^{m_i} = \omega_{g_i^{m_i}} E, \qquad 1 \le i \le p, \tag{6.12}$$

$$D(g_i)D(g_j) = \omega_{ij} \prod_{k=1}^{p} D(g_k)^{\mu_k(i,j)}, \qquad 1 \le j < i \le p, \tag{6.13}$$

where

$$\omega_{ij} = \omega_{g_i, g_j} \omega^{*}_{\prod_{i=1}^{p} g_i^{\mu_i}}. \tag{6.14}$$

We call ω_{ij} the *character factors*.

Introducing new matrices of the generators $D'(g_i)$ such that[2]

$$D'(g_i) = \frac{1}{\sqrt[m_i]{\omega_{g_i^{m_i}}}} D(g_i), \tag{6.15}$$

we simplify the defining relations (6.12) and (6.13):

$$D'(g_i)^{m_i} = E, \qquad 1 \le i \le p, \tag{6.16}$$

$$D'(g_i)D'(g_j) = \omega'_{ij} \prod_{k=1}^{p} D'(g_k)^{\mu_k(i,j)}, \qquad 1 \le j < i \le p, \tag{6.17}$$

where

$$\omega'_{ij} = \omega_{ij} \prod_{k=1}^{p} \left(\sqrt[m_k]{\omega_{g_k^{m_k}}} \right)^{\mu_k} \Big/ \left(\sqrt[m_i]{\omega_{g_i^{m_i}}} \sqrt[m_j]{\omega_{g_j^{m_j}}} \right). \tag{6.18}$$

---

[2] By the mth root we mean an arbitrary but fixed value (branch) of the complex root function.

In the general case, the character factors ω'_{ij} are related to each other by equations of the type

$$\prod_{i,j} \omega'^{\,\mu_{ij}}_{ij} = 1, \tag{6.19}$$

coming from defining relations (6.16) and (6.17). These relations allow to choose a collection of independent character factors that satisfy the equations

$$\omega'^{\,m_{ij}}_{ij} = 1 \tag{6.20}$$

and to express the remaining character factors in terms of the independent ones. Note that some of the character factors can be simply equal to unity.

The transformation (6.15) is essentially a gauge transformation, which maps the initial projective representation $\mathcal{D}$ to an associated representation $\mathcal{D}'$ with a simpler factor system. Often one can further simplify the defining relations (6.17) by applying to $\mathcal{D}'$ a subsequent gauge transformation that preserves the simple form of relations (6.16).

We illustrate this approach with an example of the group T. For brevity, here and henceforth, instead of $g_1, g_2, g_3, \ldots$ we write $a, b, c, \ldots$ and instead of $D(g_1), D(g_2), D(g_3), \ldots$ we write $A, B, C, \ldots$ Accordingly, the projective factors $\omega_{g_1,g_2}, \omega_{g_2,g_3}, \ldots$ are written as $\omega_{a,b}, \omega_{b,c}, \ldots$, and the character factors $\omega_{21}, \omega_{32}, \ldots$ as $\omega_{BA}, \omega_{CB}, \ldots$ In this notation the defining relations (6.10) and (6.11) of the group T are written as

$$\begin{aligned} &a^2 = e\,, && && \\ &b^2 = e\,, && ba = ab\,, && \\ &c^3 = e\,, && ca = bc\,, && cb = abc\,, \end{aligned} \tag{6.21}$$

and relations (6.12) and (6.13), which define the projective representations of this group, are written in the form

$$\begin{aligned} &A^2 = \omega_{a^2} E\,, && && \\ &B^2 = \omega_{b^2} E\,, && BA = \omega_{BA} AB\,, && \\ &C^3 = \omega_{c^3} E\,, && CA = \omega_{CA} BC\,, && CB = \omega_{CB} ABC\,, \end{aligned} \tag{6.22}$$

where

$$\begin{aligned} &\omega_{a^2} = \omega_{a,a}\,, && && \\ &\omega_{b^2} = \omega_{b,b}\,, && \omega_{BA} = \omega_{b,a}\omega^*_{a,b}\,, && \\ &\omega_{c^3} = \omega_{c,c}\omega_{c^2,c}\,, && \omega_{CA} = \omega_{c,a}\omega^*_{b,c}\,, && \omega_{CB} = \omega_{c,b}(\omega_{a,b}\omega_{ab,c})^*\,. \end{aligned} \tag{6.23}$$

As a result of the gauge transformation (6.15) relations (6.22) become

$$
\begin{aligned}
&A'^2 = E\,, && && \\
&B'^2 = E\,, && B'A' = \omega_{B'A'} A'B'\,, && \\
&C'^3 = E\,, && C'A' = \omega_{C'A'} B'C'\,, && C'B' = \omega_{C'B'} A'B'C'\,,
\end{aligned}
\tag{6.24}
$$

where

$$
\omega_{B'A'} = \omega_{BA} \qquad \omega_{C'A'} = \omega_{CA}\sqrt{\omega_{b^2}}\Big/\sqrt{\omega_{a^2}} \qquad \omega_{C'B'} = \omega_{CB}\sqrt{\omega_{a^2}}\,. \tag{6.25}
$$

Now we obtain equations (6.19) and (6.20) for the character factors (6.25). Multiplying from the right the second relation in the second row of (6.24) by $A'B'$, the second relation in the third row by $A'C'^2$, the third relation in the third row by $B'C'^2$ and taking all relations (6.24) into account, we have

$$
\omega^2_{B'A'} = 1 \qquad \omega^2_{C'A'} = 1 \qquad \omega^2_{C'B'}\omega_{B'A'} = 1. \tag{6.26}
$$

From the third equation, taking the first one into account, we obtain

$$
\omega_{B'A'} = \omega^2_{C'B'} \tag{6.27}
$$

and

$$
\omega^4_{C'B'} = 1\,, \tag{6.28}
$$

which means that of the three character factors (6.25) only two are independent. Finally, the second gauge transformation of the matrix B:

$$
B'' = \omega_{C'A'} B', \tag{6.29}
$$

allows us to exclude the character factor from the second relation in the third row of (6.24). The final form of the defining relations for the projective representations of the group T is as follows:

$$
\begin{aligned}
&A'^2 = E\,, && && \\
&B''^2 = E\,, && B''A' = \omega^2_{C'B'} A'B''\,, && \\
&C'^3 = E\,, && C'A' = B''C'\,, && C'B'' = \omega_{C'B'} A'B''C'\,.
\end{aligned}
\tag{6.30}
$$

Thus, using relations (6.19) and gauge transformations of the initial matrices, we were able to substantially reduce the number of character factors, and this way largely simplified the problem of constructing (and tabulating!) all possible projective representations of the group. It is precisely the gauge transformed matrices of the projective representations that we present in our tables. The method of constructing the projective representations is illustrated in Appendix B in the same example of the group T.

6.1.1.3 Double-Valued Representations

Symmetry operators that describe transformation rules of two-component spinors are determined by the equality (see details in Sect. 8.3)

$$P_{[\alpha|\mathbf{t}]}\,\varphi(\mathbf{r}) = S(\alpha)\;\varphi(\{\alpha|\mathbf{t}\}^{-1}\mathbf{r}). \tag{6.31}$$

The 2×2 matrix $S(\alpha)$ that acts on the two-component spinors has the form

$$S(\alpha) = \cos\frac{\theta}{2}\cdot E - i\,\sin\frac{\theta}{2}(\boldsymbol{\sigma}\cdot\mathbf{n}). \tag{6.32}$$

Here $-\pi < \theta \leq \pi$ is the rotation angle about the axis given by the unit vector $\mathbf{n}$, the rotation is positive when it is anticlockwise as viewed down the axis towards the origin; $\boldsymbol{\sigma} = (\sigma_x, \sigma_y, \sigma_z)$ is the vector of the Pauli matrices

$$\sigma_x = \begin{pmatrix} 0 & 1 \\ 1 & 0 \end{pmatrix}, \quad \sigma_y = \begin{pmatrix} 0 & -i \\ i & 0 \end{pmatrix}, \quad \sigma_z = \begin{pmatrix} 1 & 0 \\ 0 & -1 \end{pmatrix}. \tag{6.33}$$

From the definition (6.31) we have the multiplication law for the symmetry operators

$$P_{[\alpha|\mathbf{t}]}\,P_{[\beta|\mathbf{t}']} = \omega_{\alpha,\beta}\,P_{[\alpha\beta|\alpha\mathbf{t}'+\mathbf{t}]}, \tag{6.34}$$

where $\omega_{\alpha,\beta}$ is the numerical factor taking values ± 1. One can obtain $\omega_{\alpha,\beta}$ from comparison of matrices on both sides of equality (6.34), i.e. from comparison of the matrix product $S(\alpha)\,S(\beta)$ with the matrix $S(\alpha\beta)$.

Transformation matrices of the two-component spinors belonging to the same energy level satisfy a relation analogous to relation for the symmetry operators (6.34):

$$D([\alpha|\mathbf{t}])\,D([\beta|\mathbf{t}']) = \omega_{\alpha,\beta}\,D([\alpha\beta|\alpha\mathbf{t}' + \mathbf{t}]), \tag{6.35}$$

i.e. they form the so-called double-valued representation of the space group. The problem of constructing these representations reduces to constructing projective representations of the point groups of the $\mathbf{k}$ vector for the factor system such that each numerical factor is a product of the factor $\omega_{\alpha_m,\alpha_{m'}}$ and factor (6.5):

$$\omega'^{\mathbf{k}}_{\{\alpha_m|\tau_{\alpha_m}\},\{\alpha_{m'}|\tau_{\alpha_{m'}}\}} = \omega_{\alpha_m,\alpha_{m'}}\,\omega^{\mathbf{k}}_{\{\alpha_m|\tau_{\alpha_m}\},\{\alpha_{m'}|\tau_{\alpha_{m'}}\}}. \tag{6.36}$$

Since our tables contain all possible projective representations of the point groups, they also contain the projective representations with the factor system (6.36).

6.1.2 Guide to Tables

The Tables themselves are divided into 18 sections.

In Sect. 6.2, we present the elements of the maximal point groups O_h and D_{6h}, classified by the classes of conjugate elements (classes), and their multiplication tables.[3] In describing the elements of the group D_{6h} we used rectangular coordinate system. The bar over the element $\alpha_i\alpha_j$ in the multiplication table means that the spin factor $\omega_{\alpha_i,\alpha_j}$ is equal to minus one. An explicit form of the matrices $S(\alpha)$ is given in Tables 6.3 and 6.6.[4] Note that we number the group elements differently compared to the tables of [2].

In the seventeen next sections (by the number of nonisomorphic point groups, excluding C_1), we present the following information. In the first one or two tables of each section we describe realizations of the corresponding abstract group by elements of the maximal point groups O_h and D_{6h}, respectively. In the following table we give the defining relations of the group and its projective representations. Then we describe an explicit form of character and gauge factors. Each section is concluded by the table consisting of all projective representations $\Delta^{\mathbf{k}j}$, labeled by the order numbers. In the groups with inversion, for brevity, the representations that differ only by the sign in front of the inversion are combined and labeled by the order number with the $\pm$ sign. In the last table of each section, lower-case Greek letters denote the following:

$$\varepsilon = e^{i2\pi/3}, \qquad \gamma = e^{i\pi/3}, \qquad \sigma = e^{i\pi/4}, \qquad \varphi = e^{i\pi/12}. \tag{6.37}$$

For the point groups, their elements and classes, we use the traditional Schönflies notation[5]. In particular, the identity transformation is denoted by the symbol E (*Einheit*)[6]; the rotation through an angle $2\pi/n$ (n is an integer) is denoted by the symbol C_n, successive applications of the rotation are C_n^2, C_n^3, etc.; inversion is denoted by the symbol I.[7]

For a fixed factor system (6.5) and (6.6) determined by a vector $\mathbf{k}$, a user of our tables needs to calculate the character factors ω'_{ij} by formulas (6.18), taking (6.9) and (6.14) into account, extract the projective representations that correspond to those ω'_{ij} from the last table of one of the following sections, and multiply each matrix $D'(g_i)$ (or $D''(g_i)$) by the respective gauge factor δ_i. If one wants to obtain the matrices for all group elements, the matrices of the generators $D(g_i)$ need to be multiplied according to (6.8) and (6.9). Note that there is no need to calculate the numerical factors $\omega_{\prod_{i=1}^{p} g_i^{\mu_i}}$ using the complicated formula (6.9), because each matrix

[3] These tables are sufficient because all other point groups are subgroups of O_h or D_{6h}.

[4] There are definitions of the unitary matrix S different from ours, and they can lead to values of the projective factors $\omega_{\alpha_i,\alpha_j}$ different from ours.

[5] The correspondence between Schönflies and international symbols for all 32 point groups can be found, e.g. in [3, Appendix J].

[6] Do not confuse with the unit matrix E.

[7] In the abstract group, the inversion is still denoted by i.

$D(\prod_{i=1}^{p} g_i^{\mu_i})$ can be represented as a product of two previously calculated ones:

$$\begin{aligned} D(g_i^2) &= \omega^*_{g_i,g_i} D(g_i)D(g_i), \\ D(g_i^3) &= \omega^*_{g_i^2,g_i} D(g_i^2)D(g_i), \\ D(g_i^3 g_j) &= \omega^*_{g_i^3,g_j} D(g_i^3)D(g_j) \end{aligned} \tag{6.38}$$

and so on.

Sometimes the gauge factors δ_i include the factor $\tau_{\omega'_{ij}}$, which substantially reduces the size of the projective representation table. The reason for appearance of this factor is that the defining relations (6.16) and (6.17) sometimes remain invariant under the identical transformation, which consists in change of sign in front of ω'_{ij} and simultaneous multiplication of matrices $D(g_i)$ by the factor $\tau_{\omega'_{ij}}$. For example, the defining relations (6.30) of the group T can be expresses as

$$\begin{aligned} &(-A')^2 = E\,, \\ &(-B'')^2 = E\,, \quad (-B'')(-A') = (-\omega^2_{C'B'})(-A')(-B'')\,, \\ &C'^3 = E\,, \quad C'(-A') = (-B'')C'\,, \quad C'(-B'') = (-\omega_{C'B'})(-A')(-B'')C'\,. \end{aligned} \tag{6.39}$$

Hence the projective representations corresponding to the values $-\omega_{C'B'}$ differ from the projective representations corresponding to $\omega_{C'B'}$ only by the factor -1 in front of the matrices A' and B''.

Note that the above procedure for constructing the projective representation $\Delta^{\mathbf{k}j}$ should be used only for the $\mathbf{k}$ vectors at the boundary of the Brillouin zone in the case of a nonsymmorphic space group.

In symmorphic groups all nonprimitive translations $\boldsymbol{\tau}_\alpha$ are equal to zero and hence all the projective, character and gauge factors are equal to one. In this case, $\Delta^{\mathbf{k}j}$ are the ordinary vector representations. In the last table of each section they are presented in the very beginning.

In nonsymmorphic groups, for the $\mathbf{k}$ vectors inside the Brillouin zone ($\alpha_m\mathbf{k} = \mathbf{k}$) the projective factors (6.5) can be represented in the form

$$\omega^{\mathbf{k}}_{\{\alpha_m|\boldsymbol{\tau}_{\alpha_m}\},\{\alpha_{m'}|\boldsymbol{\tau}_{\alpha_{m'}}\}} = \frac{e^{-i\mathbf{k}\boldsymbol{\tau}_{\alpha_m}} e^{-i\mathbf{k}\boldsymbol{\tau}_{\alpha_{m'}}}}{e^{-i\mathbf{k}\boldsymbol{\tau}_{\alpha_m\alpha_{m'}}}}\,. \tag{6.40}$$

Defining new matrices $\Delta'^{\mathbf{k}j}$ by the formula

$$\Delta'^{\mathbf{k}j}(\{\alpha_m|\boldsymbol{\tau}_{\alpha_m}\}) = \frac{1}{e^{-i\mathbf{k}\boldsymbol{\tau}_{\alpha_m}}}\,\Delta^{\mathbf{k}j}(\{\alpha_m|\boldsymbol{\tau}_{\alpha_m}\}), \tag{6.41}$$

we see that they form a vector representation of the group $G^0_{\mathbf{k}}$. Therefore, the projective representation matrices $\Delta^{\mathbf{k}j}$ can be obtained from the vector representation matrices $\Delta'^{\mathbf{k}j}$ by merely multiplying with the factors $e^{-i\mathbf{k}\boldsymbol{\tau}_{\alpha_m}}$.

6.2 Elements and Multiplication Tables of the Groups O_h and D_{6h}

Elements and Multiplication Tables of the Groups O_h and D_{6h}. Spin Factors $\omega_{\alpha_i,\alpha_j}$ and Spin Matrices $S(\alpha)$ (Tables 6.1, 6.2, 6.3, 6.4, 6.5 and 6.6).

Table 6.1 Elements of the point group O_h

No. i of the element α_i	Class	Rotation axis unit vector **n**	Rotation angle θ	$\mathbf{r}' = \alpha\mathbf{r}$	No. i of the element α_i	Class	$\mathbf{r}' = \alpha\mathbf{r}$
1	E		0	(x, y, z)	25	I	$(\bar{x}, \bar{y}, \bar{z})$
2		$(1, 0, 0)$	π	$(x, \bar{y}, \bar{z})$	26		$(\bar{x}, y, z)$
3	C_4^2	$(0, 1, 0)$	π	$(\bar{x}, y, \bar{z})$	27	IC_4^2	$(x, \bar{y}, z)$
4		$(0, 0, 1)$	π	$(\bar{x}, \bar{y}, z)$	28		$(x, y, \bar{z})$
5	C_4	$(1, 0, 0)$	$\pi/2$	$(x, \bar{z}, y)$	29	IC_4	$(\bar{x}, z, \bar{y})$
6		$(1, 0, 0)$	$-\pi/2$	$(x, z, \bar{y})$	30		$(\bar{x}, \bar{z}, y)$
7		$(0, 1, 0)$	$\pi/2$	$(z, y, \bar{x})$	31		$(\bar{z}, \bar{y}, x)$
8		$(0, 1, 0)$	$-\pi/2$	$(\bar{z}, y, x)$	32		$(z, \bar{y}, \bar{x})$
9		$(0, 0, 1)$	$\pi/2$	$(\bar{y}, x, z)$	33		$(y, \bar{x}, \bar{z})$
10		$(0, 0, 1)$	$-\pi/2$	$(y, \bar{x}, z)$	34		$(\bar{y}, x, \bar{z})$
11	C_3	$1/\sqrt{3}(1, 1, 1)$	$2\pi/3$	(z, x, y)	35	IC_3	$(\bar{z}, \bar{x}, \bar{y})$
12		$1/\sqrt{3}(1, 1, 1)$	$-2\pi/3$	(y, z, x)	36		$(\bar{y}, \bar{z}, \bar{x})$
13		$1/\sqrt{3}(\bar{1}, 1, 1)$	$2\pi/3$	$(\bar{y}, z, \bar{x})$	37		$(y, \bar{z}, x)$
14		$1/\sqrt{3}(\bar{1}, 1, 1)$	$-2\pi/3$	$(\bar{z}, \bar{x}, y)$	38		$(z, x, \bar{y})$
15		$1/\sqrt{3}(1, \bar{1}, 1)$	$2\pi/3$	$(\bar{y}, \bar{z}, x)$	39		$(y, z, \bar{x})$
16		$1/\sqrt{3}(1, \bar{1}, 1)$	$-2\pi/3$	$(z, \bar{x}, \bar{y})$	40		$(\bar{z}, x, y)$
17		$1/\sqrt{3}(1, 1, \bar{1})$	$2\pi/3$	$(y, \bar{z}, \bar{x})$	41		$(\bar{y}, z, x)$
18		$1/\sqrt{3}(1, 1, \bar{1})$	$-2\pi/3$	$(\bar{z}, x, \bar{y})$	42		$(z, \bar{x}, y)$
19	C_2	$1/\sqrt{2}(1, 1, 0)$	π	$(y, x, \bar{z})$	43	IC_2	$(\bar{y}, \bar{x}, z)$
20		$1/\sqrt{2}(\bar{1}, 1, 0)$	π	$(\bar{y}, \bar{x}, \bar{z})$	44		(y, x, z)
21		$1/\sqrt{2}(1, 0, 1)$	π	$(z, \bar{y}, x)$	45		$(\bar{z}, y, \bar{x})$
22		$1/\sqrt{2}(\bar{1}, 0, 1)$	π	$(\bar{z}, \bar{y}, \bar{x})$	46		(z, y, x)
23		$1/\sqrt{2}(0, 1, 1)$	π	$(\bar{x}, z, y)$	47		$(x, \bar{z}, \bar{y})$
24		$1/\sqrt{2}(0, \bar{1}, 1)$	π	$(\bar{x}, \bar{z}, \bar{y})$	48		(x, z, y)

Table 6.2 Multiplication table of the point group O_h

i	j																							
	1	2	3	4	5	6	7	8	9	10	11	12	13	14	15	16	17	18	19	20	21	22	23	24
1	1	2	3	4	5	6	7	8	9	10	11	12	13	14	15	16	17	18	19	20	21	22	23	24
2	2	$\overline{1}$	4	$\overline{3}$	$\overline{6}$	5	21	$\overline{22}$	$\overline{20}$	19	$\overline{16}$	17	15	$\overline{18}$	$\overline{13}$	11	$\overline{12}$	14	$\overline{10}$	9	$\overline{7}$	8	24	$\overline{23}$
3	3	$\overline{4}$	$\overline{1}$	2	$\overline{24}$	23	$\overline{8}$	7	19	20	$\overline{18}$	13	$\overline{12}$	16	17	$\overline{14}$	$\overline{15}$	11	$\overline{9}$	$\overline{10}$	$\overline{22}$	21	$\overline{6}$	5
4	4	3	$\overline{2}$	$\overline{1}$	23	24	22	21	$\overline{10}$	9	$\overline{14}$	15	$\overline{17}$	11	$\overline{12}$	18	13	$\overline{16}$	20	$\overline{19}$	$\overline{8}$	$\overline{7}$	$\overline{5}$	$\overline{6}$
5	5	$\overline{6}$	23	24	2	1	11	14	15	17	21	10	9	$\overline{22}$	$\overline{20}$	7	19	8	$\overline{12}$	13	$\overline{16}$	18	4	$\overline{3}$
6	6	5	$\overline{24}$	23	1	$\overline{2}$	16	18	13	12	7	$\overline{19}$	20	8	9	$\overline{21}$	10	22	17	$\overline{15}$	11	$\overline{14}$	3	4
7	7	$\overline{22}$	$\overline{8}$	21	17	13	3	1	11	16	19	6	23	10	5	20	$\overline{24}$	9	$\overline{18}$	$\overline{14}$	2	4	$\overline{12}$	15
8	8	21	7	22	15	12	1	$\overline{3}$	18	14	9	$\overline{23}$	6	$\overline{20}$	24	10	5	$\overline{19}$	11	16	4	$\overline{2}$	13	$\overline{17}$
9	9	19	20	$\overline{10}$	11	18	13	15	4	1	23	8	22	5	21	6	7	24	3	$\overline{2}$	$\overline{12}$	$\overline{17}$	$\overline{14}$	$\overline{16}$
10	10	$\overline{20}$	19	9	14	16	17	12	1	$\overline{4}$	5	$\overline{21}$	7	$\overline{23}$	8	$\overline{24}$	$\overline{22}$	6	2	3	15	13	11	18
11	11	$\overline{18}$	$\overline{14}$	$\overline{16}$	19	9	23	5	21	7	$\overline{12}$	1	4	17	2	13	3	15	$\overline{8}$	22	$\overline{6}$	24	$\overline{10}$	$\overline{20}$
12	12	15	17	13	8	$\overline{21}$	10	$\overline{19}$	6	$\overline{23}$	1	$\overline{11}$	16	$\overline{3}$	18	$\overline{4}$	14	$\overline{2}$	5	$\overline{24}$	9	20	7	22
13	13	17	$\overline{15}$	$\overline{12}$	7	22	20	9	23	6	3	18	$\overline{14}$	1	11	$\overline{2}$	16	4	$\overline{24}$	$\overline{5}$	19	$\overline{10}$	$\overline{8}$	21
14	14	$\overline{16}$	11	18	$\overline{20}$	10	5	$\overline{23}$	8	$\overline{22}$	15	$\overline{4}$	1	$\overline{13}$	$\overline{3}$	17	2	12	21	7	24	6	9	$\overline{19}$
15	15	$\overline{12}$	13	$\overline{17}$	21	8	9	$\overline{20}$	24	5	4	14	18	2	$\overline{16}$	1	11	$\overline{3}$	23	6	$\overline{10}$	$\overline{19}$	22	$\overline{7}$
16	16	14	$\overline{18}$	11	10	20	$\overline{24}$	6	7	$\overline{21}$	17	$\overline{2}$	3	12	1	$\overline{15}$	$\overline{4}$	13	$\overline{22}$	$\overline{8}$	5	23	19	9
17	17	$\overline{13}$	$\overline{12}$	15	$\overline{22}$	7	19	10	5	$\overline{24}$	2	16	11	$\overline{4}$	14	3	$\overline{18}$	1	$\overline{6}$	23	$\overline{20}$	9	21	8
18	18	11	16	$\overline{14}$	9	$\overline{19}$	6	24	22	8	13	$\overline{3}$	$\overline{2}$	15	4	12	1	$\overline{17}$	7	$\overline{21}$	23	$\overline{5}$	20	$\overline{10}$
19	19	$\overline{9}$	$\overline{10}$	$\overline{20}$	$\overline{18}$	11	$\overline{12}$	17	2	3	$\overline{6}$	7	21	$\overline{24}$	$\overline{22}$	23	$\overline{8}$	5	$\overline{1}$	4	$\overline{13}$	15	$\overline{16}$	14
20	20	10	$\overline{9}$	19	16	$\overline{14}$	$\overline{15}$	13	3	$\overline{2}$	$\overline{24}$	22	$\overline{8}$	6	7	$\overline{5}$	$\overline{21}$	23	$\overline{4}$	$\overline{1}$	17	$\overline{12}$	$\overline{18}$	11
21	21	$\overline{8}$	22	$\overline{7}$	$\overline{12}$	15	4	2	$\overline{16}$	11	$\overline{10}$	5	24	19	$\overline{6}$	9	23	$\overline{20}$	$\overline{14}$	18	$\overline{1}$	$\overline{3}$	$\overline{17}$	$\overline{13}$
22	22	7	$\overline{21}$	$\overline{8}$	13	$\overline{17}$	$\overline{2}$	4	$\overline{14}$	18	20	24	$\overline{5}$	9	23	$\overline{19}$	6	$\overline{10}$	16	$\overline{11}$	3	$\overline{1}$	$\overline{15}$	$\overline{12}$
23	23	$\overline{24}$	$\overline{5}$	$\overline{6}$	3	4	$\overline{14}$	11	$\overline{12}$	13	$\overline{8}$	9	$\overline{10}$	7	19	22	20	21	$\overline{15}$	$\overline{17}$	$\overline{18}$	$\overline{16}$	$\overline{1}$	2
24	24	23	6	$\overline{5}$	4	$\overline{3}$	18	$\overline{16}$	$\overline{17}$	15	22	$\overline{20}$	$\overline{19}$	21	$\overline{10}$	8	9	$\overline{7}$	13	12	$\overline{14}$	$\overline{11}$	$\overline{2}$	$\overline{1}$

The table presents the products $\alpha_i\alpha_j = \alpha_k$, where $\alpha_i, \alpha_j, \alpha_k \in O$. For the elements $\alpha_l \in O_h$, $l = 25, 26, \ldots, 48$, the multiplication rules are $\alpha_{i+24}\alpha_j = \alpha_i\alpha_{j+24} = \alpha_{k+24}$ and $\alpha_{i+24}\alpha_{j+24} = \alpha_k$

Table 6.3 Spin matrices $S(\alpha)$ of the point group O

1	2	3	4
$\begin{pmatrix} 1 & 0 \\ 0 & 1 \end{pmatrix}$	$\begin{pmatrix} 0 & -i \\ -i & 0 \end{pmatrix}$	$\begin{pmatrix} 0 & -1 \\ 1 & 0 \end{pmatrix}$	$\begin{pmatrix} -i & 0 \\ 0 & i \end{pmatrix}$
5	6	7	8
$\frac{\sqrt{2}}{2}\begin{pmatrix} 1 & -i \\ -i & 1 \end{pmatrix}$	$\frac{\sqrt{2}}{2}\begin{pmatrix} 1 & i \\ i & 1 \end{pmatrix}$	$\frac{\sqrt{2}}{2}\begin{pmatrix} 1 & -1 \\ 1 & 1 \end{pmatrix}$	$\frac{\sqrt{2}}{2}\begin{pmatrix} 1 & 1 \\ -1 & 1 \end{pmatrix}$
9	10	11	12
$\frac{\sqrt{2}}{2}\begin{pmatrix} 1-i & 0 \\ 0 & 1+i \end{pmatrix}$	$\frac{\sqrt{2}}{2}\begin{pmatrix} 1+i & 0 \\ 0 & 1-i \end{pmatrix}$	$\frac{1}{2}\begin{pmatrix} 1-i & -1-i \\ 1-i & 1+i \end{pmatrix}$	$\frac{1}{2}\begin{pmatrix} 1+i & 1+i \\ -1+i & 1-i \end{pmatrix}$
13	14	15	16
$\frac{1}{2}\begin{pmatrix} 1-i & -1+i \\ 1+i & 1+i \end{pmatrix}$	$\frac{1}{2}\begin{pmatrix} 1+i & 1-i \\ -1-i & 1-i \end{pmatrix}$	$\frac{1}{2}\begin{pmatrix} 1-i & 1-i \\ -1-i & 1+i \end{pmatrix}$	$\frac{1}{2}\begin{pmatrix} 1+i & -1+i \\ 1+i & 1-i \end{pmatrix}$
17	18	19	20
$\frac{1}{2}\begin{pmatrix} 1+i & -1-i \\ 1-i & 1-i \end{pmatrix}$	$\frac{1}{2}\begin{pmatrix} 1-i & 1+i \\ -1+i & 1+i \end{pmatrix}$	$\frac{\sqrt{2}}{2}\begin{pmatrix} 0 & -1-i \\ 1-i & 0 \end{pmatrix}$	$\frac{\sqrt{2}}{2}\begin{pmatrix} 0 & -1+i \\ 1+i & 0 \end{pmatrix}$
21	22	23	24
$\frac{\sqrt{2}}{2}\begin{pmatrix} -i & -i \\ -i & i \end{pmatrix}$	$\frac{\sqrt{2}}{2}\begin{pmatrix} -i & i \\ i & i \end{pmatrix}$	$\frac{\sqrt{2}}{2}\begin{pmatrix} -i & -1 \\ 1 & i \end{pmatrix}$	$\frac{\sqrt{2}}{2}\begin{pmatrix} -i & 1 \\ -1 & i \end{pmatrix}$

Table 6.4 Elements of the point group D_{6h}

No. i of the element α_i	Class	Rotation axis unit vector $\mathbf{n}$	Rotation angle θ	$\mathbf{r}' = \alpha\mathbf{r}$	No. i of the element α_i	Class	$\mathbf{r}' = \alpha\mathbf{r}$
1	E		0	(x, y, z)	13	I	$(-x, -y, -z)$
2	C_6	$(0, 0, 1)$	$\pi/3$	$\left(\frac{1}{2}x - \frac{\sqrt{3}}{2}y, \frac{\sqrt{3}}{2}x + \frac{1}{2}y, z\right)$	14	IC_6	$\left(-\frac{1}{2}x + \frac{\sqrt{3}}{2}y, -\frac{\sqrt{3}}{2}x - \frac{1}{2}y, -z\right)$
3		$(0, 0, 1)$	$-\pi/3$	$\left(\frac{1}{2}x + \frac{\sqrt{3}}{2}y, -\frac{\sqrt{3}}{2}x + \frac{1}{2}y, z\right)$	15		$\left(-\frac{1}{2}x - \frac{\sqrt{3}}{2}y, \frac{\sqrt{3}}{2}x - \frac{1}{2}y, -z\right)$
4	C_6^2	$(0, 0, 1)$	$2\pi/3$	$\left(-\frac{1}{2}x - \frac{\sqrt{3}}{2}y, \frac{\sqrt{3}}{2}x - \frac{1}{2}y, z\right)$	16	IC_6^2	$\left(\frac{1}{2}x + \frac{\sqrt{3}}{2}y, -\frac{\sqrt{3}}{2}x + \frac{1}{2}y, -z\right)$
5		$(0, 0, 1)$	$-2\pi/3$	$\left(-\frac{1}{2}x + \frac{\sqrt{3}}{2}y, -\frac{\sqrt{3}}{2}x - \frac{1}{2}y, z\right)$	17		$\left(\frac{1}{2}x - \frac{\sqrt{3}}{2}y, \frac{\sqrt{3}}{2}x + \frac{1}{2}y, -z\right)$
6	C_6^3	$(0, 0, 1)$	π	$(-x, -y, z)$	18	IC_6^3	$(x, y, -z)$
7	C_2	$(1, 0, 0)$	π	$(x, -y, -z)$	19	IC_2	$(-x, y, z)$
8		$(\frac{1}{2}, \frac{\sqrt{3}}{2}, 0)$	π	$\left(-\frac{1}{2}x + \frac{\sqrt{3}}{2}y, \frac{\sqrt{3}}{2}x + \frac{1}{2}y, -z\right)$	20		$\left(\frac{1}{2}x - \frac{\sqrt{3}}{2}y, -\frac{\sqrt{3}}{2}x - \frac{1}{2}y, z\right)$
9		$(-\frac{1}{2}, \frac{\sqrt{3}}{2}, 0)$	π	$\left(-\frac{1}{2}x - \frac{\sqrt{3}}{2}y, -\frac{\sqrt{3}}{2}x + \frac{1}{2}y, -z\right)$	21		$\left(\frac{1}{2}x + \frac{\sqrt{3}}{2}y, \frac{\sqrt{3}}{2}x - \frac{1}{2}y, z\right)$
10	C_2'	$(\frac{\sqrt{3}}{2}, \frac{1}{2}, 0)$	π	$\left(\frac{1}{2}x + \frac{\sqrt{3}}{2}y, \frac{\sqrt{3}}{2}x - \frac{1}{2}y, -z\right)$	22	IC_2'	$\left(-\frac{1}{2}x - \frac{\sqrt{3}}{2}y, -\frac{\sqrt{3}}{2}x + \frac{1}{2}y, z\right)$
11		$(0, 1, 0)$	π	$(-x, y, -z)$	23		$(x, -y, z)$
12		$(-\frac{\sqrt{3}}{2}, \frac{1}{2}, 0)$	π	$\left(\frac{1}{2}x - \frac{\sqrt{3}}{2}y, -\frac{\sqrt{3}}{2}x - \frac{1}{2}y, -z\right)$	24		$\left(-\frac{1}{2}x + \frac{\sqrt{3}}{2}y, \frac{\sqrt{3}}{2}x + \frac{1}{2}y, z\right)$

Table 6.5 Multiplication table of the point group D_{6h}

i \ j	1	2	3	4	5	6	7	8	9	10	11	12
1	1	2	3	4	5	6	7	8	9	10	11	12
2	2	4	1	6	3	$\bar{5}$	10	11	12	8	9	$\bar{7}$
3	3	1	5	2	$\bar{6}$	4	$\overline{12}$	10	11	7	8	9
4	4	6	2	$\bar{5}$	1	$\bar{3}$	8	9	$\bar{7}$	11	12	$\overline{10}$
5	5	3	$\bar{6}$	1	$\bar{4}$	2	$\bar{9}$	7	8	$\overline{12}$	10	11
6	6	$\bar{5}$	4	$\bar{3}$	2	$\bar{1}$	11	12	$\overline{10}$	9	$\bar{7}$	$\bar{8}$
7	7	$\overline{12}$	10	$\bar{9}$	8	$\overline{11}$	$\bar{1}$	$\bar{5}$	4	$\bar{3}$	6	2
8	8	10	11	7	9	$\overline{12}$	$\bar{4}$	$\bar{1}$	$\bar{5}$	$\bar{2}$	$\bar{3}$	6
9	9	11	12	8	$\bar{7}$	10	5	$\bar{4}$	$\bar{1}$	$\bar{6}$	$\bar{2}$	$\bar{3}$
10	10	7	8	$\overline{12}$	11	$\bar{9}$	$\bar{2}$	$\bar{3}$	6	$\bar{1}$	$\bar{5}$	4
11	11	8	9	10	12	7	$\bar{6}$	$\bar{2}$	$\bar{3}$	$\bar{4}$	$\bar{1}$	$\bar{5}$
12	12	9	$\bar{7}$	11	$\overline{10}$	8	3	$\bar{6}$	$\bar{2}$	5	$\bar{4}$	$\bar{1}$

The table presents the products $\alpha_i\alpha_j = \alpha_k$, where $\alpha_i, \alpha_j, \alpha_k \in D_6$. For the elements $\alpha_l \in D_{6h}$, $l = 13, 14, \ldots, 24$, the multiplication rules are $\alpha_{i+12}\alpha_j = \alpha_i\alpha_{j+12} = \alpha_{k+12}$ and $\alpha_{i+12}\alpha_{j+12} = \alpha_k$

Table 6.6 Spin matrices $S(\alpha)$ of the point group D_6

1	2	3	4
$\begin{pmatrix} 1 & 0 \\ 0 & 1 \end{pmatrix}$	$\frac{1}{2}\begin{pmatrix} \sqrt{3}-i & 0 \\ 0 & \sqrt{3}+i \end{pmatrix}$	$\frac{1}{2}\begin{pmatrix} \sqrt{3}+i & 0 \\ 0 & \sqrt{3}-i \end{pmatrix}$	$\frac{1}{2}\begin{pmatrix} 1-i\sqrt{3} & 0 \\ 0 & 1+i\sqrt{3} \end{pmatrix}$
5	6	7	8
$\frac{1}{2}\begin{pmatrix} 1+i\sqrt{3} & 0 \\ 0 & 1-i\sqrt{3} \end{pmatrix}$	$\begin{pmatrix} -i & 0 \\ 0 & i \end{pmatrix}$	$\begin{pmatrix} 0 & -i \\ -i & 0 \end{pmatrix}$	$\frac{1}{2}\begin{pmatrix} 0 & -\sqrt{3}-i \\ \sqrt{3}-i & 0 \end{pmatrix}$
9	10	11	12
$\frac{1}{2}\begin{pmatrix} 0 & -\sqrt{3}+i \\ \sqrt{3}+i & 0 \end{pmatrix}$	$\frac{1}{2}\begin{pmatrix} 0 & -1-i\sqrt{3} \\ 1-i\sqrt{3} & 0 \end{pmatrix}$	$\begin{pmatrix} 0 & -1 \\ 1 & 0 \end{pmatrix}$	$\frac{1}{2}\begin{pmatrix} 0 & -1+i\sqrt{3} \\ 1+i\sqrt{3} & 0 \end{pmatrix}$

6.3 Groups C_2, C_i and C_s

See Tables 6.7, 6.8, 6.9, 6.10 and 6.11.

Table 6.7 Realizations of the groups in O_h

	e	a
C_2	E	C_2
	1	2
	1	3
	1	4
	1	19
	1	20
	1	21
	1	22
	1	23
	1	24
C_i	E	I
	1	25
C_s	E	IC_2
	1	26
	1	27
	1	28
	1	43
	1	44
	1	45
	1	46
	1	47
	1	48

Table 6.8 Realizations of the groups in D_{6h}

	e	a
C_2	E	C_2
	1	6
	1	7
	1	8
	1	9
	1	10
	1	11
	1	12
C_i	E	I
	1	13
C_s	E	IC_2
	1	18
	1	19
	1	20
	1	21
	1	22
	1	23
	1	24

Table 6.9 Defining relations

$a^2 = e$	$A'^2 = E$

Table 6.10 Gauge factor

δ_A
$\omega_{a,a}^{1/2}$

Table 6.11 Matrix of the generator

j	A'
1	1
2	−1

6.4 Group C_3

See Tables 6.12, 6.13, 6.14, 6.15 and 6.16.

Table 6.12 Realizations of the group in O_h

	e	a	a^2
C_3	E	C_3	C_3^2
	1	11	12
	1	13	14
	1	15	16
	1	17	18

Table 6.13 Realization of the group in D_{6h}

	e	a	a^2
C_3	E	C_3	C_3^2
	1	4	5

Table 6.14 Defining relations

$a^3 = e$	$A'^3 = E$

Table 6.15 Gauge factor

δ_A
$(\omega_{a,a}\omega_{a^2,a})^{1/3}$

Table 6.16 Matrix of the generator

j	A'
1	1
2	ε
3	ε^2

6.5 Groups C_4 and S_4

See Tables 6.17, 6.18, 6.19 and 6.20.

Table 6.17 Realizations of the groups in O_h

	e	a	a^2	a^3
C_4	E	C_4	C_4^2	C_4^3
	1	5	2	6
	1	7	3	8
	1	9	4	10
S_4	E	IC_4	C_4^2	IC_4^3
	1	29	2	30
	1	31	3	32
	1	33	4	34

Table 6.18 Defining relations

$a^4 = e$	$A'^4 = E$

Table 6.19 Gauge factor

δ_A
$(\omega_{a,a}\omega_{a^2,a}\omega_{a^3,a})^{1/4}$

Table 6.20 Matrix of the generator

j	A'
1	1
2	i
3	-1
4	$-i$

6.6 Groups D_2, C_{2h} and C_{2v}

See Tables 6.21, 6.22, 6.23, 6.24 and 6.25.

Table 6.21 Realizations of the groups in O_h

	e	a	b	ab
D_2	E	C_2	C_2'	C_2''
	1	2	3	4
	1	2	23	24
	1	3	21	22
	1	4	19	20
C_{2h}	E	C_2	I	IC_2
	1	2	25	26
	1	3	25	27
	1	4	25	28
	1	19	25	43
	1	20	25	44
	1	21	25	45
	1	22	25	46
	1	23	25	47
	1	24	25	48
C_{2v}	E	C_2	IC_2'	IC_2''
	1	2	27	28
	1	2	47	48
	1	3	26	28
	1	3	45	46
	1	4	26	27
	1	4	43	44
	1	19	28	44
	1	20	28	43
	1	21	27	46
	1	22	27	45
	1	23	26	48
	1	24	26	47

Table 6.22 Realizations of the groups in D_{6h}

	e	a	b	ab
D_2	E	C_2	C_2'	C_2''
	1	6	7	11
	1	6	8	12
	1	6	9	10
C_{2h}	E	C_2	I	IC_2
	1	6	13	18
	1	7	13	19
	1	8	13	20
	1	9	13	21
	1	10	13	22
	1	11	13	23
	1	12	13	24
C_{2v}	E	C_2	IC_2'	IC_2''
	1	6	19	23
	1	6	20	24
	1	6	21	22
	1	7	18	23
	1	8	18	24
	1	9	18	22
	1	10	18	21
	1	11	18	19
	1	12	18	20

Table 6.23 Defining relations

$a^2 = e$	$A'^2 = E$
$b^2 = e, \quad ba = ab$	$B'^2 = E, \quad B'A' = \omega_{B'A'} A'B'$
	$\omega^2_{B'A'} = 1$

Table 6.24 Character and gauge factors

$\omega_{B'A'}$	δ_A	δ_B
$\omega_{b,a}\omega^*_{a,b}$	$\omega^{1/2}_{a,a}$	$\omega^{1/2}_{b,b}$

Table 6.25 Matrices of the generators

ω_{BA}	j	A'	B'
1	1	1	1
	2	1	−1
	3	−1	1
	4	−1	−1
−1	1	$\begin{pmatrix} 1 & 0 \\ 0 & -1 \end{pmatrix}$	$\begin{pmatrix} 0 & 1 \\ 1 & 0 \end{pmatrix}$

6.7 Groups C_6, S_6 and C_{3h}

See Tables 6.26, 6.27, 6.28, 6.29 and 6.30.

Table 6.26 Realizations of the groups in O_h

	e	a	a^2	a^3	a^4	a^5
S_6	E	IC_3^2	C_3	I	C_3^2	IC_3
	1	36	11	25	12	35
	1	38	13	25	14	37
	1	40	15	25	16	39
	1	42	17	25	18	41

Table 6.27 Realizations of the groups in D_{6h}

	e	a	a^2	a^3	a^4	a^5
C_6	E	C_6	C_6^2	C_6^3	C_6^4	C_6^5
	1	2	4	6	5	3
S_6	E	IC_6^4	C_6^2	I	C_6^4	IC_6^2
	1	17	4	13	5	16
C_{3h}	E	IC_6	C_6^2	IC_6^3	C_6^4	IC_6^5
	1	14	4	18	5	15

Table 6.28 Defining relations

$a^6 = e$	$A'^6 = E$

Table 6.29 Gauge factor

δ_A
$(\omega_a, a\omega_{a^2}, a\omega_{a^3}, a\omega_{a^4}, a\omega_{a^5}, a)^{1/6}$

Table 6.30 Matrix of the generator

j	A'
1	1
2	γ
3	γ^2
4	-1
5	$-\gamma$
6	$-\gamma^2$

6.8 Groups D_3 and C_{3v}

See Tables 6.31, 6.32, 6.33, 6.34 and 6.35.

Table 6.31 Realizations of the groups in O_h

	e	a	a^2	b	ab	a^2b
D_3	E	$2C_3$		$3C_2$		
	1	11	12	20	22	24
	1	13	14	19	24	21
	1	15	16	19	23	22
	1	17	18	20	23	21
C_{3v}	E	$2C_3$		$3IC_2$		
	1	11	12	44	46	48
	1	13	14	43	48	45
	1	15	16	43	47	46
	1	17	18	44	47	45

Table 6.32 Realizations of the groups in D_{6h}

	e	a	a^2	b	ab	a^2b
D_3	E	$2C_3$		$3C_2$		
	1	4	5	7	8	9
	1	4	5	10	11	12
C_{3v}	E	$2C_3$		$3IC_2$		
	1	4	5	19	20	21
	1	4	5	22	23	24

Table 6.33 Defining relations

$a^3 = e$	$A''^3 = E$
$b^2 = e, \quad ba = a^2b$	$B'^2 = E, \quad B'A'' = A''^2B'$

Table 6.34 Gauge factors

δ_A	δ_B
$\omega^*_{b,a}\omega_{a,a}\omega_{a^2,b}$	$\omega^{1/2}_{b,b}$

Table 6.35 Matrices of the generators

j	A''	B'
1	1	1
2	1	−1
3	$\begin{pmatrix} \varepsilon & 0 \\ 0 & \varepsilon^2 \end{pmatrix}$	$\begin{pmatrix} 0 & 1 \\ 1 & 0 \end{pmatrix}$

6.9 Groups D_4, C_{4v} and D_{2d}

See Tables 6.36, 6.37, 6.38 and 6.39.

Table 6.36 Realizations of the groups in O_h

	e	a	a^3	a^2	b	a^2b	a^3b	ab
D_4	E	$2C_4$		C_4^2	$2C_2$		$2C_2'$	
	1	5	6	2	23	24	3	4
	1	7	8	3	21	22	4	2
	1	9	10	4	19	20	2	3
C_{4v}	E	$2C_4$		C_4^2	$2IC_2$		$2IC_2'$	
	1	5	6	2	47	48	27	28
	1	7	8	3	45	46	28	26
	1	9	10	4	43	44	26	27
D_{2d}	E	$2IC_4$		C_4^2	$2IC_2$		$2C_2'$	
	1	30	29	2	47	48	4	3
	1	30	29	2	28	27	24	23
	1	32	31	3	45	46	2	4
	1	32	31	3	26	28	22	21
	1	34	33	4	43	44	3	2
	1	34	33	4	27	26	20	19

Table 6.37 Defining relations

$a^4 = e$	$A'^4 = E$
$b^2 = e, \quad ba = a^3b$	$B'^2 = E, \quad B'A' = \omega_{B'A'}A'^3B'$
	$\omega_{B'A'}^4 = 1$

Table 6.38 Character and gauge factors

$\omega_{B'A'}$	δ_A	δ_B
$\omega_{b,a}(\omega_{a,a}\omega_{a^2,a}\omega_{a^3,b})^* \times ((\omega_{a,a}\omega_{a^2,a}\omega_{a^3,a})^{1/4})^2$	$(\omega_{a,a}\omega_{a^2,a}\omega_{a^3,a})^{1/4}\tau_{\omega_{B'A'}}$	$\omega_{b,b}^{1/2}$
$\tau_{\omega_{B'A'}} = \begin{cases} 1, \text{ if } \omega_{B'A'} = 1, i, \\ i, \text{ if } \omega_{B'A'} = -1, -i. \end{cases}$		

Table 6.39 Matrices of the generators

$\omega_{B'A'}$	j	A'	B'
± 1	1	1	1
	2	1	-1
	3	-1	1
	4	-1	-1
	5	$\begin{pmatrix} i & 0 \\ 0 & -i \end{pmatrix}$	$\begin{pmatrix} 0 & 1 \\ 1 & 0 \end{pmatrix}$
$\pm i$	1	$\begin{pmatrix} 1 & 0 \\ 0 & i \end{pmatrix}$	$\begin{pmatrix} 0 & 1 \\ 1 & 0 \end{pmatrix}$
	2	$\begin{pmatrix} -1 & 0 \\ 0 & -i \end{pmatrix}$	$\begin{pmatrix} 0 & 1 \\ 1 & 0 \end{pmatrix}$

6.10 Group C_{4h}

See Tables 6.40, 6.41, 6.42 and 6.43.

Table 6.40 Realization of the group in O_h

C_{4h}	as C_4 in Table 6.17	i	ai	a^2i	a^3i
		I	IC_4	IC_4^2	IC_4^3
		25	29	26	30
		25	31	27	32
		25	33	28	34

Table 6.41 Defining relations

as in Table 6.18 plus	
$i^2 = e, \quad ia = ai$	$I'^2 = E, \quad I'A' = \omega_{I'A'}A'I'$
	$\omega_{I'A'}^2 = 1$

Table 6.42 Character and gauge factors

as in Table 6.19	$\omega_{I'A'}$	δ_I
	$\omega_{i,a}\omega^*_{a,i}$	$\omega^{1/2}_{i,i}$

Table 6.43 Matrices of the generators

$\omega_{I'A'}$	j	A'	I'
1	$1^\pm$	1	± 1
	$2^\pm$	i	± 1
	$3^\pm$	-1	± 1
	$4^\pm$	$-i$	± 1
-1	1	$\begin{pmatrix} 1 & 0 \\ 0 & -1 \end{pmatrix}$	$\begin{pmatrix} 0 & 1 \\ 1 & 0 \end{pmatrix}$
	2	$\begin{pmatrix} i & 0 \\ 0 & -i \end{pmatrix}$	$\begin{pmatrix} 0 & 1 \\ 1 & 0 \end{pmatrix}$

6.11 Group D_{2h}

See Tables 6.44, 6.45, 6.46, 6.47 and 6.48.

Table 6.44 Realizations of the group in O_h

D_{2h}	as D_2 in Table 6.21	i	ai	bi	abi
		I	IC_2	IC'_2	IC''_2
		25	26	27	28
		25	26	47	48
		25	27	45	46
		25	28	43	44

Table 6.45 Realization of the group in D_{6h}

D_{2h}	as D_2 in Table 6.22	i	ai	bi	abi
		I	IC_2	IC'_2	IC''_2
		13	18	19	23
		13	18	20	24
		13	18	21	22

Table 6.46 Defining relations

as in Table 6.23 plus	
$i^2 = e, \quad ia = ai, \quad ib = bi$	$I'^2 = E, \quad I'A' = \omega_{I'A'}A'I', \quad I'B' = \omega_{I'B'}B'I'$
	$\omega^2_{I'A'} = 1, \quad \omega^2_{I'B'} = 1$

Table 6.47 Character and gauge factors

as in Table 6.24	$\omega_{I'A'}$	$\omega_{I'B'}$	δ_I
	$\omega_{i,a}\omega^*_{a,i}$	$\omega_{i,b}\omega^*_{b,i}$	$\omega^{1/2}_{i,i}$

Table 6.48 Matrices of the generators

$\omega_{B'A'}$	$\omega_{I'A'}$	$\omega_{I'B'}$	j	A'	B'	I'
1	1	1	$1^{\pm}$	1	1	± 1
			$2^{\pm}$	1	-1	± 1
			$3^{\pm}$	-1	1	± 1
			$4^{\pm}$	-1	-1	± 1
1	1	-1	1	$\begin{pmatrix} 1 & 0 \\ 0 & 1 \end{pmatrix}$	$\begin{pmatrix} 1 & 0 \\ 0 & -1 \end{pmatrix}$	$\begin{pmatrix} 0 & 1 \\ 1 & 0 \end{pmatrix}$
			2	$\begin{pmatrix} -1 & 0 \\ 0 & -1 \end{pmatrix}$	$\begin{pmatrix} 1 & 0 \\ 0 & -1 \end{pmatrix}$	$\begin{pmatrix} 0 & 1 \\ 1 & 0 \end{pmatrix}$
1	-1	1	1	$\begin{pmatrix} 1 & 0 \\ 0 & -1 \end{pmatrix}$	$\begin{pmatrix} 1 & 0 \\ 0 & 1 \end{pmatrix}$	$\begin{pmatrix} 0 & 1 \\ 1 & 0 \end{pmatrix}$
			2	$\begin{pmatrix} 1 & 0 \\ 0 & -1 \end{pmatrix}$	$\begin{pmatrix} -1 & 0 \\ 0 & -1 \end{pmatrix}$	$\begin{pmatrix} 0 & 1 \\ 1 & 0 \end{pmatrix}$
1	-1	-1	1	$\begin{pmatrix} 1 & 0 \\ 0 & -1 \end{pmatrix}$	$\begin{pmatrix} 1 & 0 \\ 0 & -1 \end{pmatrix}$	$\begin{pmatrix} 0 & 1 \\ 1 & 0 \end{pmatrix}$
			2	$\begin{pmatrix} 1 & 0 \\ 0 & -1 \end{pmatrix}$	$\begin{pmatrix} -1 & 0 \\ 0 & 1 \end{pmatrix}$	$\begin{pmatrix} 0 & 1 \\ 1 & 0 \end{pmatrix}$
-1	1	1	$1^{\pm}$	$\begin{pmatrix} 1 & 0 \\ 0 & -1 \end{pmatrix}$	$\begin{pmatrix} 0 & 1 \\ 1 & 0 \end{pmatrix}$	$\pm\begin{pmatrix} 1 & 0 \\ 0 & 1 \end{pmatrix}$
-1	1	-1	$1^{\pm}$	$\begin{pmatrix} 1 & 0 \\ 0 & -1 \end{pmatrix}$	$\begin{pmatrix} 0 & 1 \\ 1 & 0 \end{pmatrix}$	$\pm\begin{pmatrix} 1 & 0 \\ 0 & -1 \end{pmatrix}$
-1	-1	1	$1^{\pm}$	$\begin{pmatrix} 1 & 0 \\ 0 & -1 \end{pmatrix}$	$\begin{pmatrix} 0 & 1 \\ 1 & 0 \end{pmatrix}$	$\pm\begin{pmatrix} 0 & 1 \\ 1 & 0 \end{pmatrix}$
-1	-1	-1	$1^{\pm}$	$\begin{pmatrix} 1 & 0 \\ 0 & -1 \end{pmatrix}$	$\begin{pmatrix} 0 & 1 \\ 1 & 0 \end{pmatrix}$	$\pm\begin{pmatrix} 0 & i \\ -i & 0 \end{pmatrix}$

6.12 Groups D_6, D_{3d}, C_{6v} and D_{3h}

See Tables 6.49, 6.50, 6.51, 6.52 and 6.53.

Table 6.49 Realizations of the group in O_h

	e	a	a^5	a^2	a^4	a^3	b	a^2b	a^4b	ab	a^3b	a^5b
D_{3d}	E	$2IC_6^2$		$2C_6^2$		I	$3C_2$			$3IC_2'$		
	1	36	35	11	12	25	20	22	24	48	44	46
	1	38	37	13	14	25	19	24	21	45	43	48
	1	40	39	15	16	25	19	23	22	46	43	47
	1	42	41	17	18	25	20	23	21	45	44	47

Table 6.50 Realizations of the groups in D_{6h}

	e	a	a^5	a^2	a^4	a^3	b	a^2b	a^4b	ab	a^3b	a^5b
D_6	E	$2C_6$		$2C_6^2$		C_6^3	$3C_2$			$3C_2'$		
	1	2	3	4	5	6	7	8	9	10	11	12
D_{3d}	E	$2IC_6^2$		$2C_6^2$		I	$3C_2$			$3IC_2'$		
	1	17	16	4	5	13	7	8	9	21	19	20
	1	17	16	4	5	13	10	11	12	24	22	23
C_{6v}	E	$2C_6$		$2C_6^2$		C_6^3	$3IC_2$			$3IC_2'$		
	1	2	3	4	5	6	19	20	21	22	23	24
D_{3h}	E	$2IC_6$		$2C_6^2$		IC_6^3	$3C_2$			$3IC_2'$		
	1	14	15	4	5	18	7	8	9	22	23	24
	1	14	15	4	5	18	10	11	12	20	21	19

Table 6.51 Defining relations

$a^6 = e$	$A'^6 = E$
$b^2 = e, \quad ba = a^5b$	$B'^2 = E, \quad B'A' = \omega_{B'A'}A'^5B'$
	$\omega_{B'A'}^6 = 1$

Table 6.52 Character and gauge factors

$\omega_{B'A'}$	δ_A	δ_B
$\omega_{b,a}(\omega_{a,a}\omega_{a^2,a}\omega_{a^3,a}\omega_{a^4,a}\omega_{a^5,b})^*$ $\times\left((\omega_{a,a}\omega_{a^2,a}\omega_{a^3,a}\omega_{a^4,a}\omega_{a^5,a})^{1/6}\right)^4$	$\omega_{B'A'}^*$ $\times(\omega_{a,a}\omega_{a^2,a}\omega_{a^3,a}\omega_{a^4,a}\omega_{a^5,a})^{1/6}$	$\omega_{b,b}^{1/2}$

Table 6.53 Matrices of the generators

$\omega_{B'A'}$	j	A'	B'
$1, \gamma^2, \gamma^4$	1	1	1
	2	1	−1
	3	−1	1
	4	−1	−1
	5	$\begin{pmatrix} \gamma^2 & 0 \\ 0 & -\gamma \end{pmatrix}$	$\begin{pmatrix} 0 & 1 \\ 1 & 0 \end{pmatrix}$
	6	$\begin{pmatrix} -\gamma^2 & 0 \\ 0 & \gamma \end{pmatrix}$	$\begin{pmatrix} 0 & 1 \\ 1 & 0 \end{pmatrix}$
$-1, \gamma, \gamma^5$	1	$\begin{pmatrix} 1 & 0 \\ 0 & -1 \end{pmatrix}$	$\begin{pmatrix} 0 & 1 \\ 1 & 0 \end{pmatrix}$
	2	$\begin{pmatrix} \gamma^2 & 0 \\ 0 & \gamma \end{pmatrix}$	$\begin{pmatrix} 0 & 1 \\ 1 & 0 \end{pmatrix}$
	3	$\begin{pmatrix} -\gamma & 0 \\ 0 & -\gamma^2 \end{pmatrix}$	$\begin{pmatrix} 0 & 1 \\ 1 & 0 \end{pmatrix}$

6.13 Group *T*

See Tables 6.54, 6.55, 6.56 and 6.57.

Table 6.54 Realization of the group in O_h

	e	a	b	ab	c	abc	ac	bc	c^2	bc^2	abc^2	ac^2
T	E	$3C_2$			$4C_3$				$4C_3^2$			
	1	2	3	4	11	14	16	18	12	13	15	17

Table 6.55 Defining relations

$a^2 = e$	$A'^2 = E$
$b^2 = e, \quad ba = ab$	$B''^2 = E, \quad B''A' = \omega^2_{C'B'}A'B''$
$c^3 = e, \quad ca = bc, \quad cb = abc$	$C'^3 = E, \quad C'A' = B''C', \quad C'B'' = \omega_{C'B'}A'B''C'$
	$\omega^4_{C'B'} = 1$

Table 6.56 Character and gauge factors

$\omega_{C'B'}$	δ_A	δ_B	δ_C
$\omega_{c,b}(\omega_{a,b}\omega_{ab,c})^* \omega_{a,a}^{1/2}$	$\omega_{a,a}^{1/2} \tau_{\omega_{C'B'}}$	$\omega_{b,b}^{1/2} \omega_{c,a} \omega_{b,c}^* \omega_{b,b}^{1/2} (\omega_{a,a}^{1/2})^* \tau_{\omega_{C'B'}}$	$(\omega_{c,c}\omega_{c^2,c})^{1/3}$

$$\tau_{\omega_{C'B'}} = \begin{cases} 1, & \text{if } \omega_{C'B'} = 1, i, \\ -1, & \text{if } \omega_{C'B'} = -1, -i. \end{cases}$$

Table 6.57 Matrices of the generators

$\omega_{C'B'}$	j	A'	B''	C'
± 1	1	1	1	1
	2	1	1	ε
	3	1	1	ε^2
	4	$\begin{pmatrix} -1 & 0 & 0 \\ 0 & 1 & 0 \\ 0 & 0 & -1 \end{pmatrix}$	$\begin{pmatrix} 1 & 0 & 0 \\ 0 & -1 & 0 \\ 0 & 0 & -1 \end{pmatrix}$	$\begin{pmatrix} 0 & 1 & 0 \\ 0 & 0 & 1 \\ 1 & 0 & 0 \end{pmatrix}$
$\pm i$	1	$\begin{pmatrix} 1 & 0 \\ 0 & -1 \end{pmatrix}$	$\begin{pmatrix} 0 & 1 \\ 1 & 0 \end{pmatrix}$	$\frac{1}{\sqrt{2}} \begin{pmatrix} -\varphi^{11} & \varphi^5 \\ -\varphi^{11} & -\varphi^5 \end{pmatrix}$
	2	$\begin{pmatrix} 1 & 0 \\ 0 & -1 \end{pmatrix}$	$\begin{pmatrix} 0 & 1 \\ 1 & 0 \end{pmatrix}$	$\frac{1}{\sqrt{2}} \begin{pmatrix} \varphi^7 & -\varphi \\ \varphi^7 & \varphi \end{pmatrix}$
	3	$\begin{pmatrix} 1 & 0 \\ 0 & -1 \end{pmatrix}$	$\begin{pmatrix} 0 & 1 \\ 1 & 0 \end{pmatrix}$	$\frac{1}{\sqrt{2}} \begin{pmatrix} -\sigma & -\sigma^3 \\ -\sigma & \sigma^3 \end{pmatrix}$

6.14 Group C_{6h}

See Tables 6.58, 6.59, 6.60 and 6.61.

Table 6.58 Realization of the group in D_{6h}

C_{6h}	as C_6 in Table 6.27	i	ai	a^2i	a^3i	a^4i	a^5i
		I	IC_6	IC_6^2	IC_6^3	IC_6^4	IC_6^5
		13	14	16	18	17	15

Table 6.59 Defining relations

as in Table 6.28 plus	
$i^2 = e, \quad ia = ai$	$I'^2 = E, \quad I'A' = \omega_{I'A'} A'I'$
	$\omega^2_{I'A'} = 1$

Table 6.60 Character and gauge factors

as in Table 6.29	$\omega_{I'A'}$	δ_I
	$\omega_{i,a}\omega^*_{a,i}$	$\omega^{1/2}_{i,i}$

Table 6.61 Matrices of the generators

$\omega_{I'A'}$	j	A'	I'
1	$1^\pm$	1	± 1
	$2^\pm$	γ	± 1
	$3^\pm$	γ^2	± 1
	$4^\pm$	-1	± 1
	$5^\pm$	$-\gamma$	± 1
	$6^\pm$	$-\gamma^2$	± 1
-1	1	$\begin{pmatrix} 1 & 0 \\ 0 & -1 \end{pmatrix}$	$\begin{pmatrix} 0 & 1 \\ 1 & 0 \end{pmatrix}$
	2	$\begin{pmatrix} \gamma & 0 \\ 0 & -\gamma \end{pmatrix}$	$\begin{pmatrix} 0 & 1 \\ 1 & 0 \end{pmatrix}$
	3	$\begin{pmatrix} \gamma^2 & 0 \\ 0 & -\gamma^2 \end{pmatrix}$	$\begin{pmatrix} 0 & 1 \\ 1 & 0 \end{pmatrix}$

6.15 Group D_{4h}

See Tables 6.62, 6.63, 6.64 and 6.65.

Table 6.62 Realizations of the group in O_h

	as D_4	i	ai	a^3i	a^2i	bi	a^2bi	a^3bi	abi
D_{4h}	in Table 6.36	I	$2IC_4$		IC_4^2	$2IC_2$		$2IC_2'$	
		25	29	30	26	47	48	27	28
		25	31	32	27	45	46	28	26
		25	33	34	28	43	44	26	27

Table 6.63 Defining relations

as in Table 6.37 plus	
$i^2 = e, \quad ia = ai, \quad ib = bi$	$I'^2 = E, \quad I'A' = \omega_{I'A'}A'I', \quad I'B' = \omega_{I'B'}B'I'$
	$\omega_{I'A'}^2 = 1, \quad \omega_{I'B'}^2 = 1$

Table 6.64 Character and gauge factors

as in Table 6.38	$\omega_{I'A'}$	$\omega_{I'B'}$	δ_I
	$\omega_{i,a}\omega_{a,i}^*$	$\omega_{i,b}\omega_{b,i}^*$	$\omega_{i,i}^{1/2}$

Table 6.65 Matrices of the generators

$\omega_{B'A'}$	$\omega_{I'A'}$	$\omega_{I'B'}$	j	A'	B'	I'
± 1	1	1	$1^{\pm}$	1	1	± 1
			$2^{\pm}$	1	-1	± 1
			$3^{\pm}$	-1	1	± 1
			$4^{\pm}$	-1	-1	± 1
			$5^{\pm}$	$\begin{pmatrix} i & 0 \\ 0 & -i \end{pmatrix}$	$\begin{pmatrix} 0 & 1 \\ 1 & 0 \end{pmatrix}$	$\pm\begin{pmatrix} 1 & 0 \\ 0 & 1 \end{pmatrix}$
± 1	1	-1	1	$\begin{pmatrix} 1 & 0 \\ 0 & 1 \end{pmatrix}$	$\begin{pmatrix} 1 & 0 \\ 0 & -1 \end{pmatrix}$	$\begin{pmatrix} 0 & 1 \\ 1 & 0 \end{pmatrix}$
			2	$\begin{pmatrix} -1 & 0 \\ 0 & -1 \end{pmatrix}$	$\begin{pmatrix} 1 & 0 \\ 0 & -1 \end{pmatrix}$	$\begin{pmatrix} 0 & 1 \\ 1 & 0 \end{pmatrix}$
			$3^{\pm}$	$\begin{pmatrix} i & 0 \\ 0 & -i \end{pmatrix}$	$\begin{pmatrix} 0 & 1 \\ 1 & 0 \end{pmatrix}$	$\pm\begin{pmatrix} 1 & 0 \\ 0 & -1 \end{pmatrix}$
± 1	-1	1	1	$\begin{pmatrix} 1 & 0 \\ 0 & -1 \end{pmatrix}$	$\begin{pmatrix} 1 & 0 \\ 0 & 1 \end{pmatrix}$	$\begin{pmatrix} 0 & 1 \\ 1 & 0 \end{pmatrix}$
			2	$\begin{pmatrix} 1 & 0 \\ 0 & -1 \end{pmatrix}$	$\begin{pmatrix} -1 & 0 \\ 0 & -1 \end{pmatrix}$	$\begin{pmatrix} 0 & 1 \\ 1 & 0 \end{pmatrix}$
			$3^{\pm}$	$\begin{pmatrix} i & 0 \\ 0 & -i \end{pmatrix}$	$\begin{pmatrix} 0 & 1 \\ 1 & 0 \end{pmatrix}$	$\pm\begin{pmatrix} 0 & 1 \\ 1 & 0 \end{pmatrix}$
± 1	-1	-1	1	$\begin{pmatrix} 1 & 0 \\ 0 & -1 \end{pmatrix}$	$\begin{pmatrix} 1 & 0 \\ 0 & -1 \end{pmatrix}$	$\begin{pmatrix} 0 & 1 \\ 1 & 0 \end{pmatrix}$
			2	$\begin{pmatrix} 1 & 0 \\ 0 & -1 \end{pmatrix}$	$\begin{pmatrix} -1 & 0 \\ 0 & 1 \end{pmatrix}$	$\begin{pmatrix} 0 & 1 \\ 1 & 0 \end{pmatrix}$
			$3^{\pm}$	$\begin{pmatrix} i & 0 \\ 0 & -i \end{pmatrix}$	$\begin{pmatrix} 0 & 1 \\ 1 & 0 \end{pmatrix}$	$\pm\begin{pmatrix} 0 & i \\ -i & 0 \end{pmatrix}$
$\pm i$	1	1	$1^{\pm}$	$\begin{pmatrix} 1 & 0 \\ 0 & i \end{pmatrix}$	$\begin{pmatrix} 0 & 1 \\ 1 & 0 \end{pmatrix}$	$\pm\begin{pmatrix} 1 & 0 \\ 0 & 1 \end{pmatrix}$
			$2^{\pm}$	$\begin{pmatrix} -1 & 0 \\ 0 & -i \end{pmatrix}$	$\begin{pmatrix} 0 & 1 \\ 1 & 0 \end{pmatrix}$	$\pm\begin{pmatrix} 1 & 0 \\ 0 & 1 \end{pmatrix}$
$\pm i$	1	-1	$1^{\pm}$	$\begin{pmatrix} 1 & 0 \\ 0 & i \end{pmatrix}$	$\begin{pmatrix} 0 & 1 \\ 1 & 0 \end{pmatrix}$	$\pm\begin{pmatrix} 1 & 0 \\ 0 & -1 \end{pmatrix}$
			$2^{\pm}$	$\begin{pmatrix} -1 & 0 \\ 0 & -i \end{pmatrix}$	$\begin{pmatrix} 0 & 1 \\ 1 & 0 \end{pmatrix}$	$\pm\begin{pmatrix} 1 & 0 \\ 0 & -1 \end{pmatrix}$
$\pm i$	-1	1	1	$\begin{pmatrix} 1 & 0 & 0 & 0 \\ 0 & i & 0 & 0 \\ 0 & 0 & -1 & 0 \\ 0 & 0 & 0 & -i \end{pmatrix}$	$\begin{pmatrix} 0 & 1 & 0 & 0 \\ 1 & 0 & 0 & 0 \\ 0 & 0 & 0 & 1 \\ 0 & 0 & 1 & 0 \end{pmatrix}$	$\begin{pmatrix} 0 & 0 & 1 & 0 \\ 0 & 0 & 0 & 1 \\ 1 & 0 & 0 & 0 \\ 0 & 1 & 0 & 0 \end{pmatrix}$
$\pm i$	-1	-1	1	$\begin{pmatrix} 1 & 0 & 0 & 0 \\ 0 & i & 0 & 0 \\ 0 & 0 & -1 & 0 \\ 0 & 0 & 0 & -i \end{pmatrix}$	$\begin{pmatrix} 0 & 1 & 0 & 0 \\ 1 & 0 & 0 & 0 \\ 0 & 0 & 0 & -1 \\ 0 & 0 & -1 & 0 \end{pmatrix}$	$\begin{pmatrix} 0 & 0 & 1 & 0 \\ 0 & 0 & 0 & 1 \\ 1 & 0 & 0 & 0 \\ 0 & 1 & 0 & 0 \end{pmatrix}$

6.16 Groups O and T_d

See Tables 6.66, 6.67, 6.68 and 6.69.

Table 6.66 Realization of the groups in O_h

	e	a	b	ab	c^2d	ac^2d	bcd	cd	bd	ad	c	c^2
O	E	$3C_4^2$			$6C_4$							
	1	2	3	4	5	6	7	8	9	10	11	12
T_d	E	$3C_2$			$6IC_4$							
	1	2	3	4	29	30	31	32	33	34	11	12

	bc^2	abc	abc^2	ac	ac^2	bc	d	abd	$abcd$	acd	abc^2d	bc^2d
O	$8C_3$						$6C_2$					
	13	14	15	16	17	18	19	20	21	22	23	24
T_d	$8C_3$						$6IC_2$					
	13	14	15	16	17	18	43	44	45	46	47	48

Table 6.67 Defining relations

$a^2 = e$	$A'^2 = E$
$b^2 = e, \quad ba = ab$	$B''^2 = E, \quad B''A' = \omega_{C'B'}^2 A'B''$
$c^3 = e, \quad ca = bc, \quad cb = abc$	$C''^3 = E, \quad C''A' = B''C'', \quad C''B'' = \omega_{C'B'} A'B''C''$
$d^2 = e, \quad da = bd, \quad db = ad$	$D'^2 = E, \quad D'A' = B''D', \quad D'B'' = A'D'$
$dc = ac^2d$	$D'C'' = \omega_{C'B'} A'C''^2 D'$
	$\omega_{C'B'}^4 = 1$

Table 6.68 Character and gauge factors

$\omega_{C'B'}$	δ_A	δ_B
$\omega_{c,b}(\omega_{a,b}\omega_{ab,c})^*\omega_{a,a}^{1/2}$	$\omega_{a,a}^{1/2}\tau_{\omega_{C'B'}}$	$\omega_{b,b}^{1/2}\omega_{c,a}\omega_{b,c}^*\omega_{b,b}^{1/2}(\omega_{a,a}^{1/2})^*\tau_{\omega_{C'B'}}$

δ_C	δ_D
$(\omega_{c,c}\omega_{c^2,c})^{1/3}\omega_{c,b}(\omega_{a,b}\omega_{ab,c})^*\omega_{a,a}^{1/2}\left(\omega_{d,c}(\omega_{a,c}\omega_{ac,c}\omega_{ac^2,d})^*\omega_{a,a}^{1/2}(\omega_{c,c}\omega_{c^2,c})^{1/3}\right)^*$	$\omega_{d,d}^{1/2}$

$$\tau_{\omega_{C'B'}} = \begin{cases} 1, & \text{if } \omega_{C'B'} = 1,\ \mathrm{i}, \\ -1, & \text{if } \omega_{C'B'} = -1, -\mathrm{i}. \end{cases}$$

Table 6.69 Matrices of the generators

$\omega_{C'B'}$	j	A'	B''	C''	D'
± 1	1	1	1	1	1
	2	1	1	1	-1
	3	$\begin{pmatrix} 1 & 0 \\ 0 & 1 \end{pmatrix}$	$\begin{pmatrix} 1 & 0 \\ 0 & 1 \end{pmatrix}$	$\begin{pmatrix} \varepsilon & 0 \\ 0 & \varepsilon^2 \end{pmatrix}$	$\begin{pmatrix} 0 & 1 \\ 1 & 0 \end{pmatrix}$
	4	$\begin{pmatrix} -1 & 0 & 0 \\ 0 & 1 & 0 \\ 0 & 0 & -1 \end{pmatrix}$	$\begin{pmatrix} 1 & 0 & 0 \\ 0 & -1 & 0 \\ 0 & 0 & -1 \end{pmatrix}$	$\begin{pmatrix} 0 & 1 & 0 \\ 0 & 0 & 1 \\ 1 & 0 & 0 \end{pmatrix}$	$\begin{pmatrix} 0 & 1 & 0 \\ 1 & 0 & 0 \\ 0 & 0 & -1 \end{pmatrix}$
	5	$\begin{pmatrix} -1 & 0 & 0 \\ 0 & 1 & 0 \\ 0 & 0 & -1 \end{pmatrix}$	$\begin{pmatrix} 1 & 0 & 0 \\ 0 & -1 & 0 \\ 0 & 0 & -1 \end{pmatrix}$	$\begin{pmatrix} 0 & 1 & 0 \\ 0 & 0 & 1 \\ 1 & 0 & 0 \end{pmatrix}$	$\begin{pmatrix} 0 & -1 & 0 \\ -1 & 0 & 0 \\ 0 & 0 & 1 \end{pmatrix}$
$\pm i$	1	$\begin{pmatrix} 1 & 0 \\ 0 & -1 \end{pmatrix}$	$\begin{pmatrix} 0 & 1 \\ 1 & 0 \end{pmatrix}$	$\frac{1}{\sqrt{2}}\begin{pmatrix} -\sigma & -\sigma^3 \\ -\sigma & \sigma^3 \end{pmatrix}$	$\frac{1}{\sqrt{2}}\begin{pmatrix} 1 & 1 \\ 1 & -1 \end{pmatrix}$
	2	$\begin{pmatrix} 1 & 0 \\ 0 & -1 \end{pmatrix}$	$\begin{pmatrix} 0 & 1 \\ 1 & 0 \end{pmatrix}$	$\frac{1}{\sqrt{2}}\begin{pmatrix} -\sigma & -\sigma^3 \\ -\sigma & \sigma^3 \end{pmatrix}$	$\frac{1}{\sqrt{2}}\begin{pmatrix} -1 & -1 \\ -1 & 1 \end{pmatrix}$
	3	$\begin{pmatrix} 1 & 0 & 0 & 0 \\ 0 & -1 & 0 & 0 \\ 0 & 0 & 0 & 1 \\ 0 & 0 & 1 & 0 \end{pmatrix}$	$\begin{pmatrix} 0 & 1 & 0 & 0 \\ 1 & 0 & 0 & 0 \\ 0 & 0 & 1 & 0 \\ 0 & 0 & 0 & -1 \end{pmatrix}$	$\frac{1}{\sqrt{2}}\begin{pmatrix} -\varphi^{11} & \varphi^5 & 0 & 0 \\ -\varphi^{11} & -\varphi^5 & 0 & 0 \\ 0 & 0 & \varphi^7 & \varphi^7 \\ 0 & 0 & -\varphi & \varphi \end{pmatrix}$	$\begin{pmatrix} 0 & 0 & 1 & 0 \\ 0 & 0 & 0 & 1 \\ 1 & 0 & 0 & 0 \\ 0 & 1 & 0 & 0 \end{pmatrix}$

6.17 Group D_{6h}

See Tables 6.70, 6.71, 6.72 and 6.73.

Table 6.70 Realization of the group

D_{6h}	as D_6 in Table 6.50	i	ai	a^5i	a^2i	a^4i	a^3i	bi	a^2bi	a^4bi	abi	a^3bi	a^5bi
		E	$2IC_6$		$2IC_6^2$		IC_6^3	$3IC_2$			$3IC_2'$		
		13	14	15	16	17	18	19	20	21	22	23	24

Table 6.71 Defining relations

as in Table 6.51 plus	
$i^2 = e, \quad ia = ai, \quad ib = bi$	$I'^2 = E, \quad I'A' = \omega_{I'A'}A'I', \quad I'B' = \omega_{I'B'}B'I'$
	$\omega^2_{I'A'} = 1, \quad \omega^2_{I'B'} = 1$

Table 6.72 Character and gauge factors

as in Table 6.52	$\omega_{I'A'}$	$\omega_{I'B'}$	δ_I
	$\omega_{i,a}\omega^*_{a,i}$	$\omega_{i,b}\omega^*_{b,i}$	$\omega^{1/2}_{i,i}$

Table 6.73 Matrices of the generators

$\omega_{B'A'}$	$\omega_{I'A'}$	$\omega_{I'B'}$	j	A'	B'	I'
$1, \gamma^2, \gamma^4$	1	1	$1^\pm$	1	1	± 1
			$2^\pm$	1	-1	± 1
			$3^\pm$	-1	1	± 1
			$4^\pm$	-1	-1	± 1
			$5^\pm$	$\begin{pmatrix} \gamma^2 & 0 \\ 0 & -\gamma \end{pmatrix}$	$\begin{pmatrix} 0 & 1 \\ 1 & 0 \end{pmatrix}$	$\pm\begin{pmatrix} 1 & 0 \\ 0 & 1 \end{pmatrix}$
			$6^\pm$	$\begin{pmatrix} -\gamma^2 & 0 \\ 0 & \gamma \end{pmatrix}$	$\begin{pmatrix} 0 & 1 \\ 1 & 0 \end{pmatrix}$	$\pm\begin{pmatrix} 1 & 0 \\ 0 & 1 \end{pmatrix}$
		-1	1	$\begin{pmatrix} 1 & 0 \\ 0 & 1 \end{pmatrix}$	$\begin{pmatrix} 1 & 0 \\ 0 & -1 \end{pmatrix}$	$\begin{pmatrix} 0 & 1 \\ 1 & 0 \end{pmatrix}$
			2	$\begin{pmatrix} -1 & 0 \\ 0 & -1 \end{pmatrix}$	$\begin{pmatrix} 1 & 0 \\ 0 & -1 \end{pmatrix}$	$\begin{pmatrix} 0 & 1 \\ 1 & 0 \end{pmatrix}$
			$3^\pm$	$\begin{pmatrix} \gamma^2 & 0 \\ 0 & -\gamma \end{pmatrix}$	$\begin{pmatrix} 0 & 1 \\ 1 & 0 \end{pmatrix}$	$\pm\begin{pmatrix} 1 & 0 \\ 0 & -1 \end{pmatrix}$
			$4^\pm$	$\begin{pmatrix} -\gamma^2 & 0 \\ 0 & \gamma \end{pmatrix}$	$\begin{pmatrix} 0 & 1 \\ 1 & 0 \end{pmatrix}$	$\pm\begin{pmatrix} 1 & 0 \\ 0 & -1 \end{pmatrix}$
	-1	1	1	$\begin{pmatrix} 1 & 0 \\ 0 & -1 \end{pmatrix}$	$\begin{pmatrix} 1 & 0 \\ 0 & 1 \end{pmatrix}$	$\begin{pmatrix} 0 & 1 \\ 1 & 0 \end{pmatrix}$
			2	$\begin{pmatrix} 1 & 0 \\ 0 & -1 \end{pmatrix}$	$\begin{pmatrix} -1 & 0 \\ 0 & -1 \end{pmatrix}$	$\begin{pmatrix} 0 & 1 \\ 1 & 0 \end{pmatrix}$
			3	$\begin{pmatrix} \gamma^2 & 0 & 0 & 0 \\ 0 & -\gamma & 0 & 0 \\ 0 & 0 & -\gamma^2 & 0 \\ 0 & 0 & 0 & \gamma \end{pmatrix}$	$\begin{pmatrix} 0 & 1 & 0 & 0 \\ 1 & 0 & 0 & 0 \\ 0 & 0 & 0 & 1 \\ 0 & 0 & 1 & 0 \end{pmatrix}$	$\begin{pmatrix} 0 & 0 & 1 & 0 \\ 0 & 0 & 0 & 1 \\ 1 & 0 & 0 & 0 \\ 0 & 1 & 0 & 0 \end{pmatrix}$
		-1	1	$\begin{pmatrix} 1 & 0 \\ 0 & -1 \end{pmatrix}$	$\begin{pmatrix} 1 & 0 \\ 0 & -1 \end{pmatrix}$	$\begin{pmatrix} 0 & 1 \\ 1 & 0 \end{pmatrix}$
			2	$\begin{pmatrix} 1 & 0 \\ 0 & -1 \end{pmatrix}$	$\begin{pmatrix} -1 & 0 \\ 0 & 1 \end{pmatrix}$	$\begin{pmatrix} 0 & 1 \\ 1 & 0 \end{pmatrix}$
			3	$\begin{pmatrix} \gamma^2 & 0 & 0 & 0 \\ 0 & -\gamma & 0 & 0 \\ 0 & 0 & -\gamma^2 & 0 \\ 0 & 0 & 0 & \gamma \end{pmatrix}$	$\begin{pmatrix} 0 & 1 & 0 & 0 \\ 1 & 0 & 0 & 0 \\ 0 & 0 & 0 & -1 \\ 0 & 0 & -1 & 0 \end{pmatrix}$	$\begin{pmatrix} 0 & 0 & 1 & 0 \\ 0 & 0 & 0 & 1 \\ 1 & 0 & 0 & 0 \\ 0 & 1 & 0 & 0 \end{pmatrix}$

(continued)

Table 6.73 (continued)

$\omega_{B'A'}$	$\omega_{I'A'}$	$\omega_{I'B'}$	j	A'	B'	I'
$-1, \gamma, \gamma^5$	1	1	$1^{\pm}$	$\begin{pmatrix} 1 & 0 \\ 0 & -1 \end{pmatrix}$	$\begin{pmatrix} 0 & 1 \\ 1 & 0 \end{pmatrix}$	$\pm\begin{pmatrix} 1 & 0 \\ 0 & 1 \end{pmatrix}$
			$2^{\pm}$	$\begin{pmatrix} \gamma^2 & 0 \\ 0 & \gamma \end{pmatrix}$	$\begin{pmatrix} 0 & 1 \\ 1 & 0 \end{pmatrix}$	$\pm\begin{pmatrix} 1 & 0 \\ 0 & 1 \end{pmatrix}$
			$3^{\pm}$	$\begin{pmatrix} -\gamma & 0 \\ 0 & -\gamma^2 \end{pmatrix}$	$\begin{pmatrix} 0 & 1 \\ 1 & 0 \end{pmatrix}$	$\pm\begin{pmatrix} 1 & 0 \\ 0 & 1 \end{pmatrix}$
		-1	$1^{\pm}$	$\begin{pmatrix} 1 & 0 \\ 0 & -1 \end{pmatrix}$	$\begin{pmatrix} 0 & 1 \\ 1 & 0 \end{pmatrix}$	$\pm\begin{pmatrix} 1 & 0 \\ 0 & -1 \end{pmatrix}$
			$2^{\pm}$	$\begin{pmatrix} \gamma^2 & 0 \\ 0 & \gamma \end{pmatrix}$	$\begin{pmatrix} 0 & 1 \\ 1 & 0 \end{pmatrix}$	$\pm\begin{pmatrix} 1 & 0 \\ 0 & -1 \end{pmatrix}$
			$3^{\pm}$	$\begin{pmatrix} -\gamma & 0 \\ 0 & -\gamma^2 \end{pmatrix}$	$\begin{pmatrix} 0 & 1 \\ 1 & 0 \end{pmatrix}$	$\pm\begin{pmatrix} 1 & 0 \\ 0 & -1 \end{pmatrix}$
	-1	1	$1^{\pm}$	$\begin{pmatrix} 1 & 0 \\ 0 & -1 \end{pmatrix}$	$\begin{pmatrix} 0 & 1 \\ 1 & 0 \end{pmatrix}$	$\pm\begin{pmatrix} 0 & 1 \\ 1 & 0 \end{pmatrix}$
			2	$\begin{pmatrix} \gamma^2 & 0 & 0 & 0 \\ 0 & \gamma & 0 & 0 \\ 0 & 0 & -\gamma^2 & 0 \\ 0 & 0 & 0 & -\gamma \end{pmatrix}$	$\begin{pmatrix} 0 & 1 & 0 & 0 \\ 1 & 0 & 0 & 0 \\ 0 & 0 & 0 & 1 \\ 0 & 0 & 1 & 0 \end{pmatrix}$	$\begin{pmatrix} 0 & 0 & 1 & 0 \\ 0 & 0 & 0 & 1 \\ 1 & 0 & 0 & 0 \\ 0 & 1 & 0 & 0 \end{pmatrix}$
		-1	$1^{\pm}$	$\begin{pmatrix} 1 & 0 \\ 0 & -1 \end{pmatrix}$	$\begin{pmatrix} 0 & 1 \\ 1 & 0 \end{pmatrix}$	$\pm\begin{pmatrix} 0 & 1 \\ -1 & 0 \end{pmatrix}$
			2	$\begin{pmatrix} \gamma^2 & 0 & 0 & 0 \\ 0 & \gamma & 0 & 0 \\ 0 & 0 & -\gamma^2 & 0 \\ 0 & 0 & 0 & -\gamma \end{pmatrix}$	$\begin{pmatrix} 0 & 1 & 0 & 0 \\ 1 & 0 & 0 & 0 \\ 0 & 0 & 0 & -1 \\ 0 & 0 & -1 & 0 \end{pmatrix}$	$\begin{pmatrix} 0 & 0 & 1 & 0 \\ 0 & 0 & 0 & 1 \\ 1 & 0 & 0 & 0 \\ 0 & 1 & 0 & 0 \end{pmatrix}$

6.18 Group T_h

See Tables 6.74, 6.75, 6.76 and 6.77.

Table 6.74 Realization of the group in O_h

T_h	as T in Table 6.54	i	ai	bi	abi	ci	$abci$	aci	bci	c^2i	bc^2i	abc^2i	ac^2i
		I	$3IC_2$			$4IC_3$				$4IC_3^2$			
		25	26	27	28	35	38	40	42	36	37	39	41

Table 6.75 Defining relations

as in Table 6.55 plus	
$i^2 = e, \quad ia = ai, \quad ib = bi, \quad ic = ci$	$I'^2 = E, \quad I'A' = A'I', \quad I'B' = B'I', \quad I'C' = C'I'$

Table 6.76 Character and gauge factors

as in Table 6.56	δ_I
	$\omega_{i,i}^{1/2}$

Table 6.77 Matrices of the generators

as in Table 6.57 plus the unit matrix with the ± sign for the generator I'

6.19 Group O_h

See Tables 6.78, 6.79, 6.80 and 6.81.

Table 6.78 Realization of the group

O_h	as O in Table 6.66	i	ai	bi	abi	c^2di	ac^2di	$bcdi$	cdi	bdi	adi	ci	c^2i
		I	$3IC_4^2$			$6IC_4$							
		25	26	27	28	29	30	31	32	33	34	35	36

bc^2i	$abci$	abc^2i	aci	ac^2i	bci	di	$abdi$	$abcdi$	$acdi$	abc^2di	bc^2di
$8IC_3$						$6IC_2$					
37	38	39	40	41	42	43	44	45	46	47	48

Table 6.79 Defining relations

as in Table 6.67 plus	
$i^2 = e, \quad ia = ai, \quad ib = bi$	$I'^2 = E, \quad I'A' = A'I', \quad I'B'' = B''I'$
$ic = ci, \quad id = di$	$I'C'' = C''I', \quad I'D' = \omega_{I'D'} D'I'$
	$\omega_{I'D'}^2 = 1$

Table 6.80 Character and gauge factors

as in Table 6.68	$\omega_{I'D'}$	δ_I
	$\omega_{i,d}\omega_{d,i}^*$	$\omega_{i,i}^{1/2}$

Table 6.81 Matrices of the generators

$\omega_{C'B'}$	$\omega_{I'D'}$	j	A'	B''	C''	D'	I'
±1	1	$1^{\pm}$	1	1	1	1	±1
		$2^{\pm}$	1	1	1	-1	±1
		$3^{\pm}$	$\begin{pmatrix}1&0\\0&1\end{pmatrix}$	$\begin{pmatrix}1&0\\0&1\end{pmatrix}$	$\begin{pmatrix}\varepsilon&0\\0&\varepsilon^2\end{pmatrix}$	$\begin{pmatrix}0&1\\1&0\end{pmatrix}$	$\pm\begin{pmatrix}1&0\\0&1\end{pmatrix}$
		$4^{\pm}$	$\begin{pmatrix}-1&0&0\\0&1&0\\0&0&-1\end{pmatrix}$	$\begin{pmatrix}1&0&0\\0&-1&0\\0&0&-1\end{pmatrix}$	$\begin{pmatrix}0&1&0\\0&0&1\\1&0&0\end{pmatrix}$	$\begin{pmatrix}0&1&0\\1&0&0\\0&0&-1\end{pmatrix}$	$\pm\begin{pmatrix}1&0&0\\0&1&0\\0&0&1\end{pmatrix}$
		$5^{\pm}$	$\begin{pmatrix}-1&0&0\\0&1&0\\0&0&-1\end{pmatrix}$	$\begin{pmatrix}1&0&0\\0&-1&0\\0&0&-1\end{pmatrix}$	$\begin{pmatrix}0&1&0\\0&0&1\\1&0&0\end{pmatrix}$	$\begin{pmatrix}0&-1&0\\-1&0&0\\0&0&1\end{pmatrix}$	$\pm\begin{pmatrix}1&0&0\\0&1&0\\0&0&1\end{pmatrix}$
	-1	1	$\begin{pmatrix}1&0\\0&1\end{pmatrix}$	$\begin{pmatrix}1&0\\0&1\end{pmatrix}$	$\begin{pmatrix}1&0\\0&1\end{pmatrix}$	$\begin{pmatrix}1&0\\0&-1\end{pmatrix}$	$\begin{pmatrix}0&1\\1&0\end{pmatrix}$
		$2^{\pm}$	$\begin{pmatrix}1&0\\0&1\end{pmatrix}$	$\begin{pmatrix}1&0\\0&1\end{pmatrix}$	$\begin{pmatrix}\varepsilon&0\\0&\varepsilon^2\end{pmatrix}$	$\begin{pmatrix}0&1\\1&0\end{pmatrix}$	$\pm\begin{pmatrix}1&0\\0&-1\end{pmatrix}$
		3	$\begin{pmatrix}-1&0&0&0&0&0\\0&1&0&0&0&0\\0&0&-1&0&0&0\\0&0&0&-1&0&0\\0&0&0&0&1&0\\0&0&0&0&0&-1\end{pmatrix}$	$\begin{pmatrix}1&0&0&0&0&0\\0&-1&0&0&0&0\\0&0&-1&0&0&0\\0&0&0&1&0&0\\0&0&0&0&-1&0\\0&0&0&0&0&-1\end{pmatrix}$	$\begin{pmatrix}0&1&0&0&0&0\\1&0&0&0&0&0\\0&0&-1&0&0&0\\0&0&0&0&-1&0\\0&0&0&-1&0&0\\0&0&0&0&0&1\end{pmatrix}$	$\begin{pmatrix}0&1&0&0&0&0\\1&0&0&0&0&0\\0&0&-1&0&0&0\\0&0&0&0&-1&0\\0&0&0&-1&0&0\\0&0&0&0&0&1\end{pmatrix}$	$\begin{pmatrix}0&0&0&1&0&0\\0&0&0&0&1&0\\0&0&0&0&0&1\\1&0&0&0&0&0\\0&1&0&0&0&0\\0&0&1&0&0&0\end{pmatrix}$

(continued)

Table 6.81 (continued)

$\omega_{C'B'}$	$\omega_{I'D'}$	j	A'	B''	C''	D'	I'
$\pm i$	1	$1^{\pm}$	$\begin{pmatrix} 1 & 0 \\ 0 & -1 \end{pmatrix}$	$\begin{pmatrix} 0 & 1 \\ 1 & 0 \end{pmatrix}$	$\frac{1}{\sqrt{2}}\begin{pmatrix} -\sigma & -\sigma^3 \\ -\sigma & \sigma^3 \end{pmatrix}$	$\frac{1}{\sqrt{2}}\begin{pmatrix} 1 & 1 \\ 1 & -1 \end{pmatrix}$	$\pm\begin{pmatrix} 1 & 0 \\ 0 & 1 \end{pmatrix}$
		$2^{\pm}$	$\begin{pmatrix} 1 & 0 \\ 0 & -1 \end{pmatrix}$	$\begin{pmatrix} 0 & 1 \\ 1 & 0 \end{pmatrix}$	$\frac{1}{\sqrt{2}}\begin{pmatrix} -\sigma & -\sigma^3 \\ -\sigma & \sigma^3 \end{pmatrix}$	$\frac{1}{\sqrt{2}}\begin{pmatrix} -1 & -1 \\ -1 & 1 \end{pmatrix}$	$\pm\begin{pmatrix} 1 & 0 \\ 0 & 1 \end{pmatrix}$
		$3^{\pm}$	$\begin{pmatrix} 1 & 0 & 0 & 0 \\ 0 & -1 & 0 & 0 \\ 0 & 0 & 0 & 1 \\ 0 & 0 & 1 & 0 \end{pmatrix}$	$\begin{pmatrix} 0 & 1 & 0 & 0 \\ 1 & 0 & 0 & 0 \\ 0 & 0 & 1 & 0 \\ 0 & 0 & 0 & -1 \end{pmatrix}$	$\frac{1}{\sqrt{2}}\begin{pmatrix} -\varphi^{11} & \varphi^5 & 0 & 0 \\ -\varphi^{11} & -\varphi^5 & 0 & 0 \\ 0 & 0 & \varphi^7 & \varphi^7 \\ 0 & 0 & -\varphi & \varphi \end{pmatrix}$	$\begin{pmatrix} 0 & 0 & 1 & 0 \\ 0 & 0 & 0 & 1 \\ 1 & 0 & 0 & 0 \\ 0 & 1 & 0 & 0 \end{pmatrix}$	$\pm\begin{pmatrix} 1 & 0 & 0 & 0 \\ 0 & 1 & 0 & 0 \\ 0 & 0 & 1 & 0 \\ 0 & 0 & 0 & 1 \end{pmatrix}$
	-1	1	$\begin{pmatrix} 1 & 0 & 0 & 0 \\ 0 & -1 & 0 & 0 \\ 0 & 0 & 1 & 0 \\ 0 & 0 & 0 & -1 \end{pmatrix}$	$\begin{pmatrix} 0 & 1 & 0 & 0 \\ 1 & 0 & 0 & 0 \\ 0 & 0 & 0 & 1 \\ 0 & 0 & 1 & 0 \end{pmatrix}$	$\frac{1}{\sqrt{2}}\begin{pmatrix} -\sigma & -\sigma^3 & 0 & 0 \\ -\sigma & \sigma^3 & 0 & 0 \\ 0 & 0 & -\sigma & -\sigma^3 \\ 0 & 0 & -\sigma & \sigma^3 \end{pmatrix}$	$\frac{1}{\sqrt{2}}\begin{pmatrix} 1 & 1 & 0 & 0 \\ 1 & -1 & 0 & 0 \\ 0 & 0 & -1 & -1 \\ 0 & 0 & -1 & 1 \end{pmatrix}$	$\pm\begin{pmatrix} 0 & 0 & 1 & 0 \\ 0 & 0 & 0 & 1 \\ 1 & 0 & 0 & 0 \\ 0 & 1 & 0 & 0 \end{pmatrix}$
		$2^{\pm}$	$\begin{pmatrix} 1 & 0 & 0 & 0 \\ 0 & -1 & 0 & 0 \\ 0 & 0 & 0 & 1 \\ 0 & 0 & 1 & 0 \end{pmatrix}$	$\begin{pmatrix} 0 & 1 & 0 & 0 \\ 1 & 0 & 0 & 0 \\ 0 & 0 & 1 & 0 \\ 0 & 0 & 0 & -1 \end{pmatrix}$	$\frac{1}{\sqrt{2}}\begin{pmatrix} -\varphi^{11} & \varphi^5 & 0 & 0 \\ -\varphi^{11} & -\varphi^5 & 0 & 0 \\ 0 & 0 & \varphi^7 & \varphi^7 \\ 0 & 0 & -\varphi & \varphi \end{pmatrix}$	$\begin{pmatrix} 0 & 0 & 1 & 0 \\ 0 & 0 & 0 & 1 \\ 1 & 0 & 0 & 0 \\ 0 & 1 & 0 & 0 \end{pmatrix}$	$\pm\begin{pmatrix} 1 & 0 & 0 & 0 \\ 0 & 1 & 0 & 0 \\ 0 & 0 & -1 & 0 \\ 0 & 0 & 0 & -1 \end{pmatrix}$

6.20 Example of the Use of the Tables: The Space Group O_h^7

In this section we illustrate the use of our tables in the example of the space group O_h^7.

The translation subgroup of the space group O_h^7 consists of the transformations

$$\{\varepsilon|\mathbf{R}_n\}, \qquad \mathbf{R}_n = n_1\mathbf{a}_1 + n_2\mathbf{a}_2 + n_3\mathbf{a}_3, \tag{6.42}$$

where n_i are integers and $\mathbf{a}_i$ are the basic vectors,

$$\mathbf{a}_1 = a(0, 1/2, 1/2), \qquad \mathbf{a}_2 = a(1/2, 0, 1/2), \qquad \mathbf{a}_3 = a(1/2, 1/2, 0) \tag{6.43}$$

(a is the lattice constant). The point group of this space group is O_h, and, if we take a lattice site as an origin, the nonprimitive translations are given by

$$\tau_{\alpha_i} = \begin{cases} 0 & \text{for } i = 1\text{–}4,\ 11\text{–}18,\ 29\text{–}34,\ 43\text{–}48, \\ \tau = a(1/4, 1/4, 1/4) & \text{for } i = 5\text{–}10,\ 19\text{–}28,\ 35\text{–}42. \end{cases} \tag{6.44}$$

We consider the "unlucky" symmetry point W with the vector $\mathbf{k} = 2\pi/a\,(1/2, 1, 0)$.[8] The point group of this vector consists of the elements α_m with $m = 1, 2, 23, 24, 27, 28, 29, 30$, that form the group D_{2d}.

Writing each element of the group as $\prod_{i=1}^{p} g_i^{\mu_i}$ and using Table 6.36, we choose the corresponding element α_m (see Table 6.82). Using formulas (6.5) and (6.6) and taking (6.44) into account, we obtain the projective factors that enter the expressions for the character and gauge factors in Table 6.38. The results of the calculation are presented in the second row of Table 6.83. Using these results, we have

$$\omega_{B'A'} = -i\,, \tag{6.45}$$

$$\delta_A = i\,, \qquad \delta_B = e^{i\pi/4} \equiv \sigma\,. \tag{6.46}$$

Writing out the matrices A' and B' for the case $\omega_{B'A'} = \pm i$ from Table 6.39 and multiplying them by the gauge factors (6.46), we finally obtain the two projective representations for the point W (see Table 6.84).

Table 6.82 Realization of the group D_{2d} in O_h

e	a	a^3	a^2	b	a^2b	a^3b	ab
α_1	α_{30}	α_{29}	α_2	α_{28}	α_{27}	α_{24}	α_{23}

[8] Herring [4] made a mistake at the point W when he constructed the characters of the single-valued representations of the space group O_h^7. A quarter century later, Zak et al. [5] presented completely wrong characters of the double-valued representations at this point.

Table 6.83 Projective factors that enter the expressions for the character and gauge factors: for the single valued (in the second row) and for the double-valued representations (in the third row) of the point W

$\omega_{b,a}$	$\omega_{a,a}$	$\omega_{a^2,a}$	$\omega_{a^3,b}$	$\omega_{a^3,a}$	$\omega_{b,b}$
1	1	1	i	1	i
1	-1	1	i	1	$-i$

Table 6.84 Matrices of the single-valued projective representations for the generators at the point W

j	$A \equiv D(a)$	$B \equiv D(b)$
	$\Delta^{\mathbf{k}j}(\{\alpha_{30}\|\boldsymbol{\tau}_{\alpha_{30}}\})$	$\Delta^{\mathbf{k}j}(\{\alpha_{28}\|\boldsymbol{\tau}_{\alpha_{28}}\})$
1	$\begin{pmatrix} i & 0 \\ 0 & -1 \end{pmatrix}$	$\begin{pmatrix} 0 & \sigma \\ \sigma & 0 \end{pmatrix}$
2	$\begin{pmatrix} -i & 0 \\ 0 & 1 \end{pmatrix}$	$\begin{pmatrix} 0 & \sigma \\ \sigma & 0 \end{pmatrix}$

In order to construct the *double-valued* representations for the point W, we determine the projective factors that enter the expressions for the character and gauge factors using formula (6.36), i.e. each of the projective factors previously obtained by formulas (6.5) and (6.6), we multiply by the corresponding value of the factor $\omega_{\alpha_m,\alpha_{m'}}$ taken from Table 6.2. The resulting values of the projective factors are shown in the third row of Table 6.83. Using these values, we have the character and gauge factors,

$$\omega_{B'A'} = -1\,, \tag{6.47}$$

$$\delta_A = e^{i3\pi/4} \equiv \sigma^3\,, \qquad \delta_B = e^{-i\pi/4} \equiv -\sigma^3\,. \tag{6.48}$$

Now, writing out the matrices A', B' for the case $\omega_{B'A'} = \pm 1$ from Table 6.39 and multiplying them by the gauge factors (6.48), we obtain the double-valued representations for the point W (Table 6.85).

In Table 6.87 we present the matrices of the single- and double-valued projective representations $\Delta^{\mathbf{k}j}$ for *all* elements $\{\alpha_m|\boldsymbol{\tau}_{\alpha_m}\}$, which are obtained similarly to (6.38) by using Tables 6.82, 6.83, 6.84, 6.85 and 6.86.

Note that the double-valued representations 1, 2 and 3, 4 are degenerate due to the time-reversal symmetry.

Table 6.85 Matrices of the double-valued projective representations for the generators at the point W

j	$A \equiv D(a)$ $\Delta^{\mathbf{k}j}(\{\alpha_{30}\|\tau_{\alpha_{30}}\})$	$B \equiv D(b)$ $\Delta^{\mathbf{k}j}(\{\alpha_{28}\|\tau_{\alpha_{28}}\})$
1	σ^3	$-\sigma^3$
2	σ^3	σ^3
3	$-\sigma^3$	$-\sigma^3$
4	$-\sigma^3$	σ^3
5	$\begin{pmatrix} -\sigma & 0 \\ 0 & \sigma \end{pmatrix}$	$\begin{pmatrix} 0 & -\sigma^3 \\ -\sigma^3 & 0 \end{pmatrix}$

Table 6.86 Additional projective factors (on top of the ones in Table 6.83) that are neccessary to obtain representation matrices for all elements at the point W

$\omega_{a,b}$	$\omega_{a^2,b}$
$-i$	-1
$-i$	1

Table 6.87 Representations of the group of the $\mathbf{k} = 2\pi/a(1/2, 1, 0)$ vector (the point W) for the space group O_h^7

Reps.	j	$D(e)$ $\Delta^{\mathbf{k}j}(\{\alpha_1\|\boldsymbol{\tau}_{\alpha_1}\})$	$D(a)$ $\Delta^{\mathbf{k}j}(\{\alpha_{30}\|\boldsymbol{\tau}_{\alpha_{30}}\})$	$D(a^2)$ $\Delta^{\mathbf{k}j}(\{\alpha_2\|\boldsymbol{\tau}_{\alpha_2}\})$	$D(a^3)$ $\Delta^{\mathbf{k}j}(\{\alpha_{29}\|\boldsymbol{\tau}_{\alpha_{29}}\})$	$D(b)$ $\Delta^{\mathbf{k}j}(\{\alpha_{28}\|\boldsymbol{\tau}_{\alpha_{28}}\})$	$D(ab)$ $\Delta^{\mathbf{k}j}(\{\alpha_{23}\|\boldsymbol{\tau}_{\alpha_{23}}\})$	$D(a^2b)$ $\Delta^{\mathbf{k}j}(\{\alpha_{27}\|\boldsymbol{\tau}_{\alpha_{27}}\})$	$D(a^3b)$ $\Delta^{\mathbf{k}j}(\{\alpha_{24}\|\boldsymbol{\tau}_{\alpha_{24}}\})$
Single valued	1	$\begin{pmatrix} 1 & 0 \\ 0 & 1 \end{pmatrix}$	$\begin{pmatrix} i & 0 \\ 0 & -1 \end{pmatrix}$	$\begin{pmatrix} -1 & 0 \\ 0 & 1 \end{pmatrix}$	$\begin{pmatrix} -i & 0 \\ 0 & -1 \end{pmatrix}$	$\begin{pmatrix} 0 & \sigma \\ \sigma & 0 \end{pmatrix}$	$\begin{pmatrix} 0 & -\sigma \\ -\sigma^3 & 0 \end{pmatrix}$	$\begin{pmatrix} 0 & \sigma \\ -\sigma & 0 \end{pmatrix}$	$\begin{pmatrix} 0 & -\sigma \\ \sigma^3 & 0 \end{pmatrix}$
	2	$\begin{pmatrix} 1 & 0 \\ 0 & 1 \end{pmatrix}$	$\begin{pmatrix} -i & 0 \\ 0 & 1 \end{pmatrix}$	$\begin{pmatrix} -1 & 0 \\ 0 & 1 \end{pmatrix}$	$\begin{pmatrix} i & 0 \\ 0 & 1 \end{pmatrix}$	$\begin{pmatrix} 0 & \sigma \\ \sigma & 0 \end{pmatrix}$	$\begin{pmatrix} 0 & \sigma \\ \sigma^3 & 0 \end{pmatrix}$	$\begin{pmatrix} 0 & \sigma \\ -\sigma & 0 \end{pmatrix}$	$\begin{pmatrix} 0 & \sigma \\ -\sigma^3 & 0 \end{pmatrix}$
Double valued	$\left\{\begin{matrix} 1 \\ 2 \end{matrix}\right.$	1 1	σ^3 σ^3	i i	$-\sigma$ $-\sigma$	$-\sigma^3$ σ^3	-1 1	σ $-\sigma$	i $-i$
	$\left\{\begin{matrix} 3 \\ 4 \end{matrix}\right.$	1 1	$-\sigma^3$ $-\sigma^3$	i i	σ σ	$-\sigma^3$ σ^3	1 -1	σ $-\sigma$	$-i$ i
	5	$\begin{pmatrix} 1 & 0 \\ 0 & 1 \end{pmatrix}$	$\begin{pmatrix} -\sigma & 0 \\ 0 & \sigma \end{pmatrix}$	$\begin{pmatrix} -i & 0 \\ 0 & -i \end{pmatrix}$	$\begin{pmatrix} \sigma^3 & 0 \\ 0 & -\sigma^3 \end{pmatrix}$	$\begin{pmatrix} 0 & -\sigma^3 \\ -\sigma^3 & 0 \end{pmatrix}$	$\begin{pmatrix} 0 & -i \\ i & 0 \end{pmatrix}$	$\begin{pmatrix} 0 & -\sigma \\ -\sigma & 0 \end{pmatrix}$	$\begin{pmatrix} 0 & 1 \\ -1 & 0 \end{pmatrix}$

References

1. B.I. Reser, R.F. Egorov, V.K. Zvezdin, *Representations of the Space Groups. II. Tables* [in Russian] (1979), Dep. VINITI, No. 568–79
2. O.V. Kovalev, *Irreducible representations of space groups* [in Russian] (Acad. Sci. Ukrain. SSR, Kiev, 1961). Engl. transl.: Gordon and Breach, New York, 1965
3. V. Heine, *Group Theory in Quantum Mechanics* (Pergamon, Oxford, 1960). Reprinted: Dover, New York, 2009
4. C. Herring, J. Franklin Inst. **233**, 525 (1942)
5. J. Zak, A. Casher, M. Glück, Y. Gur, *The Irreducible Representations of Space Groups* (Benjamin, New York, 1969)

Chapter 7
Group Theory and Quantum Mechanics

> *The theory of the representation of groups... is exactly mathematically most important part which is necessary for an adequate description of the quantum mechanical relations. All quantum numbers, with the exception of the so-called principal quantum number, are indices characterizing representations of groups. (H. Weyl, The theory of groups and quantum mechanics, Dover, New York, 1931)*

So far we have studied *finite* groups and their representations. As an example, we considered crystallographic space groups. Now we get back to physics.

Symmetry of a quantum mechanical problem manifests itself in the invariance of the Hamiltonian, which leads to certain transformational properties of its eigenfunctions. First we consider the relation between Hilbert space transformations and group representations. This relation underlies the application of group theory to quantum mechanics. Then we sum up the use of group theoretical methods to quantum mechanics in a *general* form, without focusing on particular problems. Finally, we consider three major application domains: classification of states, splitting of states and selection rules (for details, see [1, 2]).

7.1 Representation Spaces

7.1.1 Basis Functions

Let G be a given group with elements $a \in G$. We construct an isomorphic mapping of the elements a on unitary operators P_a in Hilbert space such that

$$P_a P_b = P_{ab} \,.$$

The operators P_a should be understood as *linear* transformations of Hilbert space. Let also $\mathcal{D}$ be an l-dimensional unitary representation of the group G. Then l functions

N. B. Melnikov and B. I. Reser, *Space Group Representations*,
https://doi.org/10.1007/978-3-031-13991-8_7

f_λ ($\lambda = 1, ..., l$) of Hilbert space are called *basis functions* of the representation $\mathcal{D}$, if the following relation holds

$$P_a f_\lambda = \sum_{\mu=1}^{l} D_{\mu\lambda}(a) f_\mu \,, \tag{7.1}$$

for all $a \in G$ and $\lambda = 1, ..., l$. In this case, we say that the function f_λ *belongs to the* λ*-th row of the representation* $\mathcal{D}$. The other functions f_μ required to complete the basis for the representation are called the *partners* of the given function [3, 4].

The basis functions of an l-dimensional unitary representation $\mathcal{D}$ form an orthogonal set of functions and can be normalized to form an orthonormal basis. We will always assume that the functions f_λ are normalized.

The basis functions f_λ of the representation $\mathcal{D}$ define an l-dimensional subspace V_l of Hilbert space. We call V_l a *representation space* of the representation $\mathcal{D}$. The functions of this space will be called *representation functions* of the representation $\mathcal{D}$. In general, a representation $\mathcal{D}$ can have several representation spaces. On the opposite, a representation space determines a representation uniquely up to an equivalence relation.

From (7.1) it follows that any representation space is an *invariant* subspace of Hilbert space under the operators P_a ($a \in G$). The inverse is also true: every subspace V_l that is *invariant* under the operators P_a is a representation space of a representation of the group G.

We can formulate the following theorem. Let the representation space V_l of a representation $\mathcal{D}$ be an invariant subspace under the operators P_a of the group G. Then the invariant subspace V_l is irreducible if and only if the representation $\mathcal{D}$ is irreducible.

7.1.2 Expansion Theorem

Let $\mathcal{D}_1, ..., \mathcal{D}_r$ be a *complete* set of irreducible representations of the group G and f be a function of Hilbert space. Then one can construct representation spaces V_p ($p = 1, ..., r$) with the basis functions $f_\kappa^{(p)}$ ($\kappa = 1, ..., l_p$, where l_p is the dimension of the representation $\mathcal{D}_p$) such that

$$f = \sum_{p=1}^{r} \sum_{\kappa=1}^{l_p} c_\kappa^{(p)} f_\kappa^{(p)} . \tag{7.2}$$

In other words, any function f can be expanded in terms of representation functions of irreducible representations of the group G.

7.1.3 Reduction by Idempotent Operators

According to (7.2) any function can be presented as a linear combination of basis functions of irreducible representations of the group G. Hence we can obtain projection operators that correspond to separate rows of these representations.

We introduce the operators constructed from a given irreducible representation of the group G:

$$O^p_{\mu\nu} = \frac{l_p}{n(G)} \sum_{a \in G} D^{(p)*}_{\mu\nu}(a) P_a . \tag{7.3}$$

Here $n(G)$ is the order of the group G, and l_p is the dimension of the representation $\mathcal{D}_p$. Assuming that the operators P_a are unitary, one can prove that

$$(O^p_{\mu\nu})^\dagger = O^p_{\nu\mu}, \tag{7.4}$$

$$O^p_{\mu\nu} O^{p'}_{\mu'\nu'} = \delta_{pp'} \delta_{\nu\mu'} O^p_{\mu\nu'} . \tag{7.5}$$

Hence the operators $O^p_{\mu\mu}$ are Hermitian and idempotent[1],

$$(O^p_{\mu\mu})^\dagger = O^p_{\mu\mu}, \qquad (O^p_{\mu\mu})^2 = O^p_{\mu\mu},$$

(these are exactly the *projection operators*), and operators $O^p_{\mu\nu}$ $(\mu \neq \nu)$ are nilpotent, $(O^p_{\mu\nu})^2 = 0$.

Now it is easy to show that the projection operator $O^p_{\mu\mu}$, applied to an arbitrary function φ in Hilbert space, yields the basis function belonging to the μ-th row of the representation $\mathcal{D}_p$ and that $O^p_{\mu\nu}$, applied to the same function φ, yields all basis functions of the representation $\mathcal{D}_p$ when ν is fixed and μ goes from 1 to l_p.

Indeed, according to (7.2), any function φ can be represented as

$$\varphi = \sum_{p'=1}^{r} \sum_{\kappa=1}^{l_{p'}} c^{(p')}_\kappa f^{(p')}_\kappa,$$

where $f^{(p')}_\kappa$ is the basis function belonging to the κ-th row of the p'-th representation $\mathcal{D}_{p'}$. Then

$$O^p_{\mu\nu}\varphi = \frac{l_p}{n(G)} \sum_{p'\kappa} c^{(p')}_\kappa \sum_{a \in G} D^{(p)*}_{\mu\nu}(a) P_a\, f^{(p')}_\kappa .$$

---

[1] An operator A is called *idempotent* if $A^2 = A$ and *nilpotent* if $A^2 = 0$.

By expression (7.1), the action of the projection operator P_a is given by

$$P_a f_\kappa^{(p')} = \sum_{\lambda=1}^{l_{p'}} D_{\lambda\kappa}^{(p')}(a) f_\lambda^{(p')} .$$

Then

$$O_{\mu\nu}^p \varphi = \frac{l_p}{n(G)} \sum_{p'\kappa} c_\kappa^{(p')} \sum_\lambda \sum_{a \in G} D_{\mu\nu}^{(p)*}(a) D_{\lambda\kappa}^{(p')}(a) f_\lambda^{(p')} .$$

Using the orthogonality relations (2.8),

$$\sum_{a \in G} D_{\mu\nu}^{(p)*}(a) D_{\lambda\kappa}^{(p')}(a) = \frac{n(G)}{l_p} \delta_{\mu\lambda} \delta_{\nu\kappa} \delta'_{pp'} ,$$

we obtain

$$O_{\mu\nu}^p \varphi = c_\nu^{(p)} f_\mu^{(p)} .$$

In particular, if the expansion of φ does not have the component $f_\mu^{(p)}$, then $O_{\mu\nu}^p \varphi = 0$.

From the expansion theorem it follows that the sum of all projection operators $O_{\mu\mu}^p$ is equal to the identity operator, i.e.

$$\sum_{p=1}^{r} \sum_{\mu=1}^{l_p} O_{\mu\mu}^p = 1.$$

7.1.4 *Orthogonality of the Basis Functions*

We can now formulate a fundamental theorem that will be used for derivation of the selection rules further in this chapter.

If $f_\kappa^{(p)}$ is the basis function belonging to the κ-th row of the irreducible representation $\mathcal{D}_p$ and $g_{\kappa'}^{(p')}$ is the basis function belonging to the κ'-th row of the irreducible representation $\mathcal{D}_{p'}$ of the same group G, then

$$(f_\kappa^{(p)}, g_{\kappa'}^{(p')}) = \delta_{pp'} \delta_{\kappa\kappa'} (f_\lambda^{(p)}, g_\lambda^{(p)}) \quad (\lambda \text{ is arbitrarily}), \qquad (7.6)$$

i.e. the basis functions belonging to different irreducible representations or different rows of the same representation are orthogonal. If two basis functions belong to the same row of an irreducible representation, then their scalar product is independent of the row number, i.e. $(f_\kappa^{(p)}, g_\kappa^{(p)})$ is independent of κ.[2]

7.1.5 *Criterion for Representation Functions*

Results of Sect. 7.1.3 lead to a necessary and sufficient condition for a function $f_\kappa^{(p)}$ to belong to the κ-th row of the irreducible representation $\mathcal{D}_p$ of the group G.

A function $f_\kappa^{(p)}$ belongs to the κ-th row of the irreducible representation $\mathcal{D}_p$ of the group G if and only if

$$\sum_{a\in G} D_{\kappa\kappa}^{(p)*}(a) P_a f_\kappa^{(p)} = \frac{n(G)}{l_p} f_\kappa^{(p)}. \tag{7.7}$$

If the function $f_\kappa^{(p)}$ satisfies relation (7.7), we can introduce l_p functions

$$f_\lambda^{(p)} = \frac{l_p}{n(G)} \sum_{a\in G} D_{\lambda\kappa}^{(p)*}(a) P_a f_\kappa^{(p)} \tag{7.8}$$

($f_\kappa^{(p)}$ being one of them) that form a basis of the representation $\mathcal{D}_p$.

In conclusion, we note that (7.8) allows to obtain the entire basis of the representation $\mathcal{D}$ if we know only one of its functions f_κ belonging to the κ-th row of the representation, no matter whether the representation $\mathcal{D}$ is irreducible or not.

7.2 Classification of States

7.2.1 *Hamiltonian of a System with Symmetry*

First of all we analyze the spatial symmetry of a one-particle system with the Hamiltonian $H = T + V$. Here symmetry reveals itself in the invariance of the potential $V(\mathbf{r})$.

Any linear spatial transformation $a \equiv \{\alpha|\mathbf{t}\}$ that leaves the potential invariant,

$$V(a\mathbf{r}) = V(\mathbf{r}), \tag{7.9}$$

[2] Nothing can be said about the scalar product of the basis functions of equivalent but not the same representations.

can be associated with an operator P_a in Hilbert space commuting with the potential $V(\mathbf{r})$. The action of the operator P_a on a function $f(\mathbf{r})$ is defined as follows[3]:

$$P_a f(\mathbf{r}) = f(a^{-1}\mathbf{r}). \tag{7.10}$$

Hence the action of the operator P_a on the function $f(\mathbf{r})$ produces a new function $f'(\mathbf{r}) = f(a^{-1}\mathbf{r})$, which satisfies the relation $f'(\mathbf{r}') = f(\mathbf{r})$, where $\mathbf{r}' = a\mathbf{r}$, i.e. the value of the transformed function $f' = P_a f$ in the transformed point $\mathbf{r}' = a\mathbf{r}$ is equal to the value of the function f at the point $\mathbf{r}$.

The operators P_a defined by (7.10) form a group, which is isomorphic to the group of spatial transformations a.

The operators P_a commute with the potential if the transformation a satisfies (7.9). Furthermore, if the transformation a belongs to a space group, the operators P_a commute also with the kinetic energy T. Thus, if relation (7.9) is satisfied, the operators P_a commute with the Hamiltonian $H = T + V$. These are called *symmetry operators of the Hamiltonian.*

7.2.2 Eigenspaces and Representations

The main result here is Wigner's theorem [8] (for further discussion of Wigner's theorem, see Appendix C.1).

Theorem 1 *(Wigner's theorem). Let H be the Hamiltonian of a physical system, E_p its eigenvalues and V_p the corresponding* eigenspaces. *That means V_p consists of all eigenfunctions of the operator H corresponding to the eigenvalue E_p. Let ψ_p be an eigenfunction of the eigenspace V_p,*

$$H\psi_p = E_p\psi_p\,.$$

Let also P_a be a symmetry operator of the Hamiltonian H. Then P_a commutes with H and hence

$$HP_a\psi_p = E_p P_a\psi_p\,,$$

i.e. the function $P_a\psi_p$ also belongs to the eigenvalue E_p and lies in the eigenspace V_p. That means the space V_p is invariant *under the symmetry group of the Hamiltonian H.*

If the symmetry group of the Hamiltonian consists of only *unitary* operators[4], then each eigenspace (since it is an *invariant* subspace) will be a representation subspace of the symmetry group representation. This representation can be both reducible

---

[3] In the monograph [5] the operators $P_{\{\alpha|\mathbf{t}\}}$ are *denoted* by $\overline{\{\alpha|\mathbf{t}\}}$. In [6, 7] the symmetry operators are defined in a *different* way. We follow Wigner's definition [1].

[4] We do not consider symmetry groups with *antiunitary* operators and their *corepresentations*. The antiunitary operator of time inversion will be considered separately (see Sect. 8.4).

and irreducible. The dimension of the eigenspace is called the *degeneracy* of the corresponding eigenvalue.

The eigenvalue is said to have a *symmetry-protected or normal degeneracy* under the symmetry group of the Hamiltonian if the corresponding representation is irreducible, and it is called *accidentally degenerate* if the representation is reducible.

Now we can formulate the *irreducibility postulate*: all eigenvalues of the Hamiltonian have a symmetry-protected degeneracy under the symmetry group. According to this postulate, every eigenspace will be a representation space of the symmetry group representation.

Accidental degeneracy occurs only for certain particular potentials. From experience, it can be either completely removed or shifted to another energy eigenvalue by a small change of the potential. Therefore, in practice the accidental degeneracy is inessential.

By choosing a complete orthonormal system of eigenfunctions of H, one automatically chooses an orthonormal basis in each eigenspace V_p. Therefore, the eigenfunctions ψ of H can be labeled by triples of indices, $\psi_{pm\kappa}$, where p indicates the irreducible representation $\mathcal{D}_p$ to which the function $\psi_{pm\kappa}$ belongs; m labels the eigenspaces of H corresponding to the same representation $\mathcal{D}_p$ and can, in principle, take on an infinite number of values ($m = 1, 2, ...$); κ indicates the row of the representation $\mathcal{D}_p$ and takes on the values $1, ..., l_p$.

The eigenvalues and eigenspaces can be indexed by p and m, i.e. one can write E_{pm}, V_{pm}.

In this way we obtain a natural, based only on symmetry, classification of eigenstates and eigenfunctions of the Hamiltonian. The degeneracy of an eigenstate E_{pm} is determined by the dimension of the representation $\mathcal{D}_p$ of the symmetry group and is equal to l_p.

7.3 Splitting of Eigenvalues

We study the effect of a *small* perturbation H_1 on the eigenvalues of a unperturbed Hamiltonian H_0.

Clearly, when the operator H_0 is replaced by $H = H_0 + H_1$ each eigenvalue can either shift or split if it is degenerate. The magnitude of this splitting though is determined by the specific type of the perturbation H_1 and not just its symmetry.

To introduce the notation, we briefly recall the Schrödinger perturbation theory.

7.3.1 Schrödinger Perturbation Theory

Let us consider a particular eigenvalue E^0_{pm} of the operator H. As usual, p labels the representation $\mathcal{D}_p$ to which the eigenspace belongs, and m enumerates the various eigenspaces belonging to the representation $\mathcal{D}_p$. Since we consider a fixed eigenvalue E^0_{pm}, for brevity we will omit the indices and write simply E^0.

In the l-dimensional eigenspace V^0 of the eigenvalue E^0, we introduce an orthonormal basis

$$\psi^0_1, ..., \psi^0_l \in V^0.$$

Under the action of a small perturbation H_1 the eigenvalue E^0 can split into n eigenvalues of the operator H:

$$E^0 \to E_1, ..., E_i, ..., E_n.$$

An eigenvalue E_i generates an l_i-dimensional eigenspace V_i of the operator H. In each V_i we choose an orthonormal basis

$$\psi_{i1}, ..., \psi_{il_i} \in V_i.$$

Clearly, the sum of the dimensions of V_i is equal to the dimension of V^0,

$$\sum_{i=1}^{n} l_i = l. \tag{7.11}$$

As $H_1 \to 0$, we have[5] $E_i \to E^0$ $(i = 1, ..., n)$. We write

$$E_i = E^0 + E^1_i, \qquad \text{where} \qquad \lim_{H_1 \to 0} E^1_i = 0, \tag{7.12}$$

and analogously

$$\psi_{i\lambda} = \psi^0_{i\lambda} + \psi^1_{i\lambda}, \qquad \text{where} \qquad \lim_{H_1 \to 0} \psi^1_{i\lambda} = 0. \tag{7.13}$$

Substituting (7.12) and (7.13) in the Schrödinger equation

$$H\psi_{i\lambda} = E_i \psi_{i\lambda}$$

and taking into account that H_1, E^1_i and $\psi^1_{i\lambda}$ are small, we can equate the terms of the same order. We obtain

$$H_0 \psi^0_{i\lambda} = E^0 \psi^0_{i\lambda}, \tag{7.14}$$

$$H_1 \psi^0_{i\lambda} + (H_0 - E^0)\psi^1_{i\lambda} = E^1_i \psi^0_{i\lambda}. \tag{7.15}$$

From (7.14) we see that the functions $\psi^0_{i\lambda}$, just as the functions ψ^0_ρ $(\rho = 1, ..., l)$ introduced earlier, are eigenfunctions of the unperturbed operator H_0 corresponding

[5] We assume that the symmetry group of the operator H_1 does not change as $H_1 \to 0$.

to the eigenvalue E^0. Since $\psi^0_{i\lambda}$ are obtained from orthonormal functions $\psi_{i\lambda}$ by means of a limiting process, the functions $\psi^0_{i\lambda}$ are also orthonormal and hence form an orthonormal basis of the eigenspace V^0. The functions $\psi^0_{i\lambda}$ are called *symmetry-adapted eigenfunctions (of zeroth order)*; they do not necessarily coincide with the basis functions ψ^0_ρ. Taking the scalar product of (7.15) with ψ^0_ρ, we have

$$(\psi^0_\rho, H_1\psi^0_{i\lambda}) = E^1_i(\psi^0_\rho, \psi^0_{i\lambda}).$$

Inserting the identity operator of the eigenspace V^0 between H_1 and $\psi^0_{i\lambda}$, we come to

$$\sum_{\sigma=1}^{l}(\psi^0_\rho, H_1\psi^0_\sigma)(\psi^0_\sigma, \psi^0_{i\lambda}) = E^1_i(\psi^0_\rho, \psi^0_{i\lambda}). \tag{7.16}$$

Since ψ^0_ρ and $\psi^0_{i\lambda}$ are two orthonormal bases of the same eigenspace V^0, the symmetry-adapted eigenfunctions can be obtained from ψ^0_ρ by a unitary transformation U:

$$\psi^0_{i\lambda} = \sum_{\rho=1}^{l} U_{\rho,i\lambda}\psi^0_\rho\,. \tag{7.17}$$

Owing to (7.11) the pairs of indices $i\lambda$ ($i = 1, ..., n$; $\lambda = 1, ..., l_i$) run through the same values as the index ρ ($\rho = 1, ..., l$). Then

$$(\psi^0_\sigma, \psi^0_{i\lambda}) = U_{\sigma,i\lambda}\,,$$

and (7.16) can be written as

$$U^\dagger \mathcal{H}_1 U = E^1, \tag{7.18}$$

where

$$U = (U_{\sigma,i\lambda}), \qquad \mathcal{H}_1 = ((\psi^0_\rho, H_1\psi^0_\sigma))$$

and

$$E^1 = \left(\begin{array}{c|c|c}
\begin{matrix} E^1_1 & & 0 \\ & \ddots & \\ 0 & & E^1_1 \end{matrix} & & 0 \\
\hline
 & \ddots & \\
\hline
0 & & \begin{matrix} E^1_n & & 0 \\ & \ddots & \\ 0 & & E^1_n \end{matrix}
\end{array}\right).$$

Here E^1 is the diagonal matrix of the dimension $\sum_i l_i = l$, and hence the unitary matrix U diagonalizes the matrix $\mathcal{H}_1$. Calculating U and $\mathcal{H}_1$ and using (7.18), one obtains the corrections E_i^1 and hence the splitting of the eigenvalue E^0.

7.3.2 *Effect of Symmetry*

If G and G' are the symmetry groups of the operators H_0 and H, respectively, then in most cases G' is a subgroup of G. The perturbation usually reduces the symmetry.[6]

According to Sect. 7.3.1, the eigenspace V^0 of the operator H_0 will be perturbed into a space V, which is in general a direct sum of eigenspaces V_i of the operator H:

$$V = \bigoplus_{i=1}^{n} V_i .$$

As $H_1 \to 0$ we have $V \to V^0$ and $V_i \to V_i^0$, where the spaces V_i^0 are spanned by the functions $\psi_{i\lambda}^0$.

Since all V_i are eigenspaces of H, they are representation spaces of the group G':

$$P_a \psi_{i\lambda} = \sum_{\mu=1}^{l_i} D_{\mu\lambda}^{(i)}(a) \psi_{i\mu} \qquad (a \in G').$$

If we pass to the limit $H_1 \to 0$ in the above, we obtain

$$P_a \psi_{i\lambda}^0 = \sum_{\mu=1}^{l_i} D_{\mu\lambda}^{0(i)}(a) \psi_{i\mu}^0 \qquad (a \in G').$$

The matrices

$$D_{\mu\lambda}^{0(i)}(a) = \lim_{H_1 \to 0} D_{\mu\lambda}^{(i)}(a)$$

also form a representation of G', because the functions $\psi_{i\lambda}^0$ generate a subspace that is invariant under P_a $(a \in G')$. The representation $\mathcal{D}_i^0$ obtained from $\mathcal{D}_i$ by the limiting process is called the *limiting representation*. The characters of $\mathcal{D}_i$ and $\mathcal{D}_i^0$ differ insignificantly. According to the irreducibility criterium (2.15), the representation $\mathcal{D}_i$ is irreducible if and only if $\mathcal{D}_i^0$ is irreducible. Hence to distinguish between symmetry-protected and accidental degenerate it suffices to investigate the irreducibility of the *limiting* representations.

The symmetry-adapted eigenfunction $\psi_{i\lambda}^0$ of the operator H_0 is the basis functions of the limiting representation $\mathcal{D}_i^0$ of the symmetry group G' of the Hamiltonian H.

---

[6] But this is not always the case. For instance, the symmetry of the problem can be changed completely if the system is placed in a magnetic field.

The unperturbed functions ψ_ρ^0 are the basis functions of the representation $\mathcal{D}$ of the group G. Since the unperturbed (ψ_ρ^0) and symmetry-adapted ($\psi_{i\lambda}^0$) eigenfunctions are related by the unitary transformation U (see (7.17)) the restriction of the representation $U^\dagger \mathcal{D} U$ on G' will be a direct sum of the representations $\mathcal{D}_i^0$ of the group G'. Hence these limiting representations must also belong to the restriction of $\mathcal{D}$ on G'. Thus, any matrix U that diagonalizes $\mathcal{H}_1$ (see (7.18)) gives symmetry-adapted eigenfunctions $\psi_{i\lambda}^0$, which generate the limiting representations of the eigenvalue E^0.

Possible cases according to reducibility or irreducibility of the representations and relation between the groups ($G' = G$ or $G' \subset G$) were considered in detail by [2]. A brief resume is as follows.

An eigenvalue E^0 of the Hamiltonian H_0 corresponding to the representation $\mathcal{D}$ of the symmetry group G of H_0 in general splits under a small perturbation H_1 to eigenvalues E_i of the Hamiltonian H. The number of E_i is equal to the number of components of the restriction $\mathcal{D}^{(s)}$ of the representation $\mathcal{D}$ to the symmetry group G' of the Hamiltonian H. Moreover, E_i will be symmetry-protected degenerate eigenvalues of H. Their degeneracy is determined by the dimensions of the irreducible components of the restriction $\mathcal{D}^{(s)}$. Some of E_i can be equal (accidental degeneracy).

The magnitude of the splitting is determined by the matrix $\mathcal{H}_1$, defined in (7.18). Calculation of the matrix elements is simplified by the selection rules, which we discuss in the next section.

7.4 Selection Rules

The matrix $\mathcal{H}_1$ consists of the matrix elements of the perturbation operator H_1 between the eigenfunctions of the operator H_0 that belong to one of its eigenspaces. The matrix elements determine the transition probabilities of the system from the state ψ to the state ψ' due to the perturbation H_1. We assume ψ and ψ' to be eigenfunctions of the operator H_0. Let also ψ and ψ' be representation functions of the symmetry groups G and G' of the operators H_0 and H, respectively. The representation of the group G' is in general reducible. According to the expansion theorem (7.2), the functions ψ and ψ' can be expressed in terms of the basis functions of the irreducible representation $\mathcal{D}'_q$ of the group G':

$$\psi = \sum_{q=1}^{r'} \sum_{\kappa=1}^{l_q} c_\kappa^{(q)} f_\kappa^{(q)},$$

$$\psi' = \sum_{q=1}^{r'} \sum_{\kappa=1}^{l_q} c'^{(q)}_\kappa g_\kappa^{(q)}.$$

Then for the transition matrix element we have

$$(\psi, H_1\psi') = \sum_{q,q'=1}^{r'} \sum_{\kappa,\kappa'=1}^{l_q} c_\kappa^{(q)*} c'^{(q')}_{\kappa'} (f_\kappa^{(q)}, H_1 g_{\kappa'}^{(q')}).$$

According to Sect. 7.1.4, both functions $H_1 g_{\kappa'}^{(q')}$ and $g_{\kappa'}^{(q')}$ belong to the κ'-th row of the representation $\mathcal{D}'_{q'}$. Then, using the orthogonality relations of the basis functions (7.6), we obtain

$$(\psi, H_1\psi') = \sum_{q=1}^{r'} \left(\sum_{\kappa=1}^{l_q} c_\kappa^{(q)*} c'^{(q)}_\kappa \right) (f_\lambda^{(q)}, H_1 g_\lambda^{(q)}), \tag{7.19}$$

where $f_\lambda^{(q)}$ and $g_\lambda^{(q)}$ belong to the same representations $\mathcal{D}'_q$ (λ is arbitrary). Due to the sum

$$\sum_{\kappa=1}^{l_q} c_\kappa^{(q)*} c'^{(q)}_\kappa ,$$

the matrix element $(\psi, H_1\psi')$ vanishes if the κ-th row of the representation $\mathcal{D}'_q$ belonging to the decomposition of ψ does not occur in the decomposition of ψ' and *vice versa*.

References

1. E.P. Wigner, *Group Theory and its Application to the Quantum Mechanics of Atomic Spectra*, 2nd edn. (Academic, New York, 1959)
2. H.W. Streitwolf, *Group Theory in Solid State Physics* (MacDonald, London, 1971)
3. M. Hamermesh, *Group Theory and Its Applications to Physical Problems* (Addison-Wesley, Reading, MA, 1962). Reprinted: Dover, New York, 1989
4. M. Tinkham, *Group Theory and Quantum Mechanics* (McGraw-Hill, New York, 1964). Reprinted: Dover, New York, 2003
5. C.J. Bradley, A.P. Cracknell, *The Mathematical Theory of Symmetry in Solids. Representation Theory for Point Groups and Space Groups* (Clarendon, Oxford, 1972). Reprinted: Clarendon, Oxford, 2010
6. G.F. Koster, Solid State Phys. **5**, 173 (1957)
7. J.C. Slater, *Quantum Theory of Molecules and Solids, Vol. 4: The Self-Consistent Field for Molecules and Solids* (McGraw-Hill, New York, 1974)
8. E.P. Wigner, Z. Phys. **43**, 624 (1927)

Chapter 8
Group Theory and Solid-State Physics

Hardly any other field of natural science is as well studied and at the same time is so difficult to access from the point of view of teaching and guidance, as mathematical crystallography.
(H. Wondratschek, Acta Cryst. (1973) ***A29****, 581; transl. from German)*

After discussing general principles of group theory applications in quantum mechanics, we proceed to applications in the physics of solids. We begin with the study of an electron in a periodic field, give a mathematical formulation of the problem and apply crystal symmetry to energy band structure. We close the chapter by double-valued representations and time-reversal symmetry (for details, see [1, 2]).

8.1 Electron in a Periodic Field

8.1.1 Mathematical Formulation of the Problem

The theory of electronic band structure in crystals has been presented in many textbooks and monographs (see, e.g. [3–10]). Here we sum up the essentials, following closely [11–13].

In the one-electron approximation, electronic states in crystals are obtained from the Schrödinger equation[1]

$$\Delta\psi + (\varepsilon - V)\psi = 0\,, \tag{8.1}$$

where Δ denotes the Laplace operator in three-dimensional Euclidean space, ε is the eigenvalue, ψ is the eigenfunction, and V is the potential experienced by an electron in the crystal.

[1] Here and henceforth we use atomic units ($\hbar^2/2m = 1$), and energy is given in rydbergs.

N. B. Melnikov and B. I. Reser, *Space Group Representations*,
https://doi.org/10.1007/978-3-031-13991-8_8

The one-electron crystal potential V, no matter how it is determined, must have the full symmetry of the crystal. In particular, it must be periodic with the period of the basic vectors of the lattice $\mathbf{a}_1$, $\mathbf{a}_2$, $\mathbf{a}_3$ or, equivalently,

$$V(\mathbf{r}+\mathbf{R}_p) = V(\mathbf{r}) , \tag{8.2}$$

where $\mathbf{R}_p$ is a lattice vector. To complete the mathematical formulation of the problem it is necessary to impose boundary conditions.

In an infinite crystal it is natural to require that the wave function ψ and its first derivatives are continuous and bounded in the entire Euclidean space.[2] Then from periodicity of the potential (8.2) it follows that a solution to the Schrödinger equation (8.1) can be expressed by Bloch's theorem [16][3]

$$\psi = \psi_{n\mathbf{k}}(\mathbf{r}) = \mathrm{e}^{\mathrm{i}\mathbf{k}\mathbf{r}} u_{n\mathbf{k}}(\mathbf{r}), \tag{8.3}$$

where $u_{n\mathbf{k}}(\mathbf{r})$ is a periodic function,

$$u_{n\mathbf{k}}(\mathbf{r}+\mathbf{R}_p) = u_{n\mathbf{k}}(\mathbf{r}).$$

Here n is the index of the energy band, and $\mathbf{k}$ is the wavevector. The periodic function $u_{n\mathbf{k}}(\mathbf{r})$ can be expressed by the Fourier series

$$u_{n\mathbf{k}}(\mathbf{r}) = \sum_{\mu} c_{\mu}^{(n)}(\mathbf{k})\, \mathrm{e}^{\mathrm{i}\mathbf{K}_{\mu}\mathbf{r}} ,$$

where $\mathbf{K}_\mu$ is the reciprocal lattice vector, and hence the wave function (8.3) can be written as

$$\psi_{n\mathbf{k}}(\mathbf{r}) = \sum_{\mu} c_{\mu}^{(n)}(\mathbf{k})\, \mathrm{e}^{\mathrm{i}(\mathbf{k}+\mathbf{K}_{\mu})\mathbf{r}} .$$

It is straightforward to check that the wave function ψ and energy ε are periodic functions of the wavevector[4]:

$$\begin{aligned} \psi &= \psi_{n\mathbf{k}}(\mathbf{r}) = \psi_{n,\mathbf{k}+\mathbf{K}_{\mu}}(\mathbf{r}), \\ \varepsilon &= \varepsilon_n(\mathbf{k}) = \varepsilon_n(\mathbf{k}+\mathbf{K}_{\mu}) . \end{aligned} \tag{8.4}$$

Therefore, it is sufficient to consider the reduced wavevector $\mathbf{k}$ that takes on values in the Brillouin zone.

However, in some problems it is not quite convenient to consider an infinite crystal. Using periodicity of the potential V and Bloch's theorem (8.3) we can restrict the

---

[2] The refinements that are necessary if the potential becomes infinite at some point or has a discontinuity surface can be found in [14, 15].

[3] A derivation of Bloch's theorem under more general assumptions is given by [17, 18].

[4] The energy $\varepsilon_n(\mathbf{k})$ is an analytic, though not necessarily a single-valued, function of $\mathbf{k}$ [19].

problem to the Wigner–Seitz cell imposing the following conditions at its boundary [19, 20]:

$$\psi_{n\mathbf{k}}(\mathbf{r}_\sigma) = e^{i\mathbf{k}\mathbf{R}_\mathbf{r}} \psi_{n\mathbf{k}}(\mathbf{r}) , \qquad \frac{\partial \psi_{n\mathbf{k}}(\mathbf{r}_\sigma)}{\partial \mathbf{n}} = -e^{i\mathbf{k}\mathbf{R}_\mathbf{r}} \frac{\partial \psi_{n\mathbf{k}}(\mathbf{r})}{\partial \mathbf{n}} , \tag{8.5}$$

where $\mathbf{r}$ and $\mathbf{r}_\sigma$ are conjugate points on the Wigner–Seitz cell boundary, $\mathbf{R}_\mathbf{r}$ is the lattice translation vector between these points, $\mathbf{n}$ is the outer normal vector to the cell boundary. Henceforth we consider the eigenvalue problem (8.1) with the boundary conditions (8.5).

8.1.2 The Ritz Method

It can be proved that the eigenvalue problem (8.1) with the boundary conditions (8.5) is equivalent to the variational problem of minimizing the integral functional [21]

$$D[\psi] = \int_{\Omega_{WS}} (|\nabla \psi|^2 + V|\psi|^2)\, d\mathbf{r} \tag{8.6}$$

(Ω_{WS} is the Wigner–Seitz cell) under the additional constraint

$$\int_{\Omega_{WS}} |\psi|^2 d\mathbf{r} = 1 \tag{8.7}$$

in the class of functions satisfying the same boundary conditions (8.5).

Applying Green's theorem to the functional (8.6) and taking the boundary conditions (8.5) into account, we obtain a new functional

$$\mathcal{D}[\psi] = \int_{\Omega_{WS}} \psi^* H \psi \, d\mathbf{r} , \qquad H = -\Delta + V . \tag{8.8}$$

Finally, the constrained minimization problem (8.8) and (8.7) can be reduced to the unconstrained minimization problem of the functional

$$J[\psi] = \int_{\Omega_{WS}} \psi^* (H - \varepsilon) \psi \, d\mathbf{r} \tag{8.9}$$

in the same class of functions.

The most popular method of obtaining an approximate solution to the variational problem (8.9) is the Ritz method [22] (for details, see, e.g. [23]). The desired function is approximated by a linear combination of some coordinate functions χ_μ:

$$\psi^{(N)} = \sum_{\mu=1}^{N} c_\mu \chi_\mu \,. \tag{8.10}$$

Then the minimization of the functional (8.9) in the class of functions (8.10) reduces to the minimization of the Hermitian quadratic form

$$J(c_1, \ldots, c_N, c_1^*, \ldots, c_N^*) = \sum_{\mu,\nu=1}^{N} (H_{\mu\nu} - \varepsilon S_{\mu\nu}) c_\mu^* c_\nu \,, \tag{8.11}$$

where

$$H_{\mu\nu} = \int\limits_{\Omega_{\mathrm{WS}}} \chi_\mu^* H \chi_\nu \, d\mathbf{r} \,, \qquad S_{\mu\nu} = \int\limits_{\Omega_{\mathrm{WS}}} \chi_\mu^* \chi_\nu \, d\mathbf{r} \,. \tag{8.12}$$

Equating the partial derivatives of the function (8.11) with respect to the coefficients c_μ^*, we obtain a homogeneous system of linear equations

$$\sum_{\nu=1}^{N} (H_{\mu\nu} - \varepsilon S_{\mu\nu}) c_\nu = 0 \qquad (\mu = 1, \ldots, N). \tag{8.13}$$

This equation has a nonzero solution if and only if the determinant of the system vanishes, and we come to the *determinant equation*[5] of the form

$$\det(H_{\mu\nu} - \varepsilon S_{\mu\nu}) \equiv |H_{\mu\nu} - \varepsilon S_{\mu\nu}| = 0 \qquad (\mu, \nu = 1, \ldots, N). \tag{8.14}$$

It is easy to check that, equating the partial derivatives of the function (8.11) with respect to the coefficients c_μ and using the Hermitian character of H, we obtain the same system of equations (8.13). The approximate eigenvalues $\varepsilon^{(N)}$ are determined from the determinant equation (8.14); the corresponding coefficients $c_\mu^{(N)}$, and hence the function $\psi^{(N)}$, are obtained from the linear system (8.13). Thus, the Ritz method reduces the eigenvalue problem for the initial boundary-value problem (8.1) and (8.5) to solving the determinant equation (8.14).

Almost all existing methods of calculating the electronic energy bands of crystals differ from each other only by the choice of the coordinate functions χ_μ in (8.10). An important role in choosing the functions χ_μ is played by physical considerations, which allow to anticipate the general form the desired function ψ. A good choice of the coordinate functions can give a reasonable approximation of the eigenfunction $\psi^{(N)}$ even with a relatively small N.

[5] When $S_{\mu\nu}$ is the unit matrix (the coordinate functions χ_ν are orthonormal), Eq. (8.14) is called the characteristic (secular) equation of the matrix $H_{\mu\nu}$.

8.2 Symmetrized Coordinate Functions

At symmetry points $\mathbf{k}$ in the Brillouin zone, calculation in the Ritz method can be simplified by using the *symmetrized coordinate functions* in the expansion (8.10), i.e. functions belonging to particular rows of the small representations of $G_{\mathbf{k}}$. Then the matrices (8.12) of the determinant equation (8.14) become block-diagonal. Indeed, according to Wigner's theorem (Sect. 7.2) the matrix elements of the Hamiltonian between the functions belonging to different irreducible representations or different rows of the same representation are equal to zero, and matrix elements between the functions belonging to identical rows of the same representation are equal to each other.

The use of symmetrized coordinate functions has two important implications. First, the determinant in Eq. (8.14) factorizes, because the matrix has the block-diagonal structure corresponding to particular rows of small representations. This allows one to increase the number of the coordinate functions in the expansion (8.10) without increasing the computational load. Second, the matrix elements corresponding to different rows of the same irreducible representation are identical. Therefore, it suffices to calculate symmetrized coordinate functions belonging to only one, say the first, row of the irreducible representation. Note that the use of symmetrized coordinate functions not only simplifies the computation but also allows to classify the energy terms according to symmetry.

8.2.1 *Construction of Symmetrized Functions*

To construct the symmetrized coordinate functions, we use the projection technique presented in Sect. 7.1.3. It is important to distinguish the symmetrized functions with respect to the space group G, with respect to the group of the wavevector $G_{\mathbf{k}}$ and with respect to the *point* group of the wavevector $G^0_{\mathbf{k}}$. The symmetrized coordinate functions with respect to G are used in the studies of the phonon spectra (see, e.g. [24]). In the electron band calculations, we deal only with the symmetrized functions corresponding to $G_{\mathbf{k}}$ and $G^0_{\mathbf{k}}$ (relations between the symmetrized functions corresponding to G, $G_{\mathbf{k}}$ and $G^0_{\mathbf{k}}$ are discussed in Appendix C.2.).

Given a basis function χ_μ, we construct the symmetrized function belonging to the ρ-th row of the j-th small representation of $G_{\mathbf{k}}$ as

$$\chi^j_{\mu\rho} = O^{\mathbf{k}j}_{\rho\sigma}\,\chi_\mu\,, \tag{8.15}$$

where

$$O^{\mathbf{k}j}_{\rho\sigma} = \frac{l_j}{n(G_{\mathbf{k}})}\sum_{\{\alpha|\mathbf{t}\}\in G_{\mathbf{k}}} d^{\mathbf{k}j*}_{\rho\sigma}(\{\alpha|\mathbf{t}\})P_{\{\alpha|\mathbf{t}\}} \equiv \frac{l_j}{n(G_{\mathbf{k}})}\sum_a d^{\mathbf{k}j*}_{\rho\sigma}(a)P_a \tag{8.16}$$

(σ is fixed). We recall the notation used here:

$d^{\mathbf{k}j}$ is the j-th small representation of the group $G_{\mathbf{k}}$;
l_j is the dimension of the j-th small representation;
$n(G_{\mathbf{k}})$ is the order of the group $G_{\mathbf{k}}$;
$\{\alpha|\mathbf{t}\}$ is an element of the group $G_{\mathbf{k}}$, where α is the point transformation and $\mathbf{t} = \mathbf{R}_n + \boldsymbol{\tau}_\alpha$ is the translation vector.

There is a way to simplify the construction of the symmetrized functions in crystals. Recalling the right-coset decomposition (5.19) of $G_{\mathbf{k}}$ with respect to the invariant subgroup of translations T, we have $\{\alpha|\mathbf{t}\} = \{\varepsilon|\mathbf{R}_n\}\{\alpha_m|\boldsymbol{\tau}_{\alpha_m}\})$ for any $\{\alpha|\mathbf{t}\}$ in $G_{\mathbf{k}}$. Taking into account

$$d^{\mathbf{k}j}(\{\alpha|\mathbf{t}\}) = \mathrm{e}^{-\mathrm{i}\mathbf{k}\mathbf{R}_n}\Delta^{\mathbf{k}j}(\{\alpha_m|\boldsymbol{\tau}_{\alpha_m}\}), \tag{8.17}$$

we can write the operator (8.16) as a product of two operators,

$$O^{\mathbf{k}j}_{\rho\sigma} = \left(\frac{l_j}{n(G^0_{\mathbf{k}})}\sum_{\{\alpha_m|\boldsymbol{\tau}_{\alpha_m}\}}\Delta^{\mathbf{k}j*}_{\rho\sigma}(\{\alpha_m|\boldsymbol{\tau}_{\alpha_m}\})P_{\{\alpha_m|\boldsymbol{\tau}_{\alpha_m}\}}\right)\frac{1}{n(T)}\sum_{\{\varepsilon|\mathbf{R}_n\}}\mathrm{e}^{\mathrm{i}\mathbf{k}\mathbf{R}_n}P_{\{\varepsilon|\mathbf{R}_n\}}. \tag{8.18}$$

Here $\Delta^{\mathbf{k}j}$ is the j-th irreducible projective representation of the factor group $G_{\mathbf{k}}/T$, which is isomorphic to the point group $G^0_{\mathbf{k}}$ consisting of rotational parts α_m of the representatives $\{\alpha_m|\boldsymbol{\tau}_{\alpha_m}\}$ in the decomposition of $G_{\mathbf{k}}$ with respect to T. The action of the second operator in (8.18) on the Bloch functions

$$\psi_{\mathbf{k}}(\mathbf{r}) = \mathrm{e}^{\mathrm{i}\mathbf{k}\mathbf{r}}u_{\mathbf{k}}(\mathbf{r}) \tag{8.19}$$

leaves them unchanged. Thus, restricted to the Bloch functions, the operator (8.16) is written as

$$\tilde{O}^{\mathbf{k}j}_{\rho\sigma} = \frac{l_j}{n(G^0_{\mathbf{k}})}\sum_{\{\alpha_m|\boldsymbol{\tau}_{\alpha_m}\}}\Delta^{\mathbf{k}j*}_{\rho\sigma}(\{\alpha_m|\boldsymbol{\tau}_{\alpha_m}\})P_{\{\alpha_m|\boldsymbol{\tau}_{\alpha_m}\}}, \tag{8.20}$$

and the problem reduces to constructing the irreducible projective representations of the group $G^0_{\mathbf{k}}$. The technique of constructing those representations is presented in Chaps. 3–5, and the representations themselves are given in Chap. 6.

In general, applying the projection operator $\tilde{O}^{\mathbf{k}j}$ to the Bloch functions, we obtain linearly dependent functions. In the following chapters we show how to solve this problem in the example of plane waves and Bloch sums.

8.2.2 *Matrix Elements of the Hamiltonian*

Let us prove one useful property of the matrix elements of H between the symmetrized functions.

Since the Hamiltonian H commutes with the projection operators, the matrix element between two symmetrized functions (8.15) is written as

$$(\chi^{j'}_{\mu'\rho'}, (H-\varepsilon)\chi^{j}_{\mu\rho}) = (O^{\mathbf{k}j'}_{\rho'\sigma'}\,\chi_{\mu'}\,, (H-\varepsilon)O^{\mathbf{k}j}_{\rho\sigma}\,\chi_{\mu})$$
$$= (\chi_{\mu'}\,, (H-\varepsilon)(O^{\mathbf{k}j'}_{\rho'\sigma'})^{\dagger}\,O^{\mathbf{k}j}_{\rho\sigma}\,\chi_{\mu})\,.$$

Using properties (7.4) and (7.5) of the projection operators,

$$(O^{\mathbf{k}j}_{\rho\sigma})^{\dagger} = O^{\mathbf{k}j}_{\sigma\rho}\,,$$
$$O^{\mathbf{k}j'}_{\sigma'\rho'}\,O^{\mathbf{k}j}_{\rho\sigma} = O^{\mathbf{k}j}_{\sigma'\sigma}\,\delta_{j'j}\delta_{\rho'\rho}\,,$$

we have

$$(\chi^{j'}_{\mu'\rho'}, (H-\varepsilon)\chi^{j}_{\mu\rho}) = (\chi_{\mu'}\,, (H-\varepsilon)O^{\mathbf{k}j'}_{\sigma'\rho'}O^{\mathbf{k}j}_{\rho\sigma}\,\chi_{\mu})$$
$$= (\chi_{\mu'}\,, (H-\varepsilon)O^{\mathbf{k}j}_{\sigma'\sigma}\,\chi_{\mu})\delta_{j'j}\delta_{\rho'\rho}\,. \tag{8.21}$$

Thus, to calculate the matrix element it is sufficient to use *only one* symmetrized function, for example, the one on the right $\chi^{j}_{\mu\rho}$. The other symmetrized function $\chi^{j'}_{\mu'\rho'}$ can be replaced by the original coordinate function $\chi_{\mu'}$. Since both σ' and σ can be chosen arbitrarily, one can choose both of them equal to 1 and use $O^{\mathbf{k}j}_{11}$, so that

$$(\chi^{j'}_{\mu'\rho'}, (H-\varepsilon)\chi^{j}_{\mu\rho}) = (\chi_{\mu'}\,, (H-\varepsilon)O^{\mathbf{k}j}_{11}\,\chi_{\mu})\delta_{j'j}\delta_{\rho'\rho}\,. \tag{8.22}$$

8.3 Double-Valued Representations

Double-valued representations of the space groups were originally studied by Elliott [25]. Here we closely follow the paper by Egorov et al. [26] (for a more detailed treatment see, e.g. [2, 27]).

In Sect. 7.2 we considered symmetry of the Schrödinger equation (the quantum-mechanical equation without spin). We can summarize it as follows.

Symmetry of a quantum-mechanical system is described by the symmetry *operators*, which characterize the transformation law of the wave functions under the symmetry transformation of its coordinates. For the Schrödinger equation in a crystal, where the wave function is scalar, the symmetry operator $P_{\{\alpha|\mathbf{t}\}}$ acts as follows:

$$P_{\{\alpha|\mathbf{t}\}}\psi(\mathbf{r}) = \psi(\{\alpha|\mathbf{t}\}^{-1}\mathbf{r})\,, \tag{8.23}$$

where $\{\alpha|\mathbf{t}\}$ is the symmetry transformation of the space group. The symmetry operators $P_{\{\alpha|\mathbf{t}\}}$ commute with the Hamiltonian and satisfy the relations analogous to (5.2) and (5.3), which hold for the symmetry transformation $\{\alpha|\mathbf{t}\}$.

As was shown by Wigner [1], the matrices of the transformation coefficients of wave functions belonging to the same energy level satisfy the same multiplication rules as the symmetry operators. Hence they form a vector representation of the space group and thus can be obtained using the space group representation theory.

Now we take spin into account. A systematic treatment of spin requires relativistic effects [28, 29]. To do this it is necessary to consider the Dirac equation, whose solution is a four-component wave function Ψ. Under a certain choice of the Dirac matrices [28, Chap. 4], the symmetry operator has the form

$$P_{\{\alpha|\mathbf{t}\}}\Psi(\mathbf{r}) = \begin{pmatrix} P_{[\alpha|\mathbf{t}]}\varphi^{(1)}(\mathbf{r}) & 0 \\ 0 & P_{[\alpha|\mathbf{t}]}\varphi^{(2)}(\mathbf{r}) \end{pmatrix}, \tag{8.24}$$

where $\varphi^{(1)}$ and $\varphi^{(2)}$ are two-component spinors, which transform independently. The symmetry operator $P_{[\alpha|\mathbf{t}]}$ is defined by

$$P_{[\alpha|\mathbf{t}]}\varphi(\mathbf{r}) = S(\alpha)\ \varphi(\{\alpha|\mathbf{t}\}^{-1}\mathbf{r})\,. \tag{8.25}$$

The two-dimensional matrix $S(\alpha)$, which transforms the spinor components, has the form[6]

$$S(\alpha) = \mathrm{e}^{-\mathrm{i}(\theta/2)\boldsymbol{\sigma}\cdot\mathbf{n}} = \cos\frac{\theta}{2}\,E - \mathrm{i}\sin\frac{\theta}{2}\,(\boldsymbol{\sigma}\cdot\mathbf{n}). \tag{8.26}$$

Here θ is the rotation angle of the position vector around the axis with the unit vector $\mathbf{n}$. For the sake of definiteness, we choose the angle θ such that

$$-\pi < \theta \le \pi, \tag{8.27}$$

where a positive angle is given by a counter clockwise rotation when viewed from the endpoint of the vector $\mathbf{n}$. The vector $\boldsymbol{\sigma} = (\sigma_x, \sigma_y, \sigma_z)$ is combined of the Pauli matrices

$$\sigma_x = \begin{pmatrix} 0 & 1 \\ 1 & 0 \end{pmatrix}, \quad \sigma_y = \begin{pmatrix} 0 & -\mathrm{i} \\ \mathrm{i} & 0 \end{pmatrix}, \quad \sigma_z = \begin{pmatrix} 1 & 0 \\ 0 & -1 \end{pmatrix}. \tag{8.28}$$

The Pauli matrices are the generators of the rotations around the coordinate axes. For instance, the matrix

$$S(\alpha_z) = \mathrm{e}^{-\mathrm{i}(\theta/2)\sigma_z} = \begin{pmatrix} \mathrm{e}^{-\mathrm{i}(\theta/2)\sigma_z} & 0 \\ 0 & \mathrm{e}^{\mathrm{i}(\theta/2)\sigma_z} \end{pmatrix}$$

corresponds to the rotation about the z axis.

---

[6] Formula (8.26) is obtained for a *proper* rotation α'. In case a point group has *improper* rotations $\mathrm{i}\alpha'$, one should keep in mind that the spinors are invariant under an inversion, i.e. the matrix $S(\mathrm{i})$ is the unit matrix. Hence, for any rotation α', one can take $S(\alpha') = S(\mathrm{i}\alpha')$ and use formula (8.26) for any point transformation α.

The symmetry operators $P_{\{\alpha|t\}}$, acting on the four-component functions (8.24), commute with the Hamiltonian in the Dirac equation. Clearly, the multiplication law of these operators is determined by the multiplication law of the operators $P_{[\alpha|\mathbf{t}]}$. From the definition (8.25) it follows that

$$P_{[\alpha|\mathbf{t}]}\, P_{[\beta|\mathbf{t}']} = \omega_{\alpha,\beta}\, P_{[\alpha\beta|\alpha\mathbf{t}'+\mathbf{t}]}\,, \tag{8.29}$$

$$P_{[\alpha|\mathbf{t}]^{-1}} = P_{[\alpha^{-1}|-\alpha^{-1}\mathbf{t}]}\,, \tag{8.30}$$

where the factor $\omega_{\alpha,\beta}$ is calculated by the formula

$$\omega_{\alpha,\beta} = \begin{cases} \dfrac{\cos\frac{\theta_\alpha}{2}\cos\frac{\theta_\beta}{2} - \sin\frac{\theta_\alpha}{2}\sin\frac{\theta_\beta}{2}(\mathbf{n}_\alpha\cdot\mathbf{n}_\beta)}{\cos\frac{\theta_{\alpha\beta}}{2}}\,, & \theta_{\alpha\beta}\neq\pi, \\ \sin\frac{\theta_\alpha}{2}\cos\frac{\theta_\beta}{2}(\mathbf{n}_\alpha\cdot\mathbf{n}_{\alpha\beta}) + \cos\frac{\theta_\alpha}{2}\sin\frac{\theta_\beta}{2}(\mathbf{n}_\beta\cdot\mathbf{n}_{\alpha\beta}) & \\ +\sin\frac{\theta_\alpha}{2}\sin\frac{\theta_\beta}{2}([\mathbf{n}_\alpha\times\mathbf{n}_\beta]\cdot\mathbf{n}_{\alpha\beta}), & \theta_{\alpha\beta}=\pi, \end{cases} \tag{8.31}$$

This expression for $\omega_{\alpha,\beta}$ is obtained by comparing the matrices on the left-hand and right-hand sides of formula (8.29). Indeed, using the definition (8.25), we write formula (8.29) as

$$S(\alpha)S(\beta)\,\varphi(\{\beta|\mathbf{t}'\}^{-1}\{\alpha|\mathbf{t}\}^{-1}\mathbf{r}) = \omega_{\alpha,\beta}S(\alpha\beta)\,\varphi(\{\alpha\beta|\alpha\mathbf{t}'+\mathbf{t}\}^{-1}\mathbf{r}).$$

This is equivalent to

$$S(\alpha)S(\beta) = \omega_{\alpha,\beta}S(\alpha\beta).$$

The factor $\omega_{\alpha,\beta}$ is calculated by formula (8.31) and takes only the values ± 1 (for a proof of formula (8.31) and concrete examples of its use, see Appendix D).

Just as in the nonrelativistic case, the transformation matrices of two-component spinors, which enter the four-component wave functions belonging to the same energy level, satisfy a relation analogous to relation (8.29) for the symmetry *operators*:

$$D(P_{[\alpha|\mathbf{t}]})\, D(P_{[\beta|\mathbf{t}']}) = \omega_{\alpha,\beta}\, D(P_{[\alpha\beta|\alpha\mathbf{t}'+\mathbf{t}]})\,. \tag{8.32}$$

The matrices D that satisfy relation (8.32) form a so-called double-valued representation of the space group. From associativity of the operator multiplication (8.25) it follows that the factors $\omega_{\alpha,\beta}$ satisfy relation (4.5), and hence, in accord with the theory of Chap. 4, the matrices D form a *projective* representation of the space group.

If one follows the method of obtaining irreducible space group representations, presented in Chap. 5, it is clear that the problem of obtaining the irreducible double-valued representations of a space group can be reduced to obtaining the irreducible projective representations of the point group $G^0_{\mathbf{k}}$. The corresponding projective factors must be the products of the factors (8.31) and (5.23),

$$\omega'^{\mathbf{k}}_{\{\alpha_m|\tau_{\alpha_m}\},\{\alpha_{m'}|\tau_{\alpha_{m'}}\}} = \omega_{\alpha_m,\alpha_{m'}}\,\omega^{\mathbf{k}}_{\{\alpha_m|\tau_{\alpha_m}\},\{\alpha_{m'}|\tau_{\alpha_{m'}}\}}\,. \tag{8.33}$$

All the rest remains the same.[7] As a result, we obtain double-valued (projective) small representations $d^{\mathbf{k}j}$ of the groups of the wavevectors $G_{\mathbf{k}}$ and double-valued (projective) irreducible representations $D^{\mathbf{k}j}$ of the space group G.

8.4 Time-Reversal Symmetry

8.4.1 Time-Reversal Operator

Until now we studied only *spatial* symmetry of physical problems. But often the problems have *time-reversal* symmetry. The operator θ, which transforms the state ψ to the time-reversed state, is called the *time-reversal operator*. By its definition, the operator θ commutes with the coordinate operator and anticommutes with the momentum operator,

$$\theta^{-1}\mathbf{r}\theta = \mathbf{r}, \qquad \theta^{-1}\mathbf{p}\theta = -\mathbf{p}. \tag{8.34}$$

It is straightforward to show (see, e.g. [2]) that the time-reversal operator can be written in the form

$$\theta = \begin{cases} K, & \text{spin is not taken into account,} \\ \sigma_y K, & \text{spin is taken into account,} \end{cases} \tag{8.35}$$

where K is the complex conjugation operator.[8] Here are some properties of the time-reversal operator:

1. $\theta(a\psi_1 + b\psi_2) = a^*\theta\psi_1 + b^*\theta\psi_2$ (θ is antilinear);
2. $(\theta\psi_1, \theta\psi_2) = (\psi_2, \psi_1)$ (θ is an antiunitary);
3. $\theta^2 = 1$ in the case without spin, and $\theta^2 = -1$ in the case with spin.

8.4.2 Systems with Time-Reversal Symmetry

In the Schrödinger equation, the Hamiltonian of electrons in the crystal H clearly commutes with the time-reversal operator θ, as it follows from Eq. (8.34). Time-reversal symmetry of the Hamiltonian in this case simply reflects the fact that it is real (in the coordinate representation). In the Pauli equation—an approximation of the Dirac equation, incorporating the first nonvanishing relativistic terms—the Hamiltonian is also invariant with respect to time reversal, but in this case $\theta = \sigma_y K$. The operator θ commutes with all space symmetry operators P_a as well. Hence

---

[7] Note that the formalism of the *double groups*, introduced by Bethe [30], is not used at all here.

[8] In paper [26] formula (8.35) is written in Frei's notation: $\theta = -i\sigma_y K$ [31], where $\sigma_y = \begin{pmatrix} 0 & -1 \\ 1 & 0 \end{pmatrix}$.

we could extend the space group G by adding the operator θ and determine the irreducible representations of this group. But from the physical point of view it does not make much sense because the group contains antiunitary elements. In this case, each eigenspace of H, which is still an invariant subspace of the symmetry group, is not a representation space any more, and results of the representation theory do not have direct physical application. We could generalize the theory and consider the so-called *corepresentations* [1]. But we will proceed in an easier and more direct manner.

If the functions $\psi_1, ..., \psi_n$ form a basis of an irreducible representation $\mathcal{D}$ of a space group G, then the functions

$$\psi'_\kappa = \theta\psi_\kappa \qquad (\kappa = 1, ..., n)$$

belong to the same energy level as ψ_κ and form a basis of the irreducible representation $\mathcal{D}^*$. The functions ψ_κ and ψ'_κ define either a $2n$-dimensional or n-dimensional space, which is spanned by ψ_κ alone (for details, see, e.g. [2]). Thus, by taking time-reversal symmetry into account, we either do not get any new information about the degeneracy or the considered eigenvalue is twice degenerate as compared to what it should be according to the spatial symmetry alone. In case the spin is taken into account, the functions ψ_κ and ψ'_κ are always orthogonal to one another[9].

Since ψ_κ is the Bloch function with a wavevector $\mathbf{k}$, the wave function ψ'_κ is the Bloch function with the wavevector $-\mathbf{k}$. Both functions belong to the same energy level, and hence

$$E_{\mathbf{k}} = E_{-\mathbf{k}}. \tag{8.36}$$

This relation thus generally follows from the time-reversal symmetry. Previously we could only deduce it only for space groups with inversion ($i \in G_0$).

8.4.3 Criterion of Extra Degeneracy

8.4.3.1 Case of No Spin

When time-reversal symmetry is taken into account, it is most important to study the transformational properties of the "time-reversed" functions. Wigner has shown [1] that there is a close relation between the transformational properties a representation basis and reality property of the representation.

The Frobenius-Schur theorem (see Sect. 2.3.6) allows to decide if the irreducible representation is (a) real, (b) complex or (c) pseudoreal,

[9] When the spin is taken into account the eigenvalues of the Pauli operator are at least doubly degenerate (Kramers degeneracy). This, of course, does not mean that the *extra degeneracy* always takes place.

$$\sum_{a\in G} \chi(a^2) = \begin{cases} n(G), & \text{case a,} \\ 0, & \text{case b,} \\ -n(G), & \text{case c.} \end{cases}$$

For the induced representations this criterion can be simplified and formulated as a criterion for the small representations (Sect. 3.9). For space groups in the case of no spin, the reality criterion (3.76) takes the form[10]

$$\sum_{\beta\mathbf{k}\approx-\mathbf{k}} \chi^{\mathbf{k}\lambda}(\{\beta|\boldsymbol{\tau}_\beta\}^2) = \begin{cases} n(G^0_{\mathbf{k}}), & \text{case a: } \mathcal{D}^{\mathbf{k}\lambda} \text{ is real,} \\ 0, & \text{case b: } \mathcal{D}^{\mathbf{k}\lambda} \text{ is complex,} \\ -n(G^0_{\mathbf{k}}), & \text{case c: } \mathcal{D}^{\mathbf{k}\lambda} \text{ is pseudoreal,} \end{cases} \tag{8.37}$$

where $\chi_d^{\mathbf{k}\lambda}$ is the character of the ($\mathbf{k}\lambda$)-th small representation, $\mathcal{D}^{\mathbf{k}\lambda}$ is the regular continuation to the space group G, and summation is carried out over all elements $\{\beta|\boldsymbol{\tau}_\beta\}$ of the space group G such that $\beta\mathbf{k} \approx -\mathbf{k}$ (i.e. the two vectors differ by a reciprocal lattice vector, $\beta\mathbf{k} = -\mathbf{k} + \mathbf{K}_\mu$). Note that the element $\{\beta|\boldsymbol{\tau}_\beta\}$ can be *not* in $G_{\mathbf{k}}$, while $\{\beta|\boldsymbol{\tau}_\beta\}^2$ is always in $G_{\mathbf{k}}$. Indeed, from $\beta\mathbf{k} \approx -\mathbf{k}$ it follows that $\beta^2\mathbf{k} \approx \mathbf{k}$.

Let us prove the reality criterion for space groups (8.37). The reality criterion (3.76) holds for any finite group G with an invariant subgroup H. For space groups, we need to replace N_i/H in (3.76) by $G_{\mathbf{k}}/T$ and $n(N_i/H)$ by $n(G_{\mathbf{k}}/T)$(see Table 5.1). Since $G_{\mathbf{k}}/T$ is isomorphic to $G^0_{\mathbf{k}}$, we have $n(G_{\mathbf{k}}/T) = n(G^0_{\mathbf{k}})$. The small representations d^i are just the characters $\chi_d^{\mathbf{k}\lambda}$, so $l_i = 1$. The summation in (3.76) is carried out by the representatives in the coset decomposition of N_i with respect to H such that $\chi^{i_{gm}*}(h) = \chi^i(h)$. For space groups, that means

$$\chi^{\beta\mathbf{k}*} = \chi^{\mathbf{k}}. \tag{8.38}$$

Taking into account

$$\chi^{\mathbf{k}}(\{\varepsilon|\mathbf{R}_n\}) = \mathrm{e}^{-\mathrm{i}\mathbf{k}\mathbf{R}_n},$$

we write (8.38) as

$$\mathrm{e}^{\mathrm{i}\beta\mathbf{k}\mathbf{R}_n} = \mathrm{e}^{-\mathrm{i}\mathbf{k}\mathbf{R}_n} \quad \text{or} \quad \mathrm{e}^{\mathrm{i}(\beta\mathbf{k}+\mathbf{k})\mathbf{R}_n} = 1.$$

Thus, we obtain that the summation in (8.37) is carried out over all point group elements β such that

$$\beta\mathbf{k} \approx -\mathbf{k}.$$

Criterion (8.37), though it is written in terms of characters of the small representations $d^{\mathbf{k}\lambda}$, describes extra degeneracy of the space group representations $\mathcal{D}^{\mathbf{k}\lambda}$. We are interested in extra degeneracy of the small representations $d^{\mathbf{k}\lambda}$ (extra degeneracy of bands). To study this degeneracy it is convenient to distinguish between three

---

[10] This criterion was first obtained by Herring [32]. A more detailed proof for space groups is given in Appendix A.2.

possible categories of the vector **k**:

$$(\text{I})\,\mathbf{k} \approx -\mathbf{k}; \qquad (\text{II})\,\mathbf{k} \not\approx -\mathbf{k},\ \{\mathbf{k}\} = \{-\mathbf{k}\}; \qquad (\text{III})\,\mathbf{k} \not\approx -\mathbf{k},\ \{\mathbf{k}\} \neq \{-\mathbf{k}\}.$$

Detailed analysis of all possible variants [31] shows that extra degeneracy of the small representations $d^{\mathbf{k}\lambda}$ occurs only in four cases: Ib, Ic, IIb and IIc. Here the number corresponds to the category of the vector **k** and letter to the reality class—a, b or c—of the representation $\mathcal{D}^{\mathbf{k}\lambda}$. In cases Ic and IIc, the two *identical* representations $d^{\mathbf{k}\lambda}$ are degenerate. In cases Ib and IIb, the two different representations $d^{\mathbf{k}\lambda}$ and $d^{\mathbf{k}\mu}$ are degenerate. Moreover, in case Ib, we have

$$\chi^{\mathbf{k}\mu}(\{\alpha|\boldsymbol{\tau}_\alpha\}) = \chi^{\mathbf{k}\lambda *}(\{\alpha|\boldsymbol{\tau}_\alpha\}), \quad \{\alpha|\boldsymbol{\tau}_\alpha\} \in G_{\mathbf{k}}, \tag{8.39}$$

and the value μ is easily determined from the table of characters $\chi^{\mathbf{k}\lambda}$. In case IIb, we obtain

$$\chi^{\mathbf{k}\mu}(\{\alpha|\boldsymbol{\tau}_\alpha\}) = \chi^{\mathbf{k}\lambda *}(\{\beta|\boldsymbol{\tau}_\beta\}^{-1}\{\alpha|\boldsymbol{\tau}_\alpha\}\{\beta|\boldsymbol{\tau}_\beta\})$$
$$= \mathrm{e}^{-\mathrm{i}\mathbf{k}(\alpha\boldsymbol{\tau}_\beta + \boldsymbol{\tau}_\alpha - \boldsymbol{\tau}_\beta - \beta\boldsymbol{\tau}_{\beta^{-1}\alpha\beta})}\chi^{\mathbf{k}\lambda *}(\{\beta^{-1}\alpha\beta|\boldsymbol{\tau}_{\beta^{-1}\alpha\beta}\}), \tag{8.40}$$

where $\{\alpha|\boldsymbol{\tau}_\alpha\} \in G_{\mathbf{k}}$ and $\{\beta|\boldsymbol{\tau}_\beta\}$ is such that $\beta\mathbf{k} \approx -\mathbf{k}$. Here it is more difficult to determine the value μ: in addition to the table of characters $\chi^{\mathbf{k}\lambda}$ it is necessary to use the multiplication table of the group G^0 (both the table of characters $\chi^{\mathbf{k}\lambda}$ and multiplication table of the group G^0 can be found in, e.g. [33]). In some particular (but often encountered in practice) cases formula (8.40) can be simplified. If $\{\beta|\boldsymbol{\tau}_\beta\}$ can be chosen such that $\beta\alpha = \alpha\beta$, then

$$\chi^{\mathbf{k}\mu}(\{\alpha|\boldsymbol{\tau}_\alpha\}) = \mathrm{e}^{-\mathrm{i}\mathbf{k}(\alpha\boldsymbol{\tau}_\beta + \boldsymbol{\tau}_\alpha - \boldsymbol{\tau}_\beta - \beta\boldsymbol{\tau}_\alpha)}\chi^{\mathbf{k}\lambda *}(\{\alpha|\boldsymbol{\tau}_\alpha\}), \tag{8.41}$$

and μ is again determined only from the table of characters.[11] Furthermore, if $\{i|0\} \in G$, we can take $\{\beta|\boldsymbol{\tau}_\beta\} = \{i|0\}$, and from (8.41) it immediately follows that[12]

$$\chi^{\mathbf{k}\mu}(\{\alpha|\boldsymbol{\tau}_\alpha\}) = \mathrm{e}^{-\mathrm{i}2\mathbf{k}\boldsymbol{\tau}_\alpha}\chi^{\mathbf{k}\lambda *}(\{\alpha|\boldsymbol{\tau}_\alpha\}). \tag{8.42}$$

An algorithm for studying extra degeneracy of bands, which may occur due to time-reversal symmetry, can be summarized as follows. First, one should determine the category to which the wavevector **k** belongs. If **k** belongs to case III ($\mathbf{k} \in$ III), the answer is known immediately—there is no extra degeneracy of bands. If $\mathbf{k} \in$ I or $\mathbf{k} \in$ II, it is necessary, for each small representation $d^{\mathbf{k}\lambda}$, to calculate the sum on the left-hand side of (8.37) for determining the reality class—a, b or c—of the representation $\mathcal{D}^{\mathbf{k}\lambda}$. Further investigation is necessary only in case b, because in case a there is no extra degeneracy of the band, and in case c the two identical representations $d^{\mathbf{k}\lambda}$ are

[11] For example, the case $\{i|\boldsymbol{\tau}_i\} \in G$ can be considered this way.

[12] This formula was first obtained by Elliott [25].

degenerate. To determine the value μ of the representation $d^{\mathbf{k}\mu}$, which is degenerate with the representation $d^{\mathbf{k}\lambda}$, in case Ib it is necessary to use formula (8.39), and in case IIb—formula (8.40) or (8.41) or (8.42).

As for general points, their vectors **k** can belong to any of the three categories[13]. For the vector in categories I and II we always have case a. That means, there is no degeneracy at a general point (in the case of no spin!).

8.4.3.2 Case with Spin

In the case with spin, as we mentioned in the previous section, a small representation $d^{\mathbf{k}j}$ is an irreducible *projective* representation of the group of the wavevector $G_{\mathbf{k}}$. Hence the regular continuation $\mathcal{D}^{\mathbf{k}j}$ is an irreducible *projective* representation of a group G. For a projective representation, the Frobenius-Schur theorem is formulated as follows:

$$\sum_{a\in G} \omega_{a,a}\, \chi(a^2) = \begin{cases} n(G), & \text{case a,} \\ 0, & \text{case b,} \\ -n(G), & \text{case c,} \end{cases}$$

(for a proof, see Appendix A.1). In particular, for a *double-valued* representation of a space group, we have

$$\omega_{a,a} \equiv \omega_{\{\alpha|\tau_\alpha\},\{\alpha|\tau_\alpha\}} = \omega_{\alpha,\alpha}\,,$$

where $\omega_{\alpha,\alpha}$ is the projective factor of the point group (8.31). Then, the reality criterion (8.37) takes the form

$$\sigma = \sum_{\beta\mathbf{k}\approx-\mathbf{k}} \omega_{\beta,\beta}\, \chi_d^{\mathbf{k}j}(\{\beta|\tau_\beta\}^2) = \begin{cases} n(G^0_{\mathbf{k}}), & \text{case a,} \\ 0, & \text{case b,} \\ -n(G^0_{\mathbf{k}}), & \text{case c,} \end{cases} \tag{8.43}$$

where $\chi_d^{\mathbf{k}j}$ is the character of the $(\mathbf{k}j)$-th small representation and $\omega_{\beta,\beta}$ is the projective factor (8.31), which is related to the spin symmetry.

To classify the transformational properties of the "time-reversed" functions we introduce the following notation. Let $\mathcal{S}_n^{\{\mathbf{k}\}j}$ be the set of functions belonging to the $(\{\mathbf{k}\}j)$-th irreducible representation of a space group, and $\int_n^{\mathbf{k}j}$ is the set of functions belonging to the $(\mathbf{k}j)$-th small representation of the group of the wavevector (Frei [31] uses $\mathcal{S}$ and $\int$ from the word subspace). The index n is the energy band number, it labels different sets of functions that are transformed in the same way.

By dividing all wavevectors **k** into three categories, just as in the no spin case, one can describe the transformational properties of the "time-reversed" functions. A

[13] For example, category I will contain the vector $\mathbf{k} = (0, 0, 0)$ of the crystals with triclinic lattice such that $G^0 = C_1$; category II contains all general points of the crystals such that $i \in G^0$; and category III—most of general points of the crystals such that $i \notin G^0$.

Table 8.1 Transformational properties of the "time-reversed" functions

σ		I	II	III
No spin	With spin	$-\mathbf{k} = \mathbf{k} + \mathbf{K}_\mu$ $\{-\mathbf{k}\} = \{\mathbf{k}\}$	$-\mathbf{k} \neq \mathbf{k} + \mathbf{K}_\mu$ $\{-\mathbf{k}\} = \{\mathbf{k}\}$	$-\mathbf{k} \neq \mathbf{k} + \mathbf{K}_\mu$ $\{-\mathbf{k}\} \neq \{\mathbf{k}\}$
$n(G_{\mathbf{k}}^0)$	$-n(G_{\mathbf{k}}^0)$	$\theta \mathcal{S}_n^{\{\mathbf{k}\}j} = \mathcal{S}_n^{\{\mathbf{k}\}j}$ $\theta \int_n^{\mathbf{k}j} = \int_n^{\mathbf{k}j}$	$\theta \mathcal{S}_n^{\{\mathbf{k}\}j} = \mathcal{S}_n^{\{\mathbf{k}\}j}$ $\theta \int_n^{\mathbf{k}j} = \int_n^{-\mathbf{k}j}$	
0	0	$\theta \mathcal{S}_n^{\{\mathbf{k}\}j} = \mathcal{S}_n^{\{\mathbf{k}\}j'}$ $\theta \int_n^{\mathbf{k}j} = \int_n^{\mathbf{k}j'}$	$\theta \mathcal{S}_n^{\{\mathbf{k}\}j} = \mathcal{S}_n^{\{\mathbf{k}\}j'}$ $\theta \int_n^{\mathbf{k}j} = \int_n^{-\mathbf{k}j'}$	$\theta \mathcal{S}_n^{\{\mathbf{k}\}j} = \mathcal{S}_n^{\{-\mathbf{k}\}j'}$ $\theta \int_n^{\mathbf{k}j} = \int_n^{-\mathbf{k}j'}$
$-n(G_{\mathbf{k}}^0)$	$n(G_{\mathbf{k}}^0)$	$\theta \mathcal{S}_n^{\{\mathbf{k}\}j} = \mathcal{S}_{n'}^{\{\mathbf{k}\}j}$ $\theta \int_n^{\mathbf{k}j} = \int_{n'}^{\mathbf{k}j}$	$\theta \mathcal{S}_n^{\{\mathbf{k}\}j} = \mathcal{S}_{n'}^{\{\mathbf{k}\}j}$ $\theta \int_n^{\mathbf{k}j} = \int_{n'}^{-\mathbf{k}j}$	

detailed analysis of all possible variants is given by Frei [31]. Final results for the case with no spin and with spin are summarized in Table 8.1. These results can be interpreted as appearance of extra degeneracy of some of the energy bands. Different cases lead to degeneracy of the energy levels of different stars and different or the same representations of a group of the wavevector.

In conclusion, we note that time reversal does not give an extra classification of the energy levels, i.e. it does not lead to appearance of new quantum numbers. But, as we see, an extra degeneracy due to time reversal can occur.

References

1. E.P. Wigner, *Group Theory and its Application to the Quantum Mechanics of Atomic Spectra*, 2nd edn. (Academic, New York, 1959)
2. H.W. Streitwolf, *Group Theory in Solid State Physics* (MacDonald, London, 1971)
3. F. Seitz, *The Modern Theory of Solids* (McGraw-Hill, New York, 1940)
4. H. Jones, *Theory of Brillouin Zones and Electronic States in Crystals* (North-Holland, Amsterdam, 1960). Reprinted: LLC, Whitefish, MT, 2013
5. S. Raimes, *The Wave Mechanics of Electrons in Metals* (North-Holland, Amsterdam, 1963)
6. J. Callaway, *Energy Band Theory* (Academic, New York, 1964)
7. G.C. Fletcher, *The Electron Band Theory of Solids* (North Holland, Amsterdam, 1971)
8. W.A. Harrison, *Electronic Structure and the Properties of Solids*, 2nd edn. (Dover, New York, 1989)
9. J. Callaway, *Quantum Theory of the Solid State*, 2nd edn. (Academic, New York, 1991)
10. S.L. Altmann, *Band Theory of Solids: An Introduction from the Point of View of Symmetry* (Oxford Univ. Press, New York, 1991)
11. V.V. Dyakin, R.F. Egorov, B.I. Reser, V.P. Shirokovskii, Fiz. Metallov i Metallovedenie [in Russian] **25**, 211 (1968)
12. B.I. Reser, V.P. Shirokovskii, Fiz. Metallov i Metallovedenie [in Russian] **32**, 934 (1971)
13. B.I. Reser, *The OPW method and interpolation based on the tight binding in calculating the energy band structure of crystals* [in Russian]. Dissertation for the degree of cand. of physics and mathematical sciences, Inst. of Metal Phys., Ural Science Center, USSR Acad. of Sciences, Sverdlovsk (1972), 153 pp
14. L.D. Landau, E.M. Lifshitz, *Quantum Mechanics—Non-relativistic Theory (Course of Theoretical Physics)*, vol. 3, 3rd edn. (Pergamon, Oxford, 1991)

15. L. Schiff, *Quantum Mechanics*, 3rd edn. (McGraw-Hill, New York, 1968)
16. F. Bloch, Z. Physik **52**, 555 (1929)
17. H. Scharf, Helv. Phys. Acta **39**, 556 (1966)
18. V.V. Dyakin, Differential Equations **7**, 1520 (1971)
19. F. Odch, J.B. Keller, J. Math. Phys. **5**, 1499 (1964)
20. J.C. Slater, Phys. Rev. **45**, 794 (1934)
21. E. Schrödinger, Ann. Phys. **79**, 361 (1926)
22. W. Ritz, The Journal fur die reine und Angewandte Mathematik **135**, 1 (1909)
23. P. Gombás, *Theorie und Lösungsmethoden des Mehrteilchenproblems der Wellenmechanik* (Springer, Basel, 1950)
24. J.L. Birman, *Theory of Crystal Space Groups and Lattice Dynamics: Infra-Red and Raman Optical Processes of Insulating Crystals* (Springer, Berlin, 1974)
25. R.J. Elliott, Phys. Rev. **96**, 280 (1954)
26. R.F. Egorov, V.K. Zvezdin, B.I. Reser, *Representations of space groups* [in Russian] (1971). Dep. VINITI, No. 2865–71
27. S.L. Altmann, *Rotations, Quaternions, and Double Groups* (Oxford Univ. Press, New York, 1986)
28. S. Schweber, *An Introduction to Relativistic Quantum Field Theory* (Row, Peterson and Co., Evanston, IL, 1961)
29. V.B. Berestetskii, E.M. Lifshitz, L.P. Pitaevskii, *Quantum Electrodynamics* (Elsevier, Oxford, 1982)
30. H. Bethe, Ann. Phys. **395**, 133 (1929)
31. V. Frei, Czech. J. Phys. B **16**, 207 (1966)
32. C. Herring, Phys. Rev. **52**, 361 (1937)
33. J. Zak, A. Casher, M. Glück, Y. Gur, *The Irreducible Representations of Space Groups* (Benjamin, New York, 1969)

Chapter 9
Energy Band Calculation

For with much wisdom comes much sorrow; the more knowledge, the more grief. (Ecclesiastes 1:18)

In the previous chapter we have considered general properties of electrons in a crystal. Here we present the basics of the electronic energy band calculations and discuss in detail the use of symmetry in a simple but universal method, the orthogonal plane wave (OPW) method (see also [1–3]).

Use of symmetry in the tight-binding method will be discussed in Chap. 13 (for the use of symmetry in other energy band calculation methods, see, e.g. [4]).

9.1 Basics of Band Calculations

Almost all energy band calculation methods are based on the expansion of the wave function $\psi_{n\mathbf{k}}(\mathbf{r})$ of an electron in terms of a complete set of (coordinate) functions $\chi_i(\mathbf{r})$:

$$\psi_{n\mathbf{k}} = \sum_{i=1}^{\infty} c_i \chi_i \, . \tag{9.1}$$

To choose the coordinate functions appropriately, we first consider some characteristic features of the Bloch function

$$\psi_{n\mathbf{k}} = \frac{1}{\sqrt{\Omega_{\mathrm{WS}}}} \, \mathrm{e}^{\mathbf{ikr}} u_{n\mathbf{k}}(\mathbf{r}).$$

This function is nothing else but a plane wave exp($\mathbf{ikr}$) whose wavelength is greater than the lattice constant ($\mathbf{k}$ is reduced) and which is modulated by the periodic factor $u_{n\mathbf{k}}(\mathbf{r})$. In a neighborhood of the nuclei, the function must be almost equal to an atomic wave function and hence must in general vary rapidly due to radial and

N. B. Melnikov and B. I. Reser, *Space Group Representations*,
https://doi.org/10.1007/978-3-031-13991-8_9

angular nodes. Since the wavelength in the exponent of exp(i**kr**) is large, these rapid variations must be attributed to the periodic factor. Thus the Bloch waves behave essentially like plane waves exp(i**kr**), but behave like atomic functions near the nuclei.

These considerations on the form of the Bloch functions suggest the use of plane waves or Bloch sums as the coordinate functions χ_i in the expansion (9.1).[1] Both approaches have been implemented in methods for the calculation of band structures. In practice, the sum in (9.1) must be truncated at a finite value of i.

We can always take linear combinations of χ_i such that the resulting functions are basis function of the corresponding small representations $d^{\mathbf{k}j}$ of the group of the wavevector $G_{\mathbf{k}}$. Such linear combinations are called *symmetrized functions* (see Sect. 8.2). Then it is more convenient to carry out the expansion (9.1) over the symmetrized functions belonging to different rows of the representations $d^{\mathbf{k}j}$. Since $\psi_{n\mathbf{k}}$ belongs to a particular row of a particular representation $d^{\mathbf{k}j}$, its expansion consists of only those symmetrized functions which belong to the same row of the same representation. Thus, the use of symmetrized functions considerably simplifies the expansion (9.1) and allows to increase the number of functions in the expansion without increasing the computational load. Later, in Chap. 16 we derive symmetrized functions for plane waves and in Chap. 17 for Bloch sums.

9.2 The OPW Method

9.2.1 *Main Formulas*

Electronic states in crystals are usually divided into core, valence and conduction states. The core states correspond to the internal fully occupied atomic shells; valence states correspond to the outer (valence) atomic shells; and finally the states whose energies are above the valence states are called conduction states. Of the greatest interest in crystals are the valence states and lowest conduction states. The fact that in almost all physical problems, there is no need to calculate the core states, allows us to formulate the problem of finding valence states from the very beginning. Let us denote by ψ_c $(c = 1, 2, \ldots, n)$ the core state functions and by ψ the function of the first valence state. Then to find ψ by minimizing the functional (8.6), in addition to condition (8.7), it is necessary to satisfy also the orthogonality conditions

$$\int\limits_{\Omega_{\mathrm{WS}}} \psi_c^* \psi \, d\mathbf{r} = 0, \qquad c = 1, 2, \ldots, n. \tag{9.2}$$

---

[1] Note that the Bloch sums corresponding to bound electron states in an atom do not form a complete system of functions.

However, such a formulation of the problem implies knowledge of the exact eigenfunctions of the core states. Therefore, it is more convenient to reformulate the problem using the minimax principle [5, 6], which allows to determine the $(n+1)$th eigenvalue and $(n+1)$th eigenfunction directly, without knowledge of previous eigenvalues and eigenfunctions. The essence of the minimax principle is as follows. Let $u_1, u_2, \ldots, u_n$ be arbitrary functions satisfying the boundary conditions (8.5) and having continuous first derivatives inside the Wigner–Seitz cell, and let $\varepsilon(u_1, u_2, \ldots, u_n)$ be the minimum of the functional (8.6) under additional conditions (8.7) and

$$\int\limits_{\Omega_{\mathrm{WS}}} u_c^* \psi \, \mathrm{d}\mathbf{r} = 0, \qquad c = 1, 2, \ldots, n. \tag{9.3}$$

Then the $(n+1)$th eigenvalue ε_{n+1} is equal to the maximum of $\varepsilon(u_1, u_2, \ldots, u_n)$ for all possible variations of the system of functions $u_1, u_2, \ldots, u_n$, obeying all the requirements imposed on them above. This maximum value of the minimum of $\varepsilon(u_1, u_2, \ldots, u_n)$ is attained when $u_1, u_2, \ldots, u_n$ are the eigenfunctions of the operator (8.1).

Let us assume that we have constructed the core states eigenfunctions ψ_c with a certain precision. Then we can satisfy the condition (9.2) if we restrict the class of coordinate functions, which are used to minimize the functional (8.6), by the functions

$$\chi_\mu = f_\mu - \sum_{c=1}^{n} a_c \psi_c \,, \tag{9.4}$$

where f_μ is an arbitrary function satisfying the Bloch boundary conditions (8.5), and the coefficients

$$a_c = \int\limits_{\Omega_{\mathrm{WS}}} \psi_c^* f_\mu \, \mathrm{d}\mathbf{r} \tag{9.5}$$

are determined from the orthogonality condition of the function χ_μ to all other core state functions ψ_c. This choice of coordinate functions χ_μ allows one to hope that the calculation will converge to the investigated part of the energy bands. As for the functions f_μ, their choice is dictated by the physical nature of the problem.

Historically, the first and most practical method of calculation using the idea of orthogonalization was proposed by Herring [7]. He has chosen plane waves as the functions f_μ—the simplest functions satisfying the Bloch boundary conditions. Then, the coordinate functions are

$$\chi_{\mathbf{k}_\mu} = \varphi_{\mathbf{k}_\mu} - \sum_{c=1}^{n} a_c(\mathbf{k}_\mu) \psi_c \,, \tag{9.6}$$

where

$$\varphi_{\mathbf{k}_\mu} = \frac{1}{\sqrt{\Omega_{\mathrm{WS}}}} \, \mathrm{e}^{\mathrm{i}(\mathbf{k}+\mathbf{K}_\mu)\mathbf{r}} \equiv \frac{1}{\sqrt{\Omega_{\mathrm{WS}}}} \, \mathrm{e}^{\mathrm{i}\mathbf{k}_\mu \mathbf{r}}, \tag{9.7}$$

and the orthogonalization coefficients $a_c(\mathbf{k}_\mu)$ are calculated by the formula

$$a_c(\mathbf{k}_\mu) = \int\limits_{\Omega_{WS}} \psi_c^* \varphi_{\mathbf{k}_\mu} \, d\mathbf{r}. \tag{9.8}$$

Functions of the form (9.6) are called *orthogonalized plane waves* (OPW), and the Ritz method on these functions as applied to problem (8.6) and (8.7) is the OPW method.

From qualitative point of view, an OPW is a function that behaves like a plane wave at large distances from an atom and, at the same time, is highly oscillating, like an atomic wave function, near nuclei. Since both of these properties take place for the wave function of an electron in a crystal, the OPW are convenient coordinate functions of the problem.

Let us write the basic formulas of the OPW method following [8–10]. Substituting (9.6) into (8.12) and taking into account that

$$H\psi_c = \varepsilon_c \psi_c , \tag{9.9}$$

after straightforward calculations, for the matrix elements in the determinant equation (8.14) we obtain the following expression:

$$H_{\mu\nu} - \varepsilon S_{\mu\nu} = (\mathbf{k}_\mu^2 - \varepsilon)\delta_{\mu\nu} + V(\mathbf{k}_\mu - \mathbf{k}_\nu) + \sum_{c=1}^{n} [\varepsilon - \varepsilon_c(\mathbf{k})] a_c^*(\mathbf{k}_\mu) a_c(\mathbf{k}_\nu), \tag{9.10}$$

where $V(\mathbf{k}_\mu - \mathbf{k}_\nu)$ is the Fourier component of the crystal potential

$$V(\mathbf{k}_\mu - \mathbf{k}_\nu) = \frac{1}{\Omega_{WS}} \int\limits_{\Omega_{WS}} V(\mathbf{r}) \, e^{-i(\mathbf{k}_\mu - \mathbf{k}_\nu)\mathbf{r}} \, d\mathbf{r}. \tag{9.11}$$

In practice, direct use of (9.10) is impossible, because exact wave functions of the core states are usually not known and can only be constructed approximately.

It is usually assumed that the wave functions of the core bands can be represented in the form of individual Bloch sums, which are constructed from the atomic functions:

$$\psi_c = \Phi_{nlm}^{\mathbf{k}s}(\mathbf{r}) = \sum_p e^{i\mathbf{k}(\mathbf{R}_p + \boldsymbol{\tau}_s)} \varphi_{nlm}^s(\mathbf{r} - \mathbf{R}_p - \boldsymbol{\tau}_s), \tag{9.12}$$

where n, l, m are the atomic quantum numbers; $\boldsymbol{\tau}_s$ are the basis vectors, indicating the positions of atoms in a unit cell. In that approximation, the orthogonalization coefficients have the form

$$a_c(\mathbf{k}_\mu) = a_{nlm}^s(\mathbf{k}_\mu) = e^{-i\mathbf{k}\boldsymbol{\tau}_s} e^{i\mathbf{k}_\mu \boldsymbol{\tau}_s} \frac{1}{\sqrt{\Omega_{WS}}} \int\limits_{\infty} \varphi_{nlm}^s(\mathbf{r}) e^{i\mathbf{k}_\mu \mathbf{r}} \, d\mathbf{r}. \tag{9.13}$$

Substituting the solution of the atomic problem

$$\varphi^s_{nlm}(\mathbf{r}) = R^s_{nl}(r)\, Y_{lm}(\vartheta, \varphi) \tag{9.14}$$

into (9.13) and using the expansion of a plane wave in spherical functions

$$\mathrm{e}^{\mathrm{i}\mathbf{k}_\mu \mathbf{r}} = 4\pi \sum_{l=0}^{\infty} \sum_{m=-l}^{l} \mathrm{i}^l j_l(k_\mu r) Y_{lm}(\vartheta, \varphi) Y^*_{lm}(\vartheta_\mu, \varphi_\mu), \tag{9.15}$$

where k_μ, ϑ_μ, φ_μ are spherical coordinates of the vector $\mathbf{k}_\mu$ and $j_l(k_\mu r)$ are the spherical Bessel functions, we obtain

$$a^s_{nlm}(\mathbf{k}_\mu) = \frac{4\pi}{\sqrt{\Omega_{\mathrm{WS}}}} \mathrm{i}^l \mathrm{e}^{-\mathrm{i}\mathbf{k}\boldsymbol{\tau}_s} \mathrm{e}^{\mathrm{i}\mathbf{k}_\mu \boldsymbol{\tau}_s} A^s_{nl}(k_\mu) Y^*_{lm}(\vartheta_\mu, \varphi_\mu), \tag{9.16}$$

where

$$A^s_{nl}(k_\mu) = \int_0^\infty j_l(k_\mu r) R^s_{nl}(r) r^2 \,\mathrm{d}r. \tag{9.17}$$

Finally, one last approximation is made. Namely, it is assumed that the width of the occupied bands is so small that it can be neglected. Then one can replace ε_c in the expression for the matrix element (9.10) by the values of the corresponding "atomic" terms,[2] i.e. one can consider

$$\varepsilon_c(\mathbf{k}) = \varepsilon^s_{nl}\,. \tag{9.18}$$

Thus, substituting expressions (9.16) and (9.18) into (9.10) and using the addition theorem for the spherical functions [11]

$$P_l(\cos\vartheta_{\mu\nu}) = \frac{4\pi}{2l+1} \sum_{m=-l}^{l} Y^*_{lm}(\vartheta_\mu, \varphi_\mu) Y_{lm}(\vartheta_\nu, \varphi_\nu), \tag{9.19}$$

we obtain

$$\begin{aligned} H_{\mu\nu} - \varepsilon S_{\mu\nu} = (\mathbf{k}^2_\mu - \varepsilon)\delta_{\mu\nu} + V(\mathbf{k}_\mu - \mathbf{k}_\nu) + \frac{4\pi}{\Omega_{\mathrm{WS}}} \sum_s \mathrm{e}^{-\mathrm{i}(\mathbf{k}_\mu - \mathbf{k}_\nu)\boldsymbol{\tau}_s} \\ \times \sum_{n,l} (2l+1) P_l(\cos\vartheta_{\mu\nu})(\varepsilon - \varepsilon^s_{nl}) A^s_{nl}(k_\mu) A^s_{nl}(k_\nu) \end{aligned} \tag{9.20}$$

($\vartheta_{\mu\nu}$ is the angle between $\mathbf{k}_\mu$ and $\mathbf{k}_\nu$).

---

[2] The quotes mean that the energy values appearing in the formulas do not necessarily coincide with the corresponding values for an isolated atom and can be recalculated by taking the crystal field into account, as is done, for example, when the core shifts are taken into account.

This is the most common formulation of the OPW method[3]. We emphasize that in the actual implementation of the OPW method, one always has to choose the core eigenfunctions in the form of individual Bloch sums and assume that the crystal potential for the atomic core functions has an atom-like character, i.e. spherically symmetric. Otherwise, the calculations become immensely cumbersome. In fact, these considerations determine the range of materials to which it is advisable to apply the OPW method.

9.2.2 Symmetrized OPW

In Sect. 8.2, we considered in detail the construction of symmetrized coordinate functions in a crystal. Here we discuss an additional fact that greatly simplifies the construction of symmetrized OPW.

Construction of the symmetrized OPW can be reduced to construction of symmetrized plane waves. More specifically, applying the operator $O^{\mathbf{k}j}_{\rho\sigma}$ (σ is arbitrary) to an OPW function (9.6) constructed from the plane wave $\varphi_{\mathbf{k}_\mu}$, we obtain

$$O^{\mathbf{k}j}_{\rho\sigma}\,\chi_{\mathbf{k}_\mu} = \chi^{j}_{\mathbf{k}_\mu\rho}\,,$$

where

$$\chi^{j}_{\mathbf{k}_\mu\rho} = O^{\mathbf{k}j}_{\rho\sigma}\,\varphi_{\mathbf{k}_\mu} - \sum_{c=1}^{n}(\psi_c,\,O^{\mathbf{k}j}_{\rho\sigma}\,\varphi_{\mathbf{k}_\mu})\psi_c \tag{9.21}$$

is an OPW function constructed from the *symmetrized* plane wave $O^{\mathbf{k}j}_{\rho\sigma}\,\varphi_{\mathbf{k}_\mu}$. Indeed, taking (8.16) into account, we have

$$\sum_{c=1}^{n}(\psi_c,\varphi_{\mathbf{k}_\mu})O^{\mathbf{k}j}_{\rho\sigma}\,\psi_c = \sum_{c=1}^{n}(\psi_c,\varphi_{\mathbf{k}_\mu})\,\frac{l_j}{n(G_{\mathbf{k}})}\sum_a d^{\mathbf{k}j*}_{\rho\sigma}(a)P_a\psi_c\,.$$

Since P_a is a unitary operator, we can write

$$\sum_{c=1}^{n}(\psi_c,\varphi_{\mathbf{k}_\mu})O^{\mathbf{k}j}_{\rho\sigma}\,\psi_c = \frac{l_j}{n(G_{\mathbf{k}})}\sum_a d^{\mathbf{k}j*}_{\rho\sigma}(a)\sum_{c=1}^{n}(P_a\psi_c,\,P_a\varphi_{\mathbf{k}_\mu})\,P_a\psi_c\,. \tag{9.22}$$

The functions ψ_c form a basis of the irreducible representation $d^{\mathbf{k}j}_{\rho\sigma}$, and so from expression (7.1) it follows that

[3] In some earlier works (see, e.g. [7, 12–14]) other approximations were made, but they lead to extremely cumbersome expressions for (9.20) and are now only of historical interest.

$$P_a \psi_c = \sum_{c'=1}^{n} d^{\mathbf{k}j}_{c'c}(a)\psi_{c'} \,. \tag{9.23}$$

Using this formula and orthogonality relations (2.8), we write (9.22) as

$$\begin{aligned} &\sum_{c=1}^{n}(\psi_c, \varphi_{\mathbf{k}_\mu})O^{\mathbf{k}j}_{\rho\sigma}\,\psi_c \\ &= \sum_{c',c''=1}^{n} \frac{l_j}{n(G_{\mathbf{k}})} \sum_a d^{\mathbf{k}j*}_{\rho\sigma}(a) \sum_{c=1}^{n} d^{\mathbf{k}j*}_{cc''}(a) d^{\mathbf{k}j}_{c'c}(a)(\psi_{c''}, P_a\varphi_{\mathbf{k}_\mu})\psi_{c'} \\ &= \sum_{c=1}^{n}(\psi_c, \frac{l_j}{n(G_{\mathbf{k}})}\sum_a d^{\mathbf{k}j*}_{\rho\sigma}(a)P_a\varphi_{\mathbf{k}_\mu})\psi_c = \sum_{c=1}^{n}(\psi_c, O^{\mathbf{k}j}_{\rho\sigma}\,\varphi_{\mathbf{k}_\mu})\psi_c. \end{aligned} \tag{9.24}$$

Thus, the symmetrized OPW method requires only symmetrized plane waves.

As we showed in Sect. 8.2.2, it is sufficient to calculate symmetrized functions that belong to only one, for instance the first, row of each irreducible representation. Taking Eqs. (8.20), (9.7) and (7.10) into account, we can write the symmetrized plane wave belonging to the first row of the j-th irreducible projective representation of the group $G^0_{\mathbf{k}}$ as follows:

$$\begin{aligned} \varphi^j_{\mathbf{k}_\mu,1} = \tilde{O}^{\mathbf{k}j}_{1,1}\varphi_{\mathbf{k}_\mu} &= \frac{l_j}{n(G^0_{\mathbf{k}})\sqrt{\Omega_{\mathrm{WS}}}} \sum_{\{\alpha_m|\boldsymbol{\tau}_{\alpha m}\}} \Delta^{\mathbf{k}j*}_{1,1}(\{\alpha_m|\boldsymbol{\tau}_{\alpha_m}\})P_{\{\alpha_m|\boldsymbol{\tau}_{\alpha m}\}}\,\mathrm{e}^{\mathrm{i}\mathbf{k}_\mu\mathbf{r}} \\ &= \frac{l_j}{n(G^0_{\mathbf{k}})\sqrt{\Omega_{\mathrm{WS}}}} \sum_{\{\alpha_m|\boldsymbol{\tau}_{\alpha m}\}} \Delta^{\mathbf{k}j*}_{1,1}(\{\alpha_m|\boldsymbol{\tau}_{\alpha_m}\})\,\mathrm{e}^{-\mathrm{i}\alpha_m\mathbf{k}_\mu\boldsymbol{\tau}_{\alpha m}}\,\mathrm{e}^{\mathrm{i}\alpha_m\mathbf{k}_\mu\mathbf{r}} \\ &= \frac{l_j}{n(G^0_{\mathbf{k}})} \sum_{\{\alpha_m|\boldsymbol{\tau}_{\alpha m}\}} c^j_{\alpha_m\mathbf{k}_\mu,1}\varphi_{\alpha_m\mathbf{k}_\mu}\,, \end{aligned} \tag{9.25}$$

where

$$c^j_{\alpha_m\mathbf{k}_\mu,1} = \Delta^{\mathbf{k}j*}_{1,1}(\{\alpha_m|\boldsymbol{\tau}_{\alpha_m}\})\,\mathrm{e}^{-\mathrm{i}\alpha_m\mathbf{k}_\mu\boldsymbol{\tau}_{\alpha m}}\,, \tag{9.26}$$

$$\varphi_{\alpha_m\mathbf{k}_\mu} = \frac{1}{\sqrt{\Omega_{\mathrm{WS}}}}\,\mathrm{e}^{\mathrm{i}\alpha_m\mathbf{k}_\mu\mathbf{r}}. \tag{9.27}$$

The coefficients $c^j_{\alpha_m\mathbf{k}_\mu,1}$ are called the symmetrization coefficients. Henceforth we write the symmetrized plane wave without the factor $1/\sqrt{\Omega_{\mathrm{WS}}}$, because it is the same for all representation functions of the same representation and hence cancels in the determinant (8.14).

Now, taking Eqs. (8.22), (9.21) and (9.25) into account, we obtain the following final expression for the matrix element constructed from the symmetrized OPW:

$$(\chi^j_{\mathbf{k}_\mu,1}, (H-\varepsilon)\chi^j_{\mathbf{k}_\nu,1}) = \sum_{\{\alpha_m|\boldsymbol{\tau}_{\alpha_m}\}} c^j_{\alpha_m\mathbf{k}_\nu,1}(\chi_{\mathbf{k}_\mu}, (H-\varepsilon)\chi_{\alpha_m\mathbf{k}_\nu}) \tag{9.28}$$

$$= \sum_{\{\alpha_m|\boldsymbol{\tau}_{\alpha_m}\}} c^j_{\alpha_m\mathbf{k}_\nu,1}[(\mathbf{k}^2_\mu - \varepsilon)\delta_{\mathbf{k}_\mu,\alpha_m\mathbf{k}_\nu} + V(\mathbf{k}_\mu - \alpha_m\mathbf{k}_\nu)$$

$$+ \frac{4\pi}{\Omega_{\mathrm{WS}}}\sum_s \mathrm{e}^{-\mathrm{i}(\mathbf{k}_\mu-\alpha_m\mathbf{k}_\nu)\boldsymbol{\tau}_s}$$

$$\times \sum_{n,l}(2l+1)P_l(\cos\vartheta_{\mathbf{k}_\mu,\alpha_m\mathbf{k}_\nu})(\varepsilon - \varepsilon^s_{nl})A^s_{nl}(\mathbf{k}_\mu)A^s_{nl}(\alpha_m\mathbf{k}_\nu)],$$

which is used in practical calculations.

In order to construct the set of symmetrized plane waves belonging to the first row of the j-th representation of $G^0_{\mathbf{k}}$ using Eqs. (9.25)–(9.27), we need to take the first plane wave $\varphi_{\mathbf{k}_1}$ and sum the terms $c^j_{\alpha_m\mathbf{k}_1,1}\varphi_{\alpha_m\mathbf{k}_1}$ over all elements $\{\alpha_m|\boldsymbol{\tau}_{\alpha_m}\}$. This sum, if it is not equal to zero, will be the first symmetrized plane wave $\varphi^j_{\mathbf{k}_1,1}$. Then we must take the second plane wave $\varphi_{\mathbf{k}_2}$ and compare it with the plane waves that formed $\varphi^j_{\mathbf{k}_1,1}$; if $\varphi_{\mathbf{k}_2}$ has already appeared in $\varphi^j_{\mathbf{k}_1,1}$, we must skip the second plane wave, pass on to the third plane wave and repeat what we did with the first one, and so on. This algorithm must be repeated for each representation of each group $G^0_{\mathbf{k}}$. When there are a few plane waves, the algorithm can be easily implemented by hand. When one needs to consider hundreds of functions, the task becomes impossible.

A simple and efficient procedure of constructing the symmetrized plane waves was developed in [15]. It works for symmorphic as well as nonsymmorphic groups. This procedure is presented in Chap. 16.

9.2.3 *Crystal Potential*

9.2.3.1 Crystal Electron Density

The main quantity that determines the electronic energy bands of a solid in the one-electron approximation is the crystal potential. In a consistent theory, this potential must be calculated self-consistently according to the Hartree–Fock method. However, enormous amount of required computation does not allow the implementation of the fully self-consistent Hartree–Fock method even now. Hence additional simplifications are required. The main one is that the nonlocal (integral) operator of the exchange part of the crystal potential, depending on the considered state, is replaced by the potential function $V_{\mathrm{ex}}(\mathbf{r})$, which is the same for all states. The most common V_{ex} approximation is the Slater approximation [16],

$$V_{\mathrm{ex}}(\mathbf{r}) = -3\left[\frac{3}{\pi}\rho^{\mathrm{e}}(\mathbf{r})\right]^{\frac{1}{3}}. \tag{9.29}$$

Since the Coulomb part of the crystal potential is also expressed in terms of the crystal electron density with the help of Poisson's equation, the problem of constructing the crystal potential in the Hartree–Fock–Slater approximation is reduced to finding the crystal electron density.

It is reasonable to approximate the crystal electron density by the sum of the electron densities of neutral atoms located in lattice sites:

$$\rho^{\mathrm{e}}(\mathbf{r}) = \sum_{p,s} \rho_s^{\mathrm{e}}(\mathbf{r} - \mathbf{R}_p - \boldsymbol{\tau}_s). \tag{9.30}$$

It is easy to show that this approximation is equivalent to the assumption that the zero-order functions are the Bloch sums, that the atomic functions centered at different sites do not overlap and that filling of different **k** states is equally probable. These assumptions are quite justified for the core energy bands, but not for the valence ones. Therefore, on the whole, expression (9.30) should be considered as an *a priori* zeroth approximation for the electron density in the crystal.

9.2.3.2 Fourier Coefficients of the Coulomb Part of the Crystal Potential

The Coulomb part of the crystal potential $V_{\mathrm{C}}(\mathbf{r})$ can be obtained by solving Poisson's equation

$$\Delta V_{\mathrm{C}}(\mathbf{r}) = -8\pi\rho(\mathbf{r}) = -8\pi\left[\rho^{\mathrm{e}}(\mathbf{r}) - \sum_s Z_s\delta(\mathbf{r} - \boldsymbol{\tau}_s)\right] \tag{9.31}$$

in the Wigner–Seitz cell with periodic boundary conditions

$$V_{\mathrm{C}}(\mathbf{r}_\sigma) = V_{\mathrm{C}}(\mathbf{r}), \qquad \frac{\partial V_{\mathrm{C}}(\mathbf{r}_\sigma)}{\partial \mathbf{n}} = -\frac{\partial V_{\mathrm{C}}(\mathbf{r})}{\partial \mathbf{n}}. \tag{9.32}$$

Here Z_s is the charge of the s-th nucleus, and $\delta(\mathbf{r} - \boldsymbol{\tau}_s)$ is the Dirac delta-function. According to (9.28), the expression for the matrix element of the OPW method includes not the crystal potential itself, but its Fourier coefficients. Therefore, in equation (9.31), it is more convenient for us to use the Fourier representation:

$$V_{\mathrm{C}}(\mathbf{K}_\mu) = \frac{8\pi}{\mathbf{K}_\mu^2}\rho(\mathbf{K}_\mu), \qquad (\mathbf{K}_\mu \neq 0), \tag{9.33}$$

where

$$V_{\mathrm{C}}(\mathbf{K}_\mu) = \frac{1}{\Omega_{\mathrm{WS}}}\int_{\Omega_{\mathrm{WS}}} V_{\mathrm{C}}(\mathbf{r})\,\mathrm{e}^{-\mathrm{i}\mathbf{K}_\mu\mathbf{r}}\,\mathbf{dr}, \qquad \rho(\mathbf{K}_\mu) = \frac{1}{\Omega_{\mathrm{WS}}}\int_{\Omega_{\mathrm{WS}}} \rho(\mathbf{r})\,\mathrm{e}^{-\mathrm{i}\mathbf{K}_\mu\mathbf{r}}\,\mathbf{dr}. \tag{9.34}$$

Thus, using Poisson's equation, we have expressed the Fourier coefficients of the Coulomb part of the crystal potential $V_C(\mathbf{K}_\mu)$ directly in terms of the Fourier coefficients of the charge density $\rho(\mathbf{K}_\mu)$. Substituting the explicit expression for the density $\rho(\mathbf{r})$ into the right-hand formula of (9.34), we write (9.33) as

$$V_C(\mathbf{K}_\mu) = -\frac{8\pi}{\Omega_{WS}\mathbf{K}_\mu^2}\sum_s \mathrm{e}^{-\mathrm{i}\mathbf{K}_\mu\boldsymbol{\tau}_s}\left(Z_s - \int_\infty \rho_s^{\mathrm{e}}(\mathbf{r})\,\mathrm{e}^{-\mathrm{i}\mathbf{K}_\mu\mathbf{r}}\,\mathrm{d}\mathbf{r}\right), \qquad (\mathbf{K}_\mu \neq 0). \tag{9.35}$$

Note that assumption (9.30) is equivalent to the assumption that the Coulomb part of the crystal potential is equal to the sum of the Coulomb parts of the potentials of neutral atoms:

$$V_C(\mathbf{r}) = \sum_{p,s} V_C^s(\mathbf{r} - \mathbf{R}_p - \boldsymbol{\tau}_s)\,. \tag{9.36}$$

Indeed, in the latter case, we obtain the same expression for $V_C(\mathbf{K}_\mu)$ as under assumption (9.30).

We find the zeroth Fourier component separately, because Poisson's equation with periodic boundary conditions defines the Coulomb potential $V_C(\mathbf{r})$ only up to a constant. Taking (9.36) into account, we have

$$V_C(0) = \frac{1}{\Omega_{WS}}\int_{\Omega_{WS}} V_C(\mathbf{r})\,\mathrm{d}\mathbf{r} = \frac{1}{\Omega_{WS}}\sum_s \int_\infty V_C^s(\mathbf{r})\,\mathrm{d}\mathbf{r}\,. \tag{9.37}$$

From this, using $\rho_s^{\mathrm{e}}(\mathbf{r}) = \rho_s^{\mathrm{e}}(r)$[4] and the well-known result

$$V_C^s(\mathbf{r}) = 2\int_\infty \frac{\rho_s(r')}{|\mathbf{r}-\mathbf{r}'|}\,\mathrm{d}\mathbf{r}' = \frac{8\pi}{r}\int_0^r \rho_s(r')r'^2\,\mathrm{d}r' + 8\pi\int_r^\infty \rho_s(r')r'\,\mathrm{d}r'\,, \tag{9.38}$$

after simple calculations we obtain

$$V_C(0) = -\frac{16\pi^2}{3\Omega_{WS}}\sum_s \int_0^\infty \rho_s^{\mathrm{e}}(r)r^4\,\mathrm{d}r\,. \tag{9.39}$$

[4] For atoms with filled shells, this is always the case, and with unfilled—under the assumption of an equal probability distribution of electrons over states with different values of the quantum number m.

9.2.3.3 Fourier Coefficients of the Exchange Part of the Crystal Potential

For the exchange part of the crystal potential, as a rule, one uses the Slater approximation (9.29). However, the calculation of Fourier coefficients of this expression is a fairly difficult task [17] and is usually never done. Most often it is assumed that the exchange part of the potential can be represented as the sum of exchange potentials of neutral atoms [18]. In doing so, it turns out that the maximum error occurs for zero Fourier components $V_{\text{ex}}(0)$. Therefore, in calculations of the energy bands using this approximation, the value of $V_{\text{ex}}(0)$ has to be corrected [12, 19].

To avoid the above difficulties in calculating the Fourier coefficients of the exchange part of the crystal potential, we use the "muffin-tin" (m-t)-approximation for the electron density $\rho^{\text{e}}(\mathbf{r})$. We surround each atom s' with coordinates $\mathbf{r} = \boldsymbol{\tau}_{s'}$ by a sphere of radius $r_{s'}$, so that these spheres do not overlap. Let us select the atomic electron density in each sphere and spherically average inside the sphere the contribution from the electrons of the remaining atoms of the crystal lattice. The resulting electron density will be denoted by $\tilde{\rho}^{\text{e}}_{s'}(|\mathbf{r} - \boldsymbol{\tau}_{s'}|)$. The crystal electron density in the rest of the Wigner–Seitz cell will be assumed constant and equal ρ^{e}_c.

The constant value of the electron density ρ^{e}_c outside the (m-t)-spheres is obtained from the electro-neutrality condition of the Wigner–Seitz cell and is equal to

$$\rho^{\text{e}}_c = \frac{\sum\limits_s (Z_s - Q_s)}{\Omega_{\text{WS}} - \sum\limits_s \Omega_s} . \tag{9.40}$$

Here Q_s is the charge inside the s-th sphere,

$$Q_s = 4\pi \int\limits_0^{r_s} \tilde{\rho}^{\text{e}}_s(r') r'^2 \, \text{d}r' . \tag{9.41}$$

Derivation of rather cumbersome expressions for the Fourier coefficients of the exchange part of the crystal potential $V_{\text{ex}}(\mathbf{K}_\mu)$ in the (m-t)-approximation for the electron density is given in [10] and is omitted here for brevity. Since the Fourier coefficients of the exchange potential are, as a rule, an order of magnitude smaller than the Fourier coefficients of the Coulomb potential, it is reasonable to believe that a small error in $V_{\text{ex}}(\mathbf{K}_\mu)$, associated with the (m-t)-approximation, will not lead to a noticeable distortion of the calculation results.

Concluding the discussion of our model of the crystal potential, we note that it is the crystal potential that is usually modeled in the calculations of the energy bands. We are modeling the crystal electron density; that is, we shift the modeling onto a quantity which is directly related to the eigenfunctions of the Schrodinger equation. This allows us, if necessary, to easily carry out the subsequent refinement of the calculations, up to self-consistency.

9.2.4 Core Shifts

In Sect. 9.2.1 we have already mentioned that transition to the crystal state also changes the atomic core levels ε_c. The theoretical calculation of the magnitude of these changes is a complex and not fully resolved problem. The simplest possibility of specifying ε_c is to take into account the potential overlap of individual atoms. It is usually assumed (see, e.g. [18]) that the atomic potential changes by a constant value ΔV, which is equal to the sum of the potentials created by individual atoms at the location point of a given atom. In this case, all eigenvalues are shifted by the same amount ΔV, which is usually called the *core shift*.

Somewhat more complicated, but more correct, will be the refinement of the core states in a calculation scheme. When the core states are approximated by individual Bloch sums, the core shifts will be more consistently calculated in the following simplest version of the tight-binding method.

Let us write an expression for the energy of the c-th crystal core state, whose wave function is approximated by the Bloch sum (9.12). Since the Bloch sums (9.12) are not normalized, mean value of the energy is written as

$$\varepsilon_c(\mathbf{k}) \equiv \varepsilon_i^{s'}(\mathbf{k}) = \frac{\displaystyle\int_{\Omega_{\mathrm{WS}}} \Phi_i^{\mathbf{k}s'*}(\mathbf{r})\hat{H}\Phi_i^{\mathbf{k}s'}(\mathbf{r})\,\mathrm{d}\mathbf{r}}{\displaystyle\int_{\Omega_{\mathrm{WS}}} \Phi_i^{\mathbf{k}s'*}(\mathbf{r})\Phi_i^{\mathbf{k}s'}(\mathbf{r})\,\mathrm{d}\mathbf{r}} \tag{9.42}$$

$$= \frac{\displaystyle\sum_{n'} \mathrm{e}^{\mathrm{i}\mathbf{k}\mathbf{R}_{n'}} \int_{\infty} \varphi_i^{s'*}(\mathbf{r}-\boldsymbol{\tau}_{s'})\hat{H}\varphi_i^{s'}(\mathbf{r}-\mathbf{R}_{n'}-\boldsymbol{\tau}_{s'})\,\mathrm{d}\mathbf{r}}{\displaystyle\sum_{n'} \mathrm{e}^{\mathrm{i}\mathbf{k}\mathbf{R}_{n'}} \int_{\infty} \varphi_i^{s'*}(\mathbf{r}-\boldsymbol{\tau}_{s'})\varphi_i^{s'}(\mathbf{r}-\mathbf{R}_{n'}-\boldsymbol{\tau}_{s'})\,\mathrm{d}\mathbf{r}}\,.$$

Next, in accordance with the adopted approximation, we assume that atomic functions localized at different sites overlap weakly, i.e.

$$\int_{\infty} \varphi_i^{s'*}(\mathbf{r}-\boldsymbol{\tau}_{s'})\varphi_i^{s'}(\mathbf{r}-\mathbf{R}_{n'}-\boldsymbol{\tau}_{s'})\,\mathrm{d}\mathbf{r} \approx \delta_{\mathbf{R}_{n'},0}\,, \tag{9.43}$$

and that the crystal potential acting on the core electrons can be represented by the sum of atomic potentials:

$$V(\mathbf{r}) = \sum_{n,s} V_s(\mathbf{r}-\mathbf{R}_n-\boldsymbol{\tau}_s)\,. \tag{9.44}$$

Then, taking (9.43) and (9.44) into account and averaging over the wavevector $\mathbf{k}$, we write expression (9.42) in the form

$$\varepsilon_i^{s'} = \varepsilon_i^{s',0} + \sum_{n,s}{}' \int_\infty \varphi_i^{s'*}(\mathbf{r})\, V_s(\mathbf{r} - \mathbf{R}_n - \boldsymbol{\tau}_s + \boldsymbol{\tau}_{s'})\, \varphi_i^{s'}(\mathbf{r})\, \mathrm{d}\mathbf{r} \tag{9.45}$$

($\varepsilon_i^{s',0}$ is an atomic eigenvalue; the prime at the sum means that the term with $\mathbf{R}_n + \boldsymbol{\tau}_s - \boldsymbol{\tau}_{s'} = 0$ must be omitted). It is convenient to pass from summation over n and s to summation over the coordination spheres p and vectors of individual coordination spheres μ:

$$\varepsilon_i^{s'} = \varepsilon_i^{s',0} + \sum_{p=1}^{\infty} \sum_{\mu=1}^{n_p} \int_\infty \varphi_i^{s'*}(\mathbf{r})\, V_s(\mathbf{r} - \mathbf{R}_p^{\mu})\, \varphi_i^{s'}(\mathbf{r})\, \mathrm{d}\mathbf{r} \equiv \varepsilon_i^{s',0} + \Delta\varepsilon_i^{s'} . \tag{9.46}$$

In the accepted formulation of the OPW method, the dependence of the core eigenvalues on the magnetic quantum number m is not taken into account. Therefore, averaging the energy of the crystal core state over m and taking into account the spherical symmetry of atomic potentials, we obtain

$$\Delta\varepsilon_{nl}^{s'} = \frac{1}{2l+1} \sum_{m=-l}^{l} \Delta\varepsilon_{nlm}^{s'} = \frac{1}{2} \sum_{p=1}^{\infty} \frac{n_p}{R_p} \int_0^\infty V_s(r) r\, \mathrm{d}r \int_{|r-R_p|}^{r+R_p} R_{nl}^{s'\,2}(r') r'\, \mathrm{d}r' . \tag{9.47}$$

Note that the index s depends on p and s'.

In practical calculation of the core shifts $\Delta\varepsilon_{nl}^{s'}$ within the above approximation, it is sufficient to use an approximate analytical expression for $V_s(r)$.

It is significant that, in contrast to the commonly used approximation of core shifts by the sum of the potentials created at the location of a given atom by all other atoms, our model gives different core shifts for different states.

9.2.5 Generalized Eigenvalue Problem

Mathematically, the problem of determining the eigenfunctions and eigenvalues of the Schrödinger equation (8.1) is reduced to solving the generalized eigenvalue problem (compare with (8.13) and (8.14))

$$Hx = \varepsilon S x, \tag{9.48}$$

where H and S are in general complex Hermitian matrices, and S is also positive definite. There are several ways to solve this problem (see, e.g. [20, 21]). However, taking into account rather slow convergence of the OPW method [9] and hence necessity to use a large number of coordinate functions (several hundred), we focus on algorithms, which are suitable for solving (9.48) in the case of high-order matrices.

The easiest way is to reduce the solution of (9.48) to a twofold application of the rotation method (Givens method). One can do this with a single application of this method, if one uses the square-root method (Cholesky method) for factorizing the matrix S. However, instead of full diagonalization by the rotation method, it is better to use a technique that allows one to immediately determine only the required (small!) number of eigenvalues. In this regard, we use a method consisting of the following stages[5]:

1. The matrix S is represented in the form $LL^\dagger$, where L is the lower-triangular matrix and $L^\dagger$ is its conjugate.
2. The matrix L is inverted by the bordering method [20].
3. Problem (9.48) is reduced to the problem $Au = \varepsilon u$, where $A = L^{-1}H(L^{-1})^\dagger$ and $u = L^\dagger x$.
4. The Householder method is used to transform the matrix A into the three-diagonal matrix D [23].
5. The three-diagonal Hermitian matrix D is transformed into the three-diagonal real symmetric matrix R by the similarity transformation T ($D \to TDT^{-1} = R$).
6. The eigenvalues of the symmetric matrix R, enclosed in the interval (a_0, b_0), are calculated by the bisection method [24].

Among other methods, this method stands out for its high speed and high accuracy.

9.3 Summary

In this section we have presented step by step all the stages that make up the calculation of the energy bands by the OPW method. Although most of the necessary approximations are also present in other implementations of the OPW method, two approximations are not standard. These are the construction of the crystal potential and calculation of core shifts. We believe that our method for calculating these values is the most consistent. Construction of symmetrized OPW and solution of the generalized eigenvalue problem are the most laborious stages in the implementation of the OPW method. We hope that their detailed presentation will greatly facilitate practical calculations.

As noted above, the OPW method has rather slow convergence [9]. There are methods based on the muffin-tin approximation for the crystal potential that have much faster convergence. For example, Green's function method, commonly referred to as the Korringa–Kohn–Rostoker (KKR) method[6], has a very fast (exponential)

[5] A brief description of the method, which we use for solving (9.48), is given in [9, 10], a detailed—in [22].

[6] The method was proposed by Kohn and Rostoker [25]. Conceptually, it is close to the scattering matrix method proposed earlier by Korringa [26]. Ham and Segal [27] put the method in a form that is convenient for practical applications. The most complete presentation of both the method itself and its applications is contained in the survey [8].

convergence rate [28]. However, one has to pay the price for fast convergence with very complicated expressions for the matrix elements of the determinant equation (see [8]). In addition, in Green's function method, the matrix elements themselves depend on the unknown value of ε, and instead of solving the eigenvalue problem (8.14), one has to find the zeros of a rather complicated determinant equation. Moreover, the search for zeros is significantly complicated by the presence of false roots associated with the poles of Green's function at the points $\varepsilon = (\mathbf{k} + \mathbf{K}_\mu)^2$ and with zeros of the denominators of the diagonal terms of the determinant equation. But that's not the main point. The main thing is that there is at least one type of solids in which the muffin-tin approximation cannot be considered in any way satisfactory. These are crystals with a covalent bond, since the electron density distribution in them has a pronounced directional character. In particular, most of the semiconductors and semiconductor compounds are of this type.

References

1. H.W. Streitwolf, *Group Theory in Solid State Physics* (MacDonald, London, 1971)
2. W.A. Harrison, *Electronic Structure and the Properties of Solids*, 2nd edn. (Dover, New York, 1989)
3. J. Callaway, *Quantum Theory of the Solid State*, 2nd edn. (Academic, New York, 1991)
4. R.M. Martin, *Electronic Structure: Basic Theory and Practical Methods*, 2nd edn. (Cambridge University Press, Cambridge, 2020)
5. S.G. Mikhlin, *Variational Methods in Mathematical Physics* [in Russian], 2nd edn. (Nauka, Moscow, 1970). Engl. transl.: Pergamon, Oxford, 1964
6. R. Courant, D. Hilbert, *Methods of Mathematical Physics*, vol. 1 (Wiley, New York, 1953)
7. C. Herring, Phys. Rev. **57**, 1169 (1940)
8. V.V. Dyakin, R.F. Egorov, B.I. Reser, V.P. Shirokovskii, Fiz. Metallov i Metallovedenie [in Russian] **25**, 211 (1968)
9. B.I. Reser, V.V. Dyakin, V.P. Shirokovskii, Phys. Stat. Sol. (b) **44**, 425 (1971)
10. B.I. Reser, V.P. Shirokovskii, Fiz. Metallov i Metallovedenie **32**, 934 (1971)
11. E.U. Condon, G.H. Shortley, *The Theory of Atomic Spectra* (Cambridge University Press, Cambridge, 1959)
12. R.H. Parmenter, Phys. Rev. **86**, 552 (1952)
13. J. Callaway, Phys. Rev. **97**, 933 (1955)
14. J.R. Reitz, Solid State Phys. **1**, 1 (1955)
15. B.I. Reser, R.F. Egorov, *Construction of symmetrized plane waves* [in Russian] (1973). Dep. VINITI, No. 6658–73
16. J.C. Slater, Phys. Rev. **81**, 385 (1951)
17. F. Herman, Phys. Rev. **93**, 1214 (1954)
18. F. Herman, S. Skillman, in *Proc. Intern. Conf. Semicond. Phys.* (Czechoslovak Acad. Sci., Prague, 1960), p. 20
19. T.O. Woodruff, Phys. Rev. **103**, 1159 (1956)
20. V.V. Voevodin, *Numerical Methods of Algebra: Theory and Algorithms* [in Russian] (Nauka, Moscow, 1966)
21. J.H. Wilkinson, *The Algebraic Eigenvalue Problem*, 2nd edn. (Clarendon, Oxford, 1988)
22. B.I. Reser, *The OPW method and interpolation based on the tight binding in calculating the energy band structure of crystals* [in Russian]. Diss. for the Degree of Cand. of Phys. and Math. Sciences, Inst. of Metal Phys., Ural Science Center, USSR Acad. of Sciences, Sverdlovsk (1972), 153 pp

23. D.J. Mueller, J. Numerische Mathematik **8**, 72 (1966)
24. W. Barth, R.S. Martin, J.H. Wilkinson, Numerische Mathematik **9**, 386 (1967)
25. W. Kohn, N. Rostoker, Phys. Rev. **94**, 1111 (1954)
26. J. Korringa, Physica **13**, 392 (1957)
27. F.S. Ham, B. Segall, Phys. Rev. **124**, 1786 (1961)
28. V.V. Dyakin, V.P. Shirokovskii, U.S.S.R. Comput. Math. Math. Phys. **12**, 332 (1972)

Chapter 10
Energy Bands of Diamond

I have stressed the connection with experimental information and with the physical picture rather than the formal development of the theory. Some recent books have stressed all too much the formal side. (H. Bethe, Intermediate quantum mechanics, Benjamin, New York, 1964).

The diamond structure presents the main example of a nonsymorphic space group. This example has been analyzed in many papers and books on representation theory [1–4], and we present it here to illustrate "symmetry in action" (for details see [5]).

10.1 Crystal Structure

10.1.1 Symmetry of the Crystal

To get an idea of how atoms are arranged in crystals with the diamond structure, one can imagine two identical face-centered cubic lattices, shifted relative to each other by a quarter of the cube diagonal (see Fig. 10.1). Here we give a detailed description of this crystal structure.

The diamond lattice is a face-centered cubic (fcc) lattice with the basic vectors

$$\mathbf{a}_1 = a(0, 1, 1), \qquad \mathbf{a}_2 = a(1, 0, 1), \qquad \mathbf{a}_3 = a(1, 1, 0), \tag{10.1}$$

where $2a$ is the period of the lattice along the coordinate axes. An arbitrary vector of this lattice is

$$\mathbf{R}_n = n_1\mathbf{a}_1 + n_2\mathbf{a}_2 + n_3\mathbf{a}_3, \tag{10.2}$$

where n_i are integers. Taking (10.1) into account, we can write

N. B. Melnikov and B. I. Reser, *Space Group Representations*,
https://doi.org/10.1007/978-3-031-13991-8_10

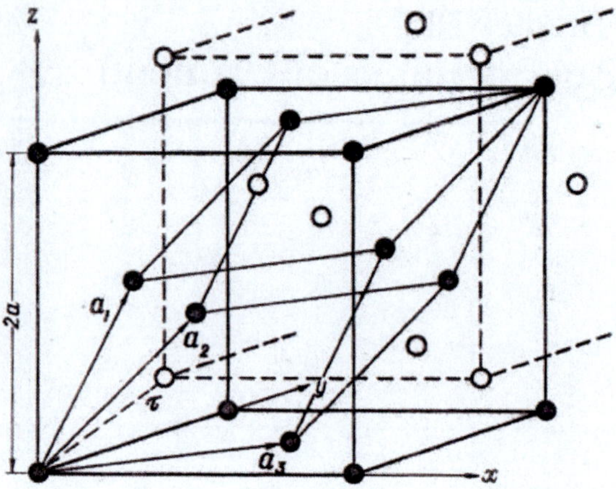

Fig. 10.1 Diamond crystal structure

$$\mathbf{R}_n = a(n_2 + n_3,\ n_1 + n_3,\ n_1 + n_2). \tag{10.3}$$

Here $(n_2 + n_3) + (n_1 + n_3) + (n_1 + n_2) = 2(n_1 + n_2 + n_3)$ is an even number. The Wigner–Seitz cell of a fcc lattice is a regular rhombic dodecahedron with volume $\Omega_{\mathrm{WS}} = 2a^3$.

The diamond structure is a structure with the basis that consists of two identical atoms. We previously denoted the position of the atoms forming the basis in the Wigner–Seitz cell by the vectors $\boldsymbol{\tau}_s$ $(s = 1, 2, \ldots, s_0)$. These vectors depend on the choice of the origin. For the diamond structure, two possibilities are usually considered: (1) the origin is located in the middle between the atoms of the basis, (2) the origin coincides with one of the atoms of the basis. We use the second choice of the origin. In this case, the basis vectors $\boldsymbol{\tau}_s$ are as follows:

$$\boldsymbol{\tau}_1 = 0, \qquad \boldsymbol{\tau}_2 = \boldsymbol{\tau} = a\left(\frac{1}{2}, \frac{1}{2}, \frac{1}{2}\right). \tag{10.4}$$

Spatial transformations that transform the diamond crystal into itself form the space group O_h^7. The translation subgroup T of this group consists of the transformations $\{\varepsilon|\mathbf{R}_n\}$.

The point transformations α of the elements of the space group O_h^7 form the point group (crystal class) O_h. Elements of the crystal class O_h are given in Table 10.1; Appendix E provides the multiplication table. Let us explain the notation in Table 10.1. A point transformation α is, as usual, a rotation, inversion, or a combination of both. The position of the symmetry axis is characterized by the coordinates of a point on this axis; rotation around the axis is performed in a counterclockwise direction as viewed from the point defining the axis position to the origin. In addition, in the description of each symmetry element, we have indicated the coordinates of the point to which the point with coordinates (x, y, z) is translated by this element.

The numbering of the elements of the crystal class is the same as in monograph [6],[1] the letter notation of the elements in the crystal class and classes of conjugate elements coincide with the notation of [7].

Since the space group O_h^7 is nonsymmorphic, a nonprimitive translation $\boldsymbol{\tau}_\alpha$ must be specified for each element of the crystal class. With our choice of the origin, the nonprimitive translations $\boldsymbol{\tau}_\alpha$ have the form shown in Table 10.1.

10.1.2 Reciprocal Lattice and Brillouin Zone

The reciprocal to the direct fcc lattice of diamond is the body-centered cubic (bcc) lattice, which is formed by the set of vectors

$$\mathbf{K}_\mu = \mu_1\mathbf{b}_1 + \mu_2\mathbf{b}_2 + \mu_3\mathbf{b}_3\,, \tag{10.5}$$

where μ_i are integers and

$$\mathbf{b}_1 = \frac{\pi}{a}(-1, 1, 1), \qquad \mathbf{b}_2 = \frac{\pi}{a}(1, -1, 1), \qquad \mathbf{b}_3 = \frac{\pi}{a}(1, 1, -1) \tag{10.6}$$

are the basic vectors of the reciprocal lattice. Taking (10.6) into account, an arbitrary reciprocal lattice vector can also be written in the form

$$\mathbf{K}_\mu = \frac{\pi}{a}(-\mu_1 + \mu_2 + \mu_3,\ \mu_1 - \mu_2 + \mu_3,\ \mu_1 + \mu_2 - \mu_3)\,, \tag{10.7}$$

where $(-\mu_1 + \mu_2 + \mu_3)$, $(\mu_1 - \mu_2 + \mu_3)$, $(\mu_1 + \mu_2 - \mu_3)$ all even or all odd. The Brillouin zone of a fcc lattice is a truncated octahedron (see Fig. 10.2) of volume $\Omega_{BZ} = 4(\pi/a)^3$.

When we study the energy bands, symmetry of the energy ε as a function of the wavevector $\mathbf{k}$:

$$\varepsilon_j(\alpha_i\mathbf{k}) = \varepsilon_j(\mathbf{k}), \qquad \alpha_i \in G_0\,, \tag{10.8}$$

allows us to consider only the vectors $\mathbf{k}$ lying in the irreducible Brillouin zone. In our case, this is the $\Gamma KWXUL$ wedge (see Fig. 10.2), which makes up 1/48 of the Brillouin zone.

In calculations, the points of the $\mathbf{k}$-space with a certain symmetry are of the greatest interest. In Fig. 10.2 all symmetry points in the Brillouin zone of the fcc lattice are shown in the BSW notation: Γ, X, L, W, K, U are the vertices of the irreducible Brillouin zone; Δ, Λ, Σ, Z, S, Q, LK, LU, KW, WU arbitrary points on the symmetry lines; ΓΛΔ, ΓΣΛ, ΓΔΣ, WUX are arbitrary points on the symmetry planes. A complete description of the symmetry points in the Brillouin zone is given

[1] Here and in some following chapters this numeration was used historically.

Table 10.1 Elements of the crystal class O_h (with nonprimitive translations $\boldsymbol{\tau}_\alpha$ for the space group O_h^7) †

No. i of the elem. α_i	Class of conjug. elem.	Rotation α_i'			
		Axis	Angle θ	$\mathbf{r}' = \alpha\mathbf{r}$	$\boldsymbol{\tau}_\alpha$
1	E		0	(x, y, z)	0
2		(1, 0, 0)	π	$(x, -y, -z)$	0
3	C_4^2	(0, 1, 0)	π	$(-x, y, -z)$	0
4		(0, 0, 1)	π	$(-x, -y, z)$	0
5	C_3	(1, 1, 1)	$-2\pi/3$	(y, z, x)	0
6		(1, 1, −1)	$2\pi/3$	$(y, -z, -x)$	0
7		(−1, 1, 1)	$2\pi/3$	$(-y, z, -x)$	0
8		(1, −1, 1)	$2\pi/3$	$(-y, -z, x)$	0
9		(1, 1, 1)	$2\pi/3$	(z, x, y)	0
10		(1, −1, 1)	$-2\pi/3$	$(z, -x, -y)$	0
11		(1, 1, −1)	$-2\pi/3$	$(-z, x, -y)$	0
12		(−1, 1, 1)	$-2\pi/3$	$(-z, -x, y)$	0
13	C_2	(−1, 1, 0)	π	$(-y, -x, -z)$	$\boldsymbol{\tau}$
14	C_4	(0, 0, 1)	$\pi/2$	$(-y, x, z)$	$\boldsymbol{\tau}$
15	C_4	(0, 0, 1)	$-\pi/2$	$(y, -x, z)$	$\boldsymbol{\tau}$
16	C_2	(1, 1, 0)	π	$(y, x, -z)$	$\boldsymbol{\tau}$
17	C_2	(0, −1, 1)	π	$(-x, -z, -y)$	$\boldsymbol{\tau}$
18	C_2	(0, 1, 1)	π	$(-x, z, y)$	$\boldsymbol{\tau}$
19	C_4	(1, 0, 0)	$\pi/2$	$(x, -z, y)$	$\boldsymbol{\tau}$
20	C_4	(1, 0, 0)	$-\pi/2$	$(x, z, -y)$	$\boldsymbol{\tau}$
21	C_2	(−1, 0, 1)	π	$(-z, -y, -x)$	$\boldsymbol{\tau}$
22	C_4	(0, 1, 0)	$-\pi/2$	$(-z, y, x)$	$\boldsymbol{\tau}$
23	C_2	(1, 0, 1)	π	$(z, -y, x)$	$\boldsymbol{\tau}$
24	C_4	(0, 1, 0)	$\pi/2$	$(z, y, -x)$	$\boldsymbol{\tau}$
25	I		0	$(-x, -y, -z)$	$\boldsymbol{\tau}$
26		(1, 0, 0)	π	$(-x, y, z)$	$\boldsymbol{\tau}$
27	IC_4^2	(0, 1, 0)	π	$(x, -y, z)$	$\boldsymbol{\tau}$
28		(0, 0, 1)	π	$(x, y, -z)$	$\boldsymbol{\tau}$
29	IC_3	(1, 1, 1)	$-2\pi/3$	$(-y, -z, -x)$	$\boldsymbol{\tau}$
30		(1, 1, −1)	$2\pi/3$	$(-y, z, x)$	$\boldsymbol{\tau}$
31		(−1, 1, 1)	$2\pi/3$	$(y, -z, x)$	$\boldsymbol{\tau}$
32		(1, −1, 1)	$2\pi/3$	$(y, z, -x)$	$\boldsymbol{\tau}$
33		(1, 1, 1)	$2\pi/3$	$(-z, -x, -y)$	$\boldsymbol{\tau}$
34		(1, −1, 1)	$-2\pi/3$	$(-z, x, y)$	$\boldsymbol{\tau}$
35		(1, 1, −1)	$-2\pi/3$	$(z, -x, y)$	$\boldsymbol{\tau}$
36		(−1, 1, 1)	$-2\pi/3$	$(z, x, -y)$	$\boldsymbol{\tau}$

(continued)

Table 10.1 (continued)

No. i of the elem. α_i	Class of conjug. elem.	Rotation α'_i			
		Axis	Angle θ	$\mathbf{r}' = \alpha\mathbf{r}$	τ_α
37	IC_2	$(-1, 1, 0)$	π	(y, x, z)	0
38	IC_4	$(0, 0, 1)$	$\pi/2$	$(y, -x, -z)$	0
39	IC_4	$(0, 0, 1)$	$-\pi/2$	$(-y, x, -z)$	0
40	IC_2	$(1, 1, 0)$	π	$(-y, -x, z)$	0
41	IC_2	$(0, -1, 1)$	π	(x, z, y)	0
42	IC_2	$(0, 1, 1)$	π	$(x, -z, -y)$	0
43	IC_4	$(1, 0, 0)$	$\pi/2$	$(-x, z, -y)$	0
44	IC_4	$(1, 0, 0)$	$-\pi/2$	$(-x, -z, y)$	0
45	IC_2	$(-1, 0, 1)$	π	(z, y, x)	0
46	IC_4	$(0, 1, 0)$	$-\pi/2$	$(z, -y, -x)$	0
47	IC_2	$(1, 0, 1)$	π	$(-z, y, -x)$	0
48	IC_4	$(0, 1, 0)$	$\pi/2$	$(-z, -y, x)$	0

$\tau = a(1/2, 1/2, 1/2)$

†The origin coincides with one of the basis atoms

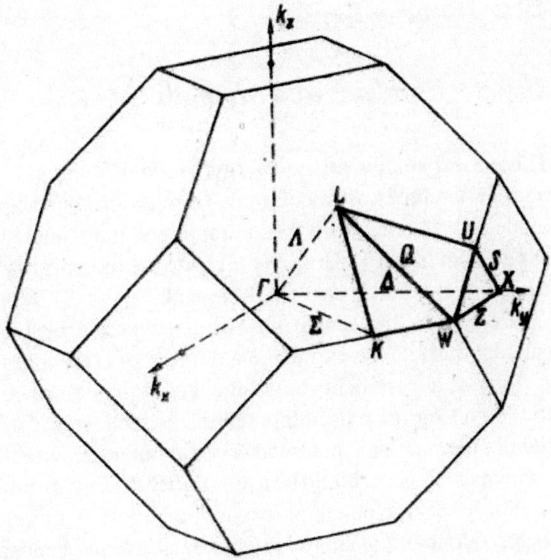

Fig. 10.2 Brillouin zone of the fcc lattice

in Table 10.2. The points of the planes LKW and LWU, like all interior points of the irreducible Brillouin zone, are in a general position; i.e. they do not have symmetry.

To construct symmetrized combinations of coordinate functions, one needs matrices of irreducible projective representations $\Delta^{\mathbf{k}j}$ of each point group $G^0_{\mathbf{k}}$ consisting of the rotational parts α_m of representatives $\{\alpha_m|\tau_{\alpha_m}\}$ in the decomposition of the group of the wavevector $G_{\mathbf{k}}$ in terms of the invariant translation subgroup T. An effective way of obtaining the matrices of these representations is presented in Chaps. 3–5. In Appendix F, we have given irreducible projective representations $\Delta^{\mathbf{k}j}$ corresponding to the space group O_h^7 for all point groups $G_{\mathbf{k}}^0$ of the Brillouin zone of the fcc lattice.

Despite the fact that the symmetry group of the diamond structure is fairly well studied (see, e.g. [1, 2]) and its description is found even in textbooks (see, e.g. [3]), we had to carry out the study of the diamond structure from the very beginning, because all published works present only tables of *characters*, and for constructing symmetrized combinations one needs the representation *matrices*.[2]

In Appendix F, we also presented all the most interesting compatibility relations, namely relations connecting irreducible representations at the vertices of the irreducible Brillouin zone and irreducible representations at points on the symmetry lines.

In Sect. 9.2.2 we have shown that construction of the symmetrized OPWs reduces to finding symmetrized plane waves. In application to the diamond structure, the construction of symmetrized plane waves has been considered in detail in [5].

10.2 Energy Bands

10.2.1 Problems and Methods

Theoretical studies of the energy bands of diamond have already been carried out both on the basis of calculations from the first principles [9, 10] and using various modeling schemes with free parameters, the values of which are determined from experiment [11–14]. However, despite the long history, the picture of the conduction bands in diamond is still controversial.

We cannot agree with the results obtained by [12, 14] on the basis of model calculations and have to make a number of critical remarks in this regard.

First of all, we believe that the technique described in [11–14] is weakly related to the concept of a pseudopotential. Indeed, in order to describe the compensated (due to the repulsive potential) crystal potential with the help of its first few Fourier components, there should be a possibility to actually introduce a sufficiently weak and smoothly varying potential (pseudopotential). In this case, the Fourier coefficients of the pseudopotential should tend to decrease and be insensitive in relation to both the method of their determination and small changes in the initial data. In the above-cited works [11–14] the situation is exactly the opposite. We explain this by the fact that the Fourier components introduced there are just some kind of mathematical parameters and do not carry any physical meaning. In this sense, their calculation (due to the ambiguity; compare the results of [11–14]) does not provide any information about

[2] Tables of all irreducible small representations of space groups are given in [6]. Unfortunately, a large number of misprints in [6] urges one to treat these data with caution [8].

Table 10.2 Symmetry points of the Brillouin zone for the fcc lattice

BSW notation	Coordinates of equivalent points in units of π/a	$G_{\mathbf{k}}^0$	No. m of the elements α_m of the group $G_{\mathbf{k}}^0$
Γ	$(0, 0, 0)$	O_h	1–48
X	$(0, 1, 0)$, $(0, -1, 0)$	D_{4h}	1–4, 21–28, 45–48
L	$\left(\frac{1}{2}, \frac{1}{2}, \frac{1}{2}\right)$, $\left(-\frac{1}{2}, -\frac{1}{2}, -\frac{1}{2}\right)$	D_{3h}	1, 5, 9, 13, 17, 21, 25, 29, 33, 37, 41, 45
W	$\left(\frac{1}{2}, 1, 0\right)$, $\left(\frac{1}{2}, -1, 0\right)$, $\left(-\frac{1}{2}, 0, 1\right)$, $\left(-\frac{1}{2}, 0, -1\right)$	D_{2d}	1, 2, 17, 18, 27, 28, 43, 44
K	$\left(\frac{3}{4}, \frac{3}{4}, 0\right)$, $\overbrace{\left(-\frac{1}{4}, -\frac{1}{4}, 1\right), \left(-\frac{1}{4}, -\frac{1}{4}, -1\right)}^{\text{U points}}$	C_{2v}	1, 16, 28, 37
U	$\left(\frac{1}{4}, 1, \frac{1}{4}\right)$, $\left(\frac{1}{4}, -1, \frac{1}{4}\right)$, $\overbrace{\left(-\frac{3}{4}, 0, -\frac{3}{4}\right)}^{\text{K point}}$	C_{2v}	1, 23, 27, 45
Δ	$(0, \mu, 0)$	C_{4v}	1, 3, 22, 24, 26, 28, 45, 47
Λ	$\left(\frac{\mu}{2}, \frac{\mu}{2}, \frac{\mu}{2}\right)$	C_{3v}	1, 5, 9, 37, 41, 45
Σ	$\left(\frac{3\mu}{4}, \frac{3\mu}{4}, 0\right)$	C_{2v}	1, 16, 28, 37
Z	$\left(\frac{\mu}{2}, 1, 0\right)$, $\left(\frac{\mu}{2}, -1, 0\right)$	C_{2v}	1, 2, 27, 28
S	$\left(\frac{\mu}{4}, 1, \frac{\mu}{4}\right)$, $\left(\frac{\mu}{4}, -1, \frac{\mu}{4}\right)$	C_{2v}	1, 23, 27, 45
Q	$\left(\frac{1}{2}, 1 - \frac{\mu}{2}, \frac{\mu}{2}\right)$, $\left(-\frac{1}{2}, -\frac{\mu}{2}, -1 + \frac{\mu}{2}\right)$	C_2	1, 17
LK	$\left(\frac{1}{2} + \frac{\mu}{4}, \frac{1}{2} + \frac{\mu}{4}, \frac{1}{2} - \frac{\mu}{2}\right)$, $\overbrace{\left(-\frac{1}{2} + \frac{\mu}{4}, -\frac{1}{2} + \frac{\mu}{4}, -\frac{1}{2} - \frac{\mu}{2}\right)}^{\text{LU point}}$	C_{1h}	1, 37
LU	$\left(\frac{1}{2} - \frac{\mu}{4}, \frac{1}{2} + \frac{\mu}{2}, \frac{1}{2} - \frac{\mu}{4}\right)$, $\overbrace{\left(-\frac{1}{2} - \frac{\mu}{4}, -\frac{1}{2} + \frac{\mu}{2}, -\frac{1}{2} - \frac{\mu}{4}\right)}^{\text{LK point}}$	C_{1h}	1, 45
KW	$\left(\frac{3}{4} - \frac{\mu}{4}, \frac{3}{4} + \frac{\mu}{4}, 0\right)$, $\overbrace{\left(-\frac{1}{4} - \frac{\mu}{4}, -\frac{1}{4} + \frac{\mu}{4}, 1\right) \left(-\frac{1}{4} - \frac{\mu}{4}, -\frac{1}{4} + \frac{\mu}{4}, -1\right)}^{\text{WU points}}$	C_{1h}	1, 28
WU	$\left(\frac{1}{2} - \frac{\mu}{4}, 1, \frac{\mu}{4}\right)$, $\left(\frac{1}{2} - \frac{\mu}{4}, -1, \frac{\mu}{4}\right)$, $\overbrace{\left(-\frac{1}{2} - \frac{\mu}{4}, 0, -1 + \frac{\mu}{4}\right)}^{\text{KW point}}$	C_{1h}	1, 27
ΓΛΔ		C_{1h}	1, 45
ΓΣΛ		C_{1h}	1, 37
ΓΔΣ		C_{1h}	1, 28
WUX		C_{1h}	1, 27

$0 < \mu < 1$

the nature of the studied bands. If we take into account the even greater sensitivity of the bands to small changes in these parameters [11] and impossibility of giving preference to any of the models by comparing the obtained results with experiment (compare Figs. 2 in [12, 14]), it is difficult to believe in the reliability of the final conclusions.

It also seems to us that the use of the so-called pseudofunctions for calculating the matrix elements of the transition probability (in calculations of the imaginary part of the dielectric constant) needs to be justified, because these pseudofunctions by no means describe the states of an electron in a crystal.

We believe that the nature of the one-electron bands of crystals can be understood only on the basis of first-principles calculations. Of course, the absolute accuracy of such calculations (associated not so much with the arithmetic accuracy of calculations, but with the accuracy of the model underlying the calculation), will not satisfy the needs of the experiment for a long time, and the use of fitting schemes, such as [9], will be dictated by necessity.

In the next section we present the calculation of the diamond bands from first principles [5], which is based on the methods for constructing the crystal potential and calculating core shifts described in Sects. 9.2.3 and 9.2.4.

10.2.2 Calculation Results

Calculations of the Fourier coefficients of the Coulomb part of the crystal potential, crystal electron density, orthogonalization coefficients and core shifts were performed with analytical atomic wave functions [15] of a neutral C atom in the $1s^2 2s^2 2p^2$ configuration (term ^1S) calculated in the Hartree–Fock approximation. The electronic states in the diamond crystal corresponding to the 1s shell were considered as core, and the states corresponding to 2s and 2p—as valence states.

The lattice constant of diamond a was taken equal to 3.363 61 a.u. [16].

We assumed the radii of the (m-t)-spheres r_s to be equal to half the distance between two atoms of the Wigner–Seitz cell, i.e. $r_s = |\boldsymbol{\tau}|/2 = a\sqrt{3}/4$.

Figure 10.3 gives a comparison of the crystal and atomic electron densities in diamond. As can be seen from the figure, the contribution from the electron densities of neighboring atoms is significant.

Calculation of the integrals in the Fourier coefficients of the exchange part of the crystal potential was carried out numerically. The charge of electrons inside the (m-t)-sphere was also calculated numerically, $Q_s = 4.266\,43$ e. The crystal electron density outside the (m-t)-spheres ρ_c^e, determined from (9.40), turned out to be equal to 0.069 03 e/(a.u.)^3. It is useful to note that most of the electron charge is concentrated inside the (m-t)-spheres, although they occupy a smaller part of the Wigner–Seitz cell: $Q_s/Z_s \approx 0.71$; $2\Omega_s/\Omega_{\mathrm{WS}} \approx 0.34$.

Atomic eigenvalue $\varepsilon_{1,0} = -22.782$ Ry was taken from the same work [15] as the atomic functions. We calculated the core shift $\Delta\varepsilon_{1,0}$ by formula (9.47) taking into account only the atoms of the first coordination sphere and using the

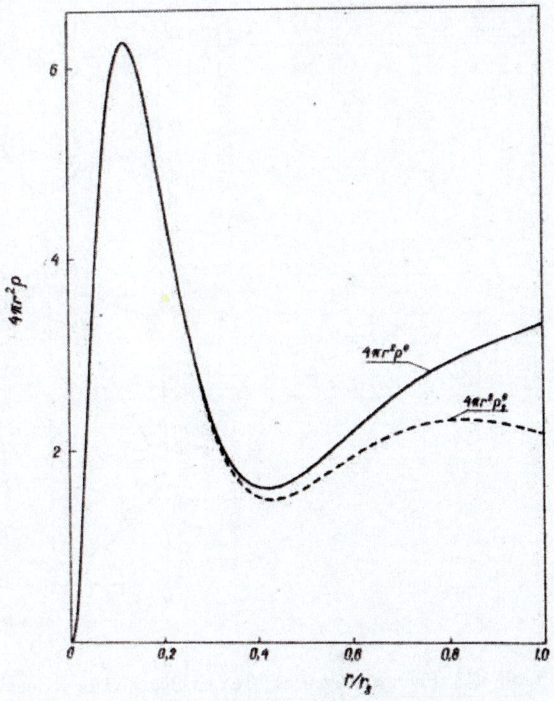

Fig. 10.3 Crystal (solid line) and atomic (dashed line) electron densities in diamond

Herman–Skillman atomic potential [17]. The value of $\Delta\varepsilon_{1,0}$ turned out to be equal to −2.450 Ry.

The energy bands were calculated at the symmetry points Γ, X, K, L, W of the Brillouin zone and along the symmetry lines Δ, Σ, Λ with a step $\pi/(4a)$, where $2a$ is the cubic lattice constant. The calculation was carried out in two variants: (a) without the core shifts taken into account and (b) with the core shifts taken into account. The results obtained in variant (b) are shown in Fig. 10.4.

The values of the characteristic energy differences for both calculation variants are given in Table 10.3. The results of Herman et al. [9] are also presented there for comparison.

All data correspond to the same number of coordinate functions (113 OPW).

10.2.3 Comparison with Experiment

Two reliably established experimental characteristics of the diamond bands are the position and value ε_d of the indirect energy gap. Namely, $\varepsilon_d = \Delta_1^m - \Gamma'_{25} = 5.47$ eV [18, 19], where Δ_1^m is the energy value corresponding to the minimum of the conduction band located at the point $\Delta(0, 3/4, 0)$. In our calculations, the minimum of

Fig. 10.4 Energy bands of diamond

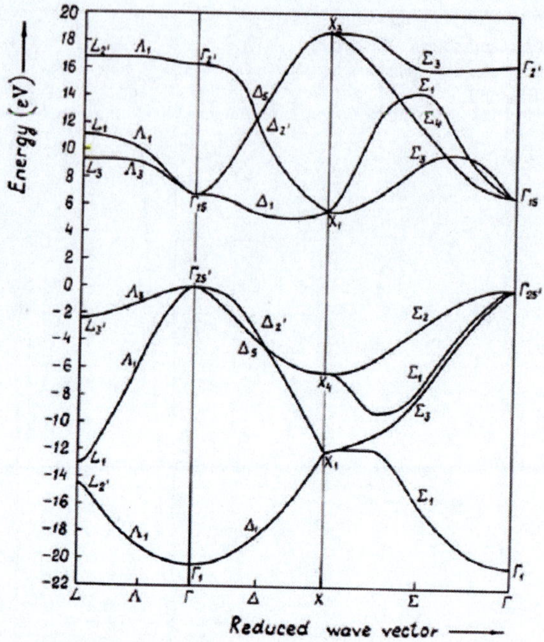

Table 10.3 Energy differences (in eV) in diamond for various models

Energy difference	a	Herman et al. [9] NRSC †	b	Herman et al. [9] PERT‡
$X_1 - \Gamma'_{25}$	4.24	4.4	5.59	6.0 ± 0.1
$\Gamma_{15} - \Gamma'_{25}$	6.75	6.7	6.75	7.1 ± 0.2
$L_1 - \Gamma'_{25}$	7.30	7.2	11.09	9.8 ± 1.0
$L_1 - L'_3$	9.64	9.5	13.43	11.9 ± 1.1
$X_1 - X_4$	10.47	10.2	11.81	11.6 ± 0.1
$\Sigma_3 - \Sigma_2^{\S}$	10.95	10.7	12.06	12.0 ± 0.3
$L_3 - L'_3$	11.67	12.3	11.67	12.8 ± 0.3
$\Gamma'_2 - \Gamma'_{25}$	10.94	10.8	16.31	14.3 ± 1.7

† Nonrelativistic self-consistent OPW calculation
‡ Empirically adjusted (perturbed) self-consistent OPW calculation
§ For $\mathbf{k} = \pi/a\,(1/2, 1/2, 0)$

the conduction band has been obtained at the same point of the Brillouin zone, and the magnitude of the indirect gap is 5.1 eV.

To compare the calculations with other experiments, we used the interpolation of the energy bands based on the tight-binding method (see Chap. 14). The eight energy bands shown in Fig. 10.4 were interpreted in the second neighbor approximation (see Sect. 14.3). A comparison of the density of states for the valence bands of diamond with experimental data on X-ray emission K-spectra is given in Sect. 14.4. The agreement with experiment is quite satisfactory.

We did not calculate the imaginary part of the dielectric constant because from our point of view, its calculation makes sense only if the dependence of the transition probability on the wavevector is correctly taken into account. However, since our data are in agreement with the results of [9], the interpretation of the optical data presented in [20] is fully applicable here as well.

10.3 Summary

We believe that our method for constructing the crystal potential and calculating core shifts made it possible to obtain quite satisfactory results for the energy bands of diamond. From our point of view, this is due to two reasons. First, apparently the crystal potential model, which is based on the representation of the crystal electron density as a superposition of atomic electron densities, approximates the crystal potential fairly well, at least for light elements. Second, in our opinion, the method for calculating the core shifts should be recognized as a successful one. The achieved success allows us to hope that the proposed model will be a good initial approximation for the energy band calculations of other crystals as well.

Finally, we note that it seems unlikely to us that a calculation that correctly reproduces the structure of the valence bands and gives the correct values of the position and width of the gap, at the same time gives such an incorrect result for the conduction bands as is actually stated in works of [12, 14].

References

1. C. Herring, J. Franklin Inst. **233**, 525 (1942)
2. W. Döring, V. Zehler, Ann. Physik **448**, 214 (1953)
3. H.W. Streitwolf, *Group Theory in Solid State Physics* (MacDonald, London, 1971)
4. J.L. Birman, *Theory of Crystal Space Groups and Lattice Dynamics: Infra-Red and Raman Optical Processes of Insulating Crystals* (Springer, Berlin, 1974)
5. B.I. Reser, *The OPW method and interpolation based on the tight binding in calculating the energy band structure of crystals* [in Russian]. Diss. for the degree of cand. of phys. and math. sciences, Inst. of Metal Phys., Ural Science Center, USSR Acad. of Sciences, Sverdlovsk (1972). 153 pp
6. O.V. Kovalev, *Irreducible Representations of Space Groups* [in Russian] (Acad. Sci. Ukrain. SSR, Kiev, 1961). Engl. transl.: Gordon and Breach, New York, 1965
7. L.P. Bouckaert, R. Smoluchowski, E. Wigner, Phys. Rev. **50**, 58 (1936)
8. J.L. Warren, Rev. Mod. Phys. **40**, 38 (1968)
9. F. Herman, R.L. Kortum, C.D. Kuglin, Internat. J. Quantum Chem. **1S**, 533 (1967)
10. R. Keown, Phys. Rev. **150**, 568 (1966)
11. W. Saslow, T.K. Bergstresser, M.L. Cohen, Phys. Rev. Lett. **16**, 354 (1966)
12. L.R. Saravia, D. Brust, Phys. Rev. **170**, 683 (1968)
13. W. van Haeringen, H.G. Junginger, Solid State Commun. **7**, 1135 (1969)
14. L.A. Hemstreet, C.Y. Fong, M.L. Cohen, Phys. Rev. B **2**, 2054 (1970)
15. E. Clementi, C.C.J. Roothaan, M. Yoshimine, Phys. Rev. **127**, 1618 (1962)

16. K. Lonsdale (ed.), *International Tables for X-Ray Crystallography*, vol. 3 (Kynoch Press, Birmingham, 1962)
17. F. Herman, S. Skillman, *Atomic Structure Calculations* (Prentice-Hall, Englewood Cliffs, N.J., 1963)
18. C.D. Clark, P.J. Dean, P.V. Harris, Proc. Roy. Soc. (London) **A277**, 312 (1964)
19. P.J. Dean, E.C. Lightowlers, D.R. Wight, Phys. Rev. **140**, A352 (1965)
20. R.A. Roberts, W.C. Walker, Phys. Rev. **161**, 730 (1967)

Chapter 11
Electronic States in Atoms and Solids

As far as I see, all a priori statements in physics have their origin in symmetry. (H. Weyl, Symmetry, Princeton Univ. Press, Princeton, NJ, 1952)

In problems that require studying the energy bands and wave functions of electrons in a crystal, it is often necessary to establish a correspondence between the states of an electron in a crystal and states of an electron in free atoms that make up this crystal.

Similarly to the problem of constructing irreducible representations of the group of the wavevector, the problem of identifying electronic states is simple for crystals with one atom per unit cell and difficult for crystals with a basis, especially when the space group is nonsymmorphic (see, e.g. [1, 2]).

From the group-theoretical point of view, the state of an electron in a crystal is characterized by an irreducible representation $d^{\mathbf{k}j}$ of the group of the wavevector $G_{\mathbf{k}}$, where $\mathbf{k}$ is the reduced wavevector, which labels different representations of the Abelian translation subgroup T, and j is the index of the irreducible projective representation of the factor group $G_{\mathbf{k}}/T$. The state of an electron in a free atom is characterized by an irreducible representation $\mathcal{D}^{l\pm}$ of the full rotation group. The correspondence between the states of an electron in a crystal and in a free atom can be established using the representation theory.

11.1 Correspondence Between Atomic and Crystal States

11.1.1 Background Material

Let us first recall some definitions and notation that will be widely used in this chapter (see also Chaps. 5 and 7).

N. B. Melnikov and B. I. Reser, *Space Group Representations*,
https://doi.org/10.1007/978-3-031-13991-8_11

11.1.1.1 Notation and Properties of the Space Transformations

Any spatial transformation is usually written in the form $\{\alpha|\mathbf{t}\}$ and its action on an arbitrary vector $\mathbf{r}$ is given by the equality

$$\{\alpha|\mathbf{t}\}\mathbf{r} = \alpha\mathbf{r} + \mathbf{t}. \tag{11.1}$$

Here α is the point transformation corresponding to the rotation α', inversion i, or their combination $i\alpha'$, and $\mathbf{t}$ is the translation vector. In this notation, $\{\varepsilon|0\}$ is the identity transformation, $\{\varepsilon|\mathbf{t}\}$ represents translations, and $\{\alpha|0\}$ represents rotations (proper or improper). It is easy to check that

$$\{\alpha|\mathbf{t}\}\{\beta|\mathbf{t}'\} = \{\alpha\beta|\alpha\mathbf{t}' + \mathbf{t}\}, \tag{11.2}$$
$$\{\alpha|\mathbf{t}\}^{-1} = \{\alpha^{-1}| - \alpha^{-1}\mathbf{t}\}. \tag{11.3}$$

The transformation $\{\alpha|\mathbf{t}\}$ in the $\mathbf{r}$-space induces the operators $P_{\{\alpha|\mathbf{t}\}}$ in the space of functions of $\mathbf{r}$. The action of $P_{\{\alpha|\mathbf{t}\}}$ on an arbitrary function $f(\mathbf{r})$ is determined by the relation

$$P_{\{\alpha|\mathbf{t}\}} f(\mathbf{r}) = f(\{\alpha|\mathbf{t}\}^{-1}\mathbf{r}). \tag{11.4}$$

Based on this definition, it is easy to verify that the introduced operators have properties similar to (11.2) and (11.3):

$$P_{\{\alpha|\mathbf{t}\}} P_{\{\beta|\mathbf{t}'\}} = P_{\{\alpha\beta|\alpha\mathbf{t}'+\mathbf{t}\}}, \tag{11.5}$$
$$P_{\{\alpha|\mathbf{t}\}^{-1}} = P_{\{\alpha^{-1}|-\alpha^{-1}\mathbf{t}\}}. \tag{11.6}$$

11.1.1.2 Irreducible Representations of the Group of the Wavevector $G_{\mathbf{k}}$

The spatial transformations

$$\{\alpha|\mathbf{R}_p + \boldsymbol{\tau}_\alpha\} = \{\varepsilon|\mathbf{R}_p\}\{\alpha|\boldsymbol{\tau}_\alpha\} \tag{11.7}$$

($\mathbf{R}_p$ are the Bravais lattice vectors, $\boldsymbol{\tau}_\alpha$ are nonprimitive translations), whose rotational parts leave $\mathbf{k}$ unchanged up to a reciprocal lattice vector:

$$\alpha\mathbf{k} = \mathbf{k} + \mathbf{K}_{\alpha,\mathbf{k}}, \tag{11.8}$$

form the group of the wavevector $G_{\mathbf{k}}$.

The matrices of irreducible representations of the group $G_{\mathbf{k}}$ are written in the form

$$d^{\mathbf{k}j}(\{\varepsilon|\mathbf{R}_p\}\{\alpha|\boldsymbol{\tau}_\alpha\}) = \mathrm{e}^{-\mathrm{i}\mathbf{k}\mathbf{R}_p}\,\Delta^{\mathbf{k}j}(\{\alpha|\boldsymbol{\tau}_\alpha\}), \tag{11.9}$$

where $\Delta^{\mathbf{k}j}$ is an irreducible projective representation of the factor group $G_{\mathbf{k}}/T$, which is isomorphic to the point group $G^0_{\mathbf{k}}$ consisting of the rotational parts α of elements of the group $G_{\mathbf{k}}$.

The irreducible projective representations $\Delta^{\mathbf{k}j}$ correspond to the system of projective factors

$$\omega^{\mathbf{k}}_{\{\alpha|\boldsymbol{\tau}_\alpha\},\{\beta|\boldsymbol{\tau}_\beta\}} = \mathrm{e}^{-\mathrm{i}\mathbf{k}(\alpha\boldsymbol{\tau}_\beta+\boldsymbol{\tau}_\alpha-\boldsymbol{\tau}_{\alpha\beta})}, \tag{11.10}$$

i.e. the matrices $\Delta^{\mathbf{k}j}$ satisfy the relations

$$\Delta^{\mathbf{k}j}(\{\alpha|\boldsymbol{\tau}_\alpha\})\,\Delta^{\mathbf{k}j}(\{\beta|\boldsymbol{\tau}_\beta\}) = \mathrm{e}^{-\mathrm{i}\mathbf{k}(\alpha\boldsymbol{\tau}_\beta+\boldsymbol{\tau}_\alpha-\boldsymbol{\tau}_{\alpha\beta})}\Delta^{\mathbf{k}j}(\{\alpha\beta|\boldsymbol{\tau}_{\alpha\beta}\})\,. \tag{11.11}$$

11.1.2 Reducible Representation

We consider the initial set of atomic functions

$$\varphi^s_{nlm}(\mathbf{r}-\boldsymbol{\tau}^s_t) = R^s_{nl}(|\mathbf{r}-\boldsymbol{\tau}^s_t|)\,Y_{lm}(\mathbf{r}-\boldsymbol{\tau}^s_t), \tag{11.12}$$

where n, l, m are the atomic quantum numbers and $\boldsymbol{\tau}^s_t$ is the basis vector indicating the position of the t-th atom of the s-th sort in the unit cell of the crystal. Linear combinations of atomic functions compatible with the translational symmetry of the crystal have the form

$$\begin{aligned}\Phi^{\mathbf{k}st}_{nlm}(\mathbf{r}) &= \sum_p \mathrm{e}^{\mathrm{i}\mathbf{k}(\mathbf{R}_p+\boldsymbol{\tau}^s_t)}\varphi^s_{nlm}(\mathbf{r}-\mathbf{R}_p-\boldsymbol{\tau}^s_t) \qquad (11.13)\\ &= \sum_p \mathrm{e}^{\mathrm{i}\mathbf{k}\mathbf{Q}^{st}_p}\,P_{\{\varepsilon|\mathbf{Q}^{st}_p\}}\,\varphi^s_{nlm}(\mathbf{r}) \qquad (\mathbf{Q}^{st}_p = \mathbf{R}_p+\boldsymbol{\tau}^s_t)\,.\end{aligned}$$

Let us show that the set of functions $\Phi^{\mathbf{k}st}_{nlm}(\mathbf{r})$ for fixed $\mathbf{k}, s, n, l$ forms a basis of some, generally reducible, projective representation $\mathcal{D}^{\mathbf{k}sl}$ of the factor group $G_{\mathbf{k}}/T$ corresponding to the system of projective factors (11.10). To do this, we act on the functions $\Phi^{\mathbf{k}st}_{nlm}$ by the operator $P_{\{\alpha|\boldsymbol{\tau}_\alpha\}}$[1]:

$$\begin{aligned}P_{\{\alpha|\boldsymbol{\tau}_\alpha\}}\,\Phi^{\mathbf{k}t}_{lm}(\mathbf{r}) &= \sum_p \mathrm{e}^{\mathrm{i}\mathbf{k}\mathbf{Q}^t_p}\,P_{\{\alpha|\boldsymbol{\tau}_\alpha\}}P_{\{\varepsilon|\mathbf{Q}^t_p\}}\,\varphi_{lm}(\mathbf{r}) \qquad (11.14)\\ &= \sum_p \mathrm{e}^{\mathrm{i}\mathbf{k}\mathbf{Q}^t_p}\,P_{\{\varepsilon|\alpha\mathbf{Q}^t_p+\boldsymbol{\tau}_\alpha\}}\,P_{\{\alpha|0\}}\,\varphi_{lm}(\mathbf{r}).\end{aligned}$$

Using the fact that the atomic functions (11.12) are the basis functions of the irreducible representation $\mathcal{D}^{l\pm}$ of the full rotation group (plus sign for even l, minus sign for odd ones), i.e. satisfy the relation

[1] In what follows, the fixed indices s and n are omitted for brevity.

$$P_{\{\alpha|0\}}\,\varphi_{lm}(\mathbf{r}) = \sum_{m'} D^{l\pm}_{m'm}(\alpha)\,\varphi_{lm'}(\mathbf{r}), \tag{11.15}$$

we obtain

$$P_{\{\alpha|\boldsymbol{\tau}_\alpha\}}\,\Phi^{\mathbf{k}t}_{lm}(\mathbf{r}) = \sum_{m'} D^{l\pm}_{m'm}(\alpha) \sum_{p} \mathrm{e}^{\mathrm{i}\mathbf{k}\mathbf{Q}^t_p}\, P_{\{\varepsilon|\alpha\mathbf{Q}^t_p+\boldsymbol{\tau}_\alpha\}}\,\varphi_{lm'}(\mathbf{r}). \tag{11.16}$$

The vector $\alpha\mathbf{Q}^t_p + \boldsymbol{\tau}_\alpha$ can be represented as $\{\alpha|\boldsymbol{\tau}_\alpha\}\mathbf{Q}^t_p$ and, since $\{\alpha|\boldsymbol{\tau}_\alpha\}$ is an element of the crystal symmetry group,

$$\alpha\mathbf{Q}^t_p + \boldsymbol{\tau}_\alpha = \{\alpha|\boldsymbol{\tau}_\alpha\}\mathbf{Q}^t_p = \mathbf{Q}^v_u \tag{11.17}$$

is a vector characterizing the position of a v-th atom of a u-th cell. From equality (11.17) written in the form

$$\mathbf{R}_u + \boldsymbol{\tau}_v = \alpha\mathbf{R}_p + \alpha\boldsymbol{\tau}_t + \boldsymbol{\tau}_\alpha = \alpha\mathbf{R}_p - \mathbf{R}_{\alpha,t} + \alpha\boldsymbol{\tau}_t + \boldsymbol{\tau}_\alpha + \mathbf{R}_{\alpha,t}\,, \tag{11.18}$$

the values of u and v are determined directly:

$$\mathbf{R}_u = \alpha\mathbf{R}_p - \mathbf{R}_{\alpha,t}\,, \tag{11.19}$$

$$\boldsymbol{\tau}_v = \alpha\boldsymbol{\tau}_t + \boldsymbol{\tau}_\alpha + \mathbf{R}_{\alpha,t}\,, \tag{11.20}$$

where $\mathbf{R}_{\alpha,t}$ is a translation vector chosen so that the vector on the right-hand side of equality (11.20) is in the Wigner–Seitz cell (i.e. is a basis vector). Since α transforms the crystal lattice into itself, in expression (11.16) one can go over from summation over p to summation over u. Taking (11.17) into account, we write[2]

$$\begin{aligned}\sum_{p} \mathrm{e}^{\mathrm{i}\mathbf{k}\mathbf{Q}^t_p}\, P_{\{\varepsilon|\alpha\mathbf{Q}^t_p+\boldsymbol{\tau}_\alpha\}}\,\varphi_{lm'}(\mathbf{r}) &= \sum_{u} \mathrm{e}^{\mathrm{i}\mathbf{k}\alpha^{-1}(\mathbf{Q}^v_u-\boldsymbol{\tau}_\alpha)}\, P_{\{\varepsilon|\mathbf{Q}^v_u\}}\,\varphi_{lm'}(\mathbf{r}) \\ &= \sum_{u} \mathrm{e}^{\mathrm{i}(\mathbf{k}+\mathbf{K}_\alpha)(\mathbf{Q}^v_u-\boldsymbol{\tau}_\alpha)}\, P_{\{\varepsilon|\mathbf{Q}^v_u\}}\,\varphi_{lm'}(\mathbf{r}) \\ &= \mathrm{e}^{\mathrm{i}(\mathbf{K}_\alpha\boldsymbol{\tau}_v-(\mathbf{k}+\mathbf{K}_\alpha)\boldsymbol{\tau}_\alpha)} \sum_{u} \mathrm{e}^{\mathrm{i}\mathbf{k}\mathbf{Q}^v_u}\, P_{\{\varepsilon|\mathbf{Q}^v_u\}}\,\varphi_{lm'}(\mathbf{r}) \\ &= \mathrm{e}^{\mathrm{i}(\mathbf{K}_\alpha\boldsymbol{\tau}_v-(\mathbf{k}+\mathbf{K}_\alpha)\boldsymbol{\tau}_\alpha)}\, \Phi^{\mathbf{k}v}_{lm'}(\mathbf{r})\,.\end{aligned} \tag{11.21}$$

Substituting (11.21) into (11.16), we obtain

$$\begin{aligned}P_{\{\alpha|\boldsymbol{\tau}_\alpha\}}\,\Phi^{\mathbf{k}t}_{lm}(\mathbf{r}) &= \sum_{m'} D^{l\pm}_{m'm}(\alpha)\,\mathrm{e}^{\mathrm{i}(\mathbf{K}_\alpha\boldsymbol{\tau}_v-(\mathbf{k}+\mathbf{K}_\alpha)\boldsymbol{\tau}_\alpha)}\,\Phi^{\mathbf{k}v}_{lm'}(\mathbf{r}) \\ &= \sum_{m't'} D^{l\pm}_{m'm}(\alpha)\,\mathrm{e}^{\mathrm{i}(\mathbf{K}_\alpha\boldsymbol{\tau}_{t'}-(\mathbf{k}+\mathbf{K}_\alpha)\boldsymbol{\tau}_\alpha)}\,\delta_{t',v(\alpha,t)}\,\Phi^{\mathbf{k}t'}_{lm'}(\mathbf{r})\,.\end{aligned} \tag{11.22}$$

[2] The arguments α, t of the indices u and v are sometimes omitted for brevity.

We introduce the notation

$$D^{\mathbf{k}l}_{m't',mt}(\{\alpha|\boldsymbol{\tau}_\alpha\}) \equiv D^{l\pm}_{m'm}(\alpha)\,\mathrm{e}^{\mathrm{i}(\mathbf{K}_\alpha \boldsymbol{\tau}_{t'} - (\mathbf{k}+\mathbf{K}_\alpha)\boldsymbol{\tau}_\alpha)}\,\delta_{t',v(\alpha,t)}\,. \tag{11.23}$$

Let us prove that $\mathcal{D}^{\mathbf{k}l}$ is a projective representation of the factor group $G_{\mathbf{k}}/T$ corresponding to the system of projective factors (11.10):

$$\begin{aligned}
&\sum_{mt} D^{\mathbf{k}l}_{m't',mt}(\{\alpha|\boldsymbol{\tau}_\alpha\})\, D^{\mathbf{k}l}_{mt,m''t''}(\beta|\boldsymbol{\tau}_\beta\}) \\
&= \sum_{m} D^{l\pm}_{m'm}(\alpha)\, D^{l\pm}_{mm''}(\beta) \sum_{t} \mathrm{e}^{\mathrm{i}(\mathbf{K}_\alpha \boldsymbol{\tau}_{t'} - (\mathbf{k}+\mathbf{K}_\alpha)\boldsymbol{\tau}_\alpha)}\,\delta_{t',v(\alpha,t)} \\
&\times \mathrm{e}^{\mathrm{i}(\mathbf{K}_\beta \boldsymbol{\tau}_t - (\mathbf{k}+\mathbf{K}_\beta)\boldsymbol{\tau}_\beta)}\,\delta_{t,v(\beta,t'')}\,.
\end{aligned} \tag{11.24}$$

The delta-symbol[3]

$$\delta_{t',v(\alpha,t)} = \tilde{\delta}_{\boldsymbol{\tau}_{t'},\alpha\boldsymbol{\tau}_t+\boldsymbol{\tau}_\alpha} = \tilde{\delta}_{\alpha^{-1}(\boldsymbol{\tau}_{t'}-\boldsymbol{\tau}_\alpha),\boldsymbol{\tau}_t} \tag{11.25}$$

removes the sum over t, and expression (11.24) takes the form

$$\begin{aligned}
&D^{l\pm}_{m'm''}(\alpha\beta)\,\mathrm{e}^{\mathrm{i}(\mathbf{K}_\alpha \boldsymbol{\tau}_{t'} - (\mathbf{k}+\mathbf{K}_\alpha)\boldsymbol{\tau}_\alpha + \mathbf{K}_\beta \alpha^{-1}(\boldsymbol{\tau}_{t'}-\boldsymbol{\tau}_\alpha) - \alpha(\mathbf{k}+\mathbf{K}_\beta)\alpha\boldsymbol{\tau}_\beta}\,\tilde{\delta}_{\alpha^{-1}(\boldsymbol{\tau}_{t'}-\boldsymbol{\tau}_\alpha),\beta\boldsymbol{\tau}_{t''}+\boldsymbol{\tau}_\beta} \\
&= D^{l\pm}_{m'm''}(\alpha\beta)\,\mathrm{e}^{\mathrm{i}((\alpha\mathbf{K}_\beta+\mathbf{K}_\alpha)\boldsymbol{\tau}_{t'} - (\mathbf{k}+\alpha\mathbf{K}_\beta+\mathbf{K}_\alpha)(\alpha\boldsymbol{\tau}_\beta+\boldsymbol{\tau}_\alpha))}\,\tilde{\delta}_{\boldsymbol{\tau}_{t'},\alpha\beta\boldsymbol{\tau}_{t''}+\alpha\boldsymbol{\tau}_\beta+\boldsymbol{\tau}_\alpha}\,.
\end{aligned} \tag{11.26}$$

Using (11.8), it is easy to check that

$$\alpha\mathbf{K}_\beta + \mathbf{K}_\alpha = \mathbf{K}_{\alpha\beta}\,. \tag{11.27}$$

Taking into account this equality and the fact that $\alpha\boldsymbol{\tau}_\beta + \boldsymbol{\tau}_\alpha - \boldsymbol{\tau}_{\alpha\beta}$ is a lattice vector, we obtain

$$\begin{aligned}
&\sum_{mt} D^{\mathbf{k}l}_{m't',mt}(\{\alpha|\boldsymbol{\tau}_\alpha\})\, D^{\mathbf{k}l}_{mt,m''t''}(\beta|\boldsymbol{\tau}_\beta\}) \\
&= D^{l\pm}_{m'm''}(\alpha\beta)\,\mathrm{e}^{\mathrm{i}(\mathbf{K}_{\alpha\beta}\boldsymbol{\tau}_{t'} - (\mathbf{k}+\mathbf{K}_{\alpha\beta})(\boldsymbol{\tau}_{\alpha\beta}+\alpha\boldsymbol{\tau}_\beta+\boldsymbol{\tau}_\alpha-\boldsymbol{\tau}_{\alpha\beta}))}\,\tilde{\delta}_{\boldsymbol{\tau}_{t'},\alpha\beta\boldsymbol{\tau}_{t''}+\boldsymbol{\tau}_{\alpha\beta}} \\
&= \mathrm{e}^{-\mathrm{i}\mathbf{k}(\alpha\boldsymbol{\tau}_\beta+\boldsymbol{\tau}_\alpha-\boldsymbol{\tau}_{\alpha\beta})}\, D^{l\pm}_{m'm''}(\alpha\beta)\,\mathrm{e}^{\mathrm{i}(\mathbf{K}_{\alpha\beta}\boldsymbol{\tau}_{t'} - (\mathbf{k}+\mathbf{K}_{\alpha\beta})\boldsymbol{\tau}_{\alpha\beta})}\,\delta_{t',v(\alpha\beta,t'')} \\
&= \mathrm{e}^{-\mathrm{i}\mathbf{k}(\alpha\boldsymbol{\tau}_\beta+\boldsymbol{\tau}_\alpha-\boldsymbol{\tau}_{\alpha\beta})}\, D^{\mathbf{k}l}_{m't',m''t''}(\{\alpha\beta|\boldsymbol{\tau}_{\alpha\beta}\})\,.
\end{aligned} \tag{11.28}$$

Thus, matrices (11.23) form a projective representation, and the functions $\Phi^{\mathbf{k}t}_{lm}(\mathbf{r})$ satisfying relation (11.22),

[3] The tilde over the delta-symbol $\delta_{\mathbf{a},\mathbf{b}}$ means that equality (and inequality) of vectors $\mathbf{a}$ and $\mathbf{b}$ is understood up to a lattice vector.

$$P_{\{\alpha|\boldsymbol{\tau}_\alpha\}}\,\Phi^{\mathbf{k}t}_{lm}(\mathbf{r}) = \sum_{m't'} D^{\mathbf{k}l}_{m't',mt}(\{\alpha|\boldsymbol{\tau}_\alpha\})\,\Phi^{\mathbf{k}t'}_{lm'}(\mathbf{r}), \tag{11.29}$$

form a basis of this representation.

11.1.3 Decomposition into Irreducible Components

In the general case, each representation $\mathcal{D}^{\mathbf{k}l}$ decomposes into irreducible projective components. The *multiplicity of occurrence* $a^{\mathbf{k}j}$ of the irreducible projective representation $\Delta^{\mathbf{k}j}$, i.e. the number of times it occurs in the representation $\mathcal{D}^{\mathbf{k}l}$, is calculated by the formula

$$a^{\mathbf{k}j} = \frac{1}{n(G^0_{\mathbf{k}})} \sum_{\{\alpha\}} \chi^{\mathbf{k}l}(\{\alpha\})\,\chi^{\mathbf{k}j*}(\{\alpha\}), \tag{11.30}$$

where $\{\alpha\}$ is a shorthand notation for $\{\alpha|\boldsymbol{\tau}_\alpha\}$; $\chi^{\mathbf{k}l}(\{\alpha\})$ and $\chi^{\mathbf{k}j}(\{\alpha\})$ are the characters of the matrices $D^{\mathbf{k}l}(\{\alpha\})$ and $\Delta^{\mathbf{k}j}(\{\alpha\})$, respectively; $n(G^0_{\mathbf{k}})$ is the order of the group $G^0_{\mathbf{k}}$. Denoting by $c^{\mathbf{k}}_i$ the number of elements α in i-th class of the group $G^0_{\mathbf{k}}$, we can write the latter formula in a more convenient for calculations form (see formula (2.13))

$$a^{\mathbf{k}j} = \frac{1}{n(G^0_{\mathbf{k}})} \sum_{i} c^{\mathbf{k}}_i\,\chi^{\mathbf{k}l}(\{\alpha_i\})\,\chi^{\mathbf{k}j*}(\{\alpha_i\}), \tag{11.31}$$

because the characters of the elements $\{\alpha_i\}$ of the same class are all equal. To determine the multiplicities of occurrence $a^{\mathbf{k}j}$ by formula (11.31), it suffices to calculate the characters of the matrices of the reducible representation $\mathcal{D}^{\mathbf{k}l}$ only for the elements $\{\alpha_i\}$ one from each class.

Recalling formula (11.23), we obtain an expression for the character of the reducible representation $\mathcal{D}^{\mathbf{k}l}$:

$$\begin{aligned}
\chi^{\mathbf{k}l}(\{\alpha\}) &= \sum_{m,t} D^{l\pm}_{mm}(\alpha)\,\mathrm{e}^{\mathrm{i}(\mathbf{K}_{\alpha,\mathbf{k}}\boldsymbol{\tau}_t-(\mathbf{k}+\mathbf{K}_{\alpha,\mathbf{k}})\boldsymbol{\tau}_\alpha)}\,\delta_{t,v(\alpha,t)} \\
&= \sum_{m} D^{l\pm}_{mm}(\alpha)\,\mathrm{e}^{-\mathrm{i}(\mathbf{k}+\mathbf{K}_{\alpha,\mathbf{k}})\boldsymbol{\tau}_\alpha} \sum_t \mathrm{e}^{\mathrm{i}\mathbf{K}_{\alpha,\mathbf{k}}\boldsymbol{\tau}_t}\,\delta_{t,v(\alpha,t)} \\
&= \chi^{l\pm}(\alpha)\,\mathrm{e}^{-\mathrm{i}(\mathbf{k}+\mathbf{K}_{\alpha,\mathbf{k}})\boldsymbol{\tau}_\alpha} \sum_t \mathrm{e}^{\mathrm{i}\mathbf{K}_{\alpha,\mathbf{k}}\boldsymbol{\tau}_t}\,\delta_{t,v(\alpha,t)}\,.
\end{aligned} \tag{11.32}$$

Here $\chi^{l\pm}(\alpha)$ is the character of the irreducible representation of the full rotation group, calculated by the formulas

Table 11.1 Characters $\chi^l(\alpha')$ of the irreducible representations of the rotation group for rotations through the angles θ corresponding to the crystallographic point groups

$\theta \setminus l$	0	1	2	3	4	5
$\pm 2\pi/2$	1	−1				
$\pm 2\pi/3$	1	0	−1			
$\pm 2\pi/4$	1	1	−1	−1		
$\pm 2\pi/6$	1	2	1	−1	−2	−1

$$\chi^{l\pm}(\alpha') = \chi^l(\alpha'), \tag{11.33}$$

$$\chi^{l\pm}(\mathrm{i}\alpha') = (-1)^l \, \chi^l(\alpha'), \tag{11.34}$$

where (see, e.g. [3])

$$\chi^l(\alpha') = \sum_{m=-l}^{l} \mathrm{e}^{im\theta} = \frac{\sin(l + \frac{1}{2})\theta}{\sin \frac{1}{2}\theta} \tag{11.35}$$

(θ is the rotation angle corresponding to the transformation α', it is chosen in the range $-\pi < \theta \leq \pi$, and positive angles are measured counterclockwise as viewed down the axis toward the origin).

It follows directly from formula (11.32) that the character $\chi^{\mathbf{k}l}(\{\alpha_1\}) \equiv \chi^{\mathbf{k}l}(\{\varepsilon|0\})$ is equal to the product of the number of atomic functions with a fixed value of l and number of atoms of fixed type s in the Wigner–Seitz cell:

$$\chi^{\mathbf{k}l}(\{\alpha_1\}) = (2l + 1)\, t_s \,. \tag{11.36}$$

The characters $\chi^{\mathbf{k}l}$ of the remaining elements $\{\alpha_i\}$ of the reducible representation $\mathcal{D}^{\mathbf{k}l}$, as it follows from (11.32)–(11.34), depend on the specific values of the characters $\chi^l(\alpha_i)$ of the representation of the rotation group, wavevector $\mathbf{k}$, nonprimitive translations $\boldsymbol{\tau}_\alpha$, and basis vectors $\boldsymbol{\tau}_t$. It is sufficient to consider only the computation of the characters $\chi^l(\alpha_i')$.

Crystallographic point groups contain rotations only by the angles $\theta = \pm 2\pi/n$, where $n = 2, 3, 4, 6$. Calculating the characters χ^l for these elements by formula (11.35),

$$\chi^l(\pm 2\pi/n) = \frac{\sin((l + \frac{1}{2})(2\pi/n))}{\sin(\pi/n)}\,, \tag{11.37}$$

we see that it is sufficient to determine χ^l for $0 \leq l < n$ (the case $l = mn + p$, where m is an integer and $p < n$, repeats the results for $l = p$). All the necessary values of the characters χ^l are presented in Table 11.1.

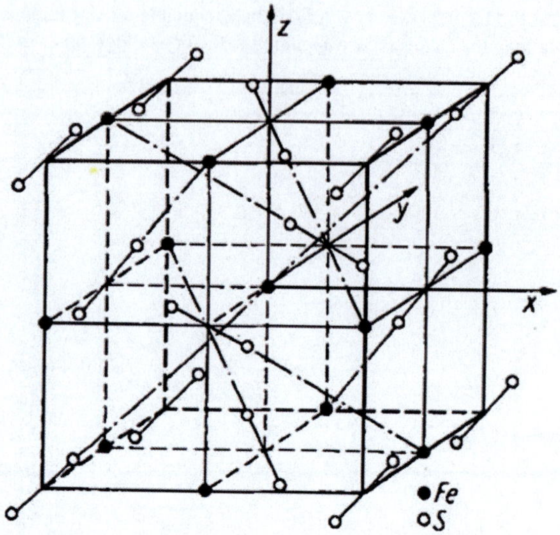

Fig. 11.1 Space structure of the pyrite cell

11.2 An Example: The Pyrite Structure

Let us apply the proposed general method for comparing atomic and crystal states of electrons to the pyrite structure (FeS_2) . Crystals with the pyrite structure are mainly compounds of transition metals with elements of IVB–VB groups. An idea of the pyrite structure can be gained by considering it as the NaCl structure with atoms of "iron" and pairs of atoms of "sulfur" (Fig. 11.1).[4]

The choice of the pyrite structure is due to two reasons. First, a number of crystals have the pyrite structure, many of them are interesting for their special electrical, magnetic and optical properties [4, 5]. Earth abundance makes metal pyrites attractive materials for various potential applications (see, e.g. [6]). Furthermore, FeS_2 may have played an important role in the formation, evolution and composition of the Earth's core and of the cores of other planets. Therefore, its physical and chemical properties have been studied at high temperatures and pressures (see, e.g. [7, 8]). Second, this structure allows us to illustrate our method of identifying electronic states using a rather complicated example with several atoms per unit cell and complex symmetry elements (for group-theoretical analysis of the pyrite structure, see, e.g. [2, 9, 10]).

[4] The sulfur ions are located very close to each other: the distance in pairs is 2.05 Å (instead of 3.5 Å—the double ionic radius). Pairs with their axes are oriented along the diagonals of small cubes, moreover, in such a way that they do not intersect with each other.

11.2.1 Crystal Structure and Space Group T_h^6

The set of all spatial transformations that leave the pyrite structure unchanged forms the space group T_h^6 [11, 12]. The group T_h^6, like any space group, is completely determined by its translation subgroup, point group and nonprimitive translations.

(a) The subgroup of translations of the space group T_h^6 consists of the transformations

$$\{\varepsilon|\mathbf{R}_n\}, \qquad \mathbf{R}_n = a(n_1, n_2, n_3), \tag{11.38}$$

where a is the lattice constant and n_i are integers. The set of vectors $\mathbf{R}_n$ defines the simple cubic (SC) Bravais lattice. The Wigner–Seitz cell of the SC lattice is a cube of side a.

(b) Point transformations α of elements of the space group T_h^6 form a point group (crystal class) $T_h = T \times C_i$. Elements of this group are listed in Table 11.2. The numbering of the elements is the same as in monograph [13], the letter notation of the classes coincide with the notation of [14] (hereafter referred to as BSW). In Table 11.2, the point transformations α, always representing proper rotations α' or rotations with inversion $i\alpha'$, are specified by indicating the axis of rotation and angle of rotation θ in the range $-\pi < \theta \le \pi$; positive angles are measured in the counterclockwise direction as viewed down the axis toward the origin. In addition, for each element α Table 11.2 gives its action on the components x, y, z of an arbitrary vector.

(c) If the origin coincides with the center of an "iron" atom, the nonprimitive translations $\boldsymbol{\tau}_\alpha$ have the form shown in Table 11.2.

The pyrite structure is the one with a basis. The position of the atoms forming the basis is indicated by the vectors $\boldsymbol{\tau}_t^s$, given in Table 11.3. In our notation, the "iron" atoms are of the sort $s = 1$ and "sulfur" atoms are of the sort $s = 2$.

11.2.2 Atomic and Crystal States

Before calculating the characters of the reducible representation $\chi^{\mathbf{k}sl}$, we note the following. According to (11.32), the contribution to the character $\chi^{\mathbf{k}sl}(\{\alpha_i\})$ comes only from the basis atoms $\boldsymbol{\tau}_t^s$ that do not change (up to a lattice vector) their position under the transformation $\{\alpha_i\}$. Therefore, for all α and t in formula (11.20), we determine the values of the index v coinciding with t (Table 11.4).

The set of vectors $\mathbf{K}_\mu = 2\pi/a(\mu_1, \mu_2, \mu_3)$, where μ_i are integers, forms the reciprocal lattice of the SC lattice. The Brillouin zone of the reciprocal lattice is a cube of side $2\pi/a$. Let us consider the symmetry points Γ, X, M and R [14], described in Table 11.5. The characters of the irreducible representations of the groups of the wavevectors corresponding to these symmetry points can be obtained using the representation matrices given in monograph [13] or in [9], specially devoted to the space

Table 11.2 The elements of the point group T_h (with the nonprimitive translations $\boldsymbol{\tau}_\alpha$ of the space group T_h^6)

No. i of the elem. α_i	class of conj. elems.	rotation α_i' axis	angle θ	$\mathbf{r}' = \alpha\mathbf{r}$	$\boldsymbol{\tau}_\alpha$ in units of a
1	E		0	(x, y, z)	$(0, 0, 0)$
2		$(1, 0, 0)$	π	$(x, -y, -z)$	$(1/2, 1/2, 0)$
3	C_2	$(0, 1, 0)$	π	$(-x, y, -z)$	$(0, 1/2, 1/2)$
4		$(0, 0, 1)$	π	$(-x, -y, z)$	$(1/2, 0, 1/2)$
5	C_3^2	$(1, 1, 1)$	$-2\pi/3$	(y, z, x)	$(0, 0, 0)$
6		$(1, 1, -1)$	$2\pi/3$	$(y, -z, -x)$	$(1/2, 1/2, 0)$
7		$(-1, 1, 1)$	$2\pi/3$	$(-y, z, -x)$	$(0, 1/2, 1/2)$
8		$(1, -1, 1)$	$2\pi/3$	$(-y, -z, x)$	$(1/2, 0, 1/2)$
9	C_3	$(1, 1, 1)$	$2\pi/3$	(z, x, y)	$(0, 0, 0)$
10		$(1, -1, 1)$	$-2\pi/3$	$(z, -x, -y)$	$(1/2, 1/2, 0)$
11		$(1, 1, -1)$	$-2\pi/3$	$(-z, x, -y)$	$(0, 1/2, 1/2)$
12		$(-1, 1, 1)$	$-2\pi/3$	$(-z, -x, y)$	$(1/2, 0, 1/2)$
25	I		0	$(-x, -y, -z)$	$(0, 0, 0)$
26		$(1, 0, 0)$	π	$(-x, y, z)$	$(1/2, 1/2, 0)$
27	IC_2	$(0, 1, 0)$	π	$(x, -y, z)$	$(0, 1/2, 1/2)$
28		$(0, 0, 1)$	π	$(x, y, -z)$	$(1/2, 0, 1/2)$
29	IC_3^2	$(1, 1, 1)$	$-2\pi/3$	$(-y, -z, -x)$	$(0, 0, 0)$
30		$(1, 1, -1)$	$2\pi/3$	$(-y, z, x)$	$(1/2, 1/2, 0)$
31		$(-1, 1, 1)$	$2\pi/3$	$(y, -z, x)$	$(0, 1/2, 1/2)$
32		$(1, -1, 1)$	$2\pi/3$	$(y, z, -x)$	$(1/2, 0, 1/2)$
33	IC_3	$(1, 1, 1)$	$2\pi/3$	$(-z, -x, -y)$	$(0, 0, 0)$
34		$(1, -1, 1)$	$-2\pi/3$	$(-z, x, y)$	$(1/2, 1/2, 0)$
35		$(1, 1, -1)$	$-2\pi/3$	$(z, -x, y)$	$(0, 1/2, 1/2)$
36		$(-1, 1, 1)$	$-2\pi/3$	$(z, x, -y)$	$(1/2, 0, 1/2)$

group T_h^6. Unfortunately, [9] gives only matrices of the generators, and monograph [13] contains a large number of misprints and inaccuracies.[5] Here we use the matrices of irreducible representations of the groups of the wavevectors obtained by the method described in Chaps. 3–6.[6] The characters of these representations for points Γ, X, M and R are given in Tables 11.6 and 11.7, in the notation of [9]. In accordance with (11.31), we present one element $\{\alpha_i\}$ from each class[7]. The representations

[5] Some of them are mentioned in [15].

[6] Our representation matrices are in agreement with the matrices from [9]. As for the monograph [13], in the table T190, which is used to obtain all representations at the point R, there is an inaccuracy: the representation matrix of the elements h_{11} and h_{35} must be $\gamma^*\begin{pmatrix} i & -i \\ -1 & -1 \end{pmatrix}$.

[7] For the distribution of the point group elements among the classes, see, e.g. [16].

Table 11.3 Basis vectors τ_t^s of the pyrite structure

Sort s	No. t	Basis vector τ_t^s in units a[†]
1	1	(0, 0, 0)
	2	(1/2, 1/2, 0)
	3	(0, 1/2, 1/2)
	4	(1/2, 0, 1/2)
2	1	(u, u, u)
	2	$(-1/2+u, 1/2-u, -u)$
	3	$(-u, -1/2+u, 1/2-u)$
	4	$(1/2-u, -u, -1/2+u)$
	5	$(-u, -u, -u)$
	6	$(1/2-u, -1/2+u, u)$
	7	$(u, 1/2-u, -1/2+u)$
	8	$(-1/2+u, u, 1/2-u)$

[†]The parameter u is close to 0.40

Table 11.4 Numbers of the basis atoms that do not change their position under the action of $\{\alpha_i\}$ for the pyrite structure (the values of the index $v(\alpha, t)$ coinciding with t)

Element α	$v(\alpha, t)$		Element α	$v(\alpha, t)$	
	$s=1$	$s=2$		$s=1$	$s=2$
α_1	1–4	1–8	α_{25}	1–4	–
α_2	–	–	α_{26}	–	–
α_3	–	–	α_{27}	–	–
α_4	–	–	α_{28}	–	–
α_5	1	1, 5	α_{29}	1	–
α_6	4	4, 8	α_{30}	4	–
α_7	2	2, 6	α_{31}	2	–
α_8	3	3, 7	α_{32}	3	–
α_9	1	1, 5	α_{33}	1	–
α_{10}	3	3, 7	α_{34}	3	–
α_{11}	4	4, 8	α_{35}	4	–
α_{12}	2	2, 6	α_{36}	2	–

paired by curly braces are degenerate with respect to the time-reversal symmetry, i.e. they correspond to the same energy value if time-reversal symmetry is taken into account.

Now let us focus on the calculation of the character of the reducible representation $\chi^{\mathbf{k}l}$ for the elements $\{\alpha_i\}$. Since the characters $\chi^{\mathbf{k}l}$ are used in expression (11.31), it is necessary to determine them only for the elements $\{\alpha_i\}$ such that the character $\chi^{\mathbf{k}j}(\{\alpha_i\})$ is nonzero for at least one irreducible representation. At points X and M, for example, apart from the identity element, there is only one such element: $\{\alpha_{28}\}$ and $\{\alpha_2\}$, respectively. Before calculating the characters of the reducible representation

Table 11.5 Symmetry points of the Brillouin zone for the simple cubic lattice

Symmetry point in the notation of [9]	Coordinates of the **k**-point in units $2\pi/a$	Element α satisfying $\alpha\mathbf{k} = \mathbf{k} + \mathbf{K}_\alpha$	Point group $G_\mathbf{k}^0$
Γ	(0, 0, 0)	$\alpha_1 - \alpha_{12}$, $\alpha_{25} - \alpha_{36}$	T_h
X	(0, 1/2, 0)	$\alpha_1 - \alpha_4$, $\alpha_{25} - \alpha_{28}$	D_{2h}
M	(1/2, 1/2, 0)	$\alpha_1 - \alpha_4$, $\alpha_{25} - \alpha_{28}$	D_{2h}
R	(1/2, 1/2, 1/2)	$\alpha_1 - \alpha_{12}$, $\alpha_{25} - \alpha_{36}$	T_h

Table 11.6 Characters of irreducible representations of the groups of the wavevectors for the points Γ and R

	$\{\alpha_1\}$	$\{\alpha_2\}$	$\{\alpha_5\}$	$\{\alpha_9\}$	$\{\alpha_{25}\}$	$\{\alpha_{26}\}$	$\{\alpha_{29}\}$	$\{\alpha_{33}\}$
Γ_1	1	1	1	1	1	1	1	1
Γ_2	1	1	ε^2	ε	1	1	ε^2	ε
Γ_3	1	1	ε	ε^2	1	1	ε	ε^2
Γ_4	3	−1	0	0	3	−1	0	0
Γ_5	1	1	1	1	−1	−1	−1	−1
Γ_6	1	1	ε^2	ε	−1	−1	$-\varepsilon^2$	$-\varepsilon$
Γ_7	1	1	ε	ε^2	−1	−1	$-\varepsilon$	$-\varepsilon^2$
Γ_8	3	−1	0	0	−3	1	0	0
R_1	2	0	−1	−1	2	0	−1	−1
R_2	2	0	$-\varepsilon^2$	$-\varepsilon$	2	0	$-\varepsilon^2$	$-\varepsilon$
R_3	2	0	$-\varepsilon$	$-\varepsilon^2$	2	0	$-\varepsilon$	$-\varepsilon^2$
R_4	2	0	−1	−1	−2	0	1	1
R_5	2	0	$-\varepsilon^2$	$-\varepsilon$	−2	0	ε^2	ε
R_6	2	0	$-\varepsilon$	$-\varepsilon^2$	−2	0	ε	ε^2

$\varepsilon = \exp(i2\pi/3)$

$\chi^{\mathbf{k}l}$ by formula (11.32) for the remaining elements $\{\alpha_i\}$, it is necessary to make sure that at least one of the basis atoms does not change its position under the action of this $\{\alpha_i\}$ (Table 11.4) and that the character of the irreducible representation of the rotation group χ^l for α_i' is not equal to zero (Table 11.1). At the points X and M, for example, it generally remains to determine only the character $\chi^{\mathbf{k}l}$ for the identity element (see (11.36)).

Having calculated the characters of the reducible representation $\chi^{\mathbf{k}l}$ for all necessary elements $\{\alpha_i\}$ by formula (11.31), we determine the multiplicities of occurrence $a^{\mathbf{k}j}$ of each irreducible representation of the group of the wavevector **k**. The calculation results are summarized in Table 11.8.

The comparison of the electronic states at the points of the Brillouin zone located on the symmetry lines and symmetry planes can be carried out in a completely analogous way. However, it is easier, having performed the comparison at the points Γ,X,

Table 11.7 Characters of irreducible representations of the groups of the wavevectors for the points X and M

	$\{\alpha_1\}$	$\{\alpha_2\}$	$\{\alpha_3\}$	$\{\alpha_4\}$	$\{\alpha_{25}\}$	$\{\alpha_{26}\}$	$\{\alpha_{27}\}$	$\{\alpha_{28}\}$
X_1	2	0	0	0	0	0	0	-2
X_2	2	0	0	0	0	0	0	2
M_1	2	$-2i$	0	0	0	0	0	0
M_2	2	$2i$	0	0	0	0	0	0

Table 11.8 Correspondence between atomic and crystal electronic states (multiplicity of occurrence $a^{\mathbf{k}j}$) for the pyrite structure †

	Atom of "iron"				Atom of "sulfur"			
	s	p	d	f	s	p	d	f
Γ_1	1		1		1	1	1	3
Γ_2			2			1	2	2
Γ_3			2			1	2	2
Γ_4	1		5		1	3	5	7
Γ_5		1		3	1	1	1	3
Γ_6		1		2		1	2	2
Γ_7		1		2		1	2	2
Γ_8		3		7	1	3	5	7
X_1	1	3	5	7	2	6	10	14
X_2	1	3	5	7	2	6	10	14
M_1	1	3	5	7	2	6	10	14
M_2	1	3	5	7	2	6	10	14
R_1			4			2	4	4
R_2	1		3		1	2	3	5
R_3	1		3		1	2	3	5
R_4		1		4		2	4	4
R_5		1		5	1	2	3	5
R_6		1		5	1	2	3	5

†In all empty positions $a^{\mathbf{k}j} = 0$

M and R, to use then the compatibility relations that exist between the irreducible representations at the symmetry points Γ, X, M and R and irreducible representations on the lines and planes adjacent to these points (for a concrete example, see Sect. 12.3).

11.3 Summary

As can be seen from formula (11.32), the comparison of electronic states in crystals with one atom per unit cell is really not difficult. In this case, expression (11.32) for the character of the reduced representation takes the form

$$\chi^{\mathbf{k}l}(\{\alpha\}) = \chi^{l\pm}(\alpha)\,, \tag{11.39}$$

and the irreducible projective representations $\Delta^{\mathbf{k}j}$ of the factor group $G_{\mathbf{k}}/T$ become the usual vector representations of crystallographic point groups. The compatibility tables of the irreducible representations $\mathcal{D}^{l\pm}$ of the full rotation group with the irreducible representations of point groups can be found, for example, in the work by [17].

In complex crystal structures with two or more atoms per unit cell,f it becomes very difficult to answer the question, which atomic states a given crystal state is associated with. The method described above is universal and solves this problem in a simple way. It should only be emphasized that when atomic states with different l and s contribute to the crystal state, and this happens quite often in crystals with a basis (see Table 11.8), the *value* of individual contributions can be judged only on the basis of quantum mechanical calculations.

Finally, we note that the choice of the origin in crystals with a basis is ambiguous. The choice of the origin determines the values of nonprimitive translations $\boldsymbol{\tau}_\alpha$ and basis vectors $\boldsymbol{\tau}^s_t$. However, it is easy to prove (see [2, Appendix]), that the results of the comparison of electronic states do not depend on the choice of the origin.[8]

References

1. A.K. Sedaghat, A.P. Cracknell, Physica **71**, 615 (1974)
2. B.I. Reser, Y.S. Mitrokhin, *Correspondence between the classifications of electronic states according to symmetry in atoms and crystals* [in Russian] (1976). Dep. VINITI, No. 4142–76
3. M. Hamermesh, *Group Theory and Its Applications to Physical Problems* (Addison-Wesley, Reading, MA, 1962). Reprinted: Dover, New York, 1989
4. F. Hulliger, E. Mooser, J. Phys. Chem. Solids **26**, 429 (1965)
5. G. Oertzen, R.T. Jones, A.R. Gerson, Phys. Chem. Minerals **32**, 255 (2005)
6. M.S. Faber, M.A. Lukowski, Q. Ding, N.S. Kaiser, S. Jin, J. Phys. Chem. C **118**, 21347 (2014)
7. J. Muscat, A. Hung, S. Russo, I. Yarovsky, Phys. Rev. B **65**, 054107 (2002)
8. M. Blanchard, M. Alfredsson, J. Brodholt, G.D. Price, K. Wright, C.R.A. Catlow, J. Phys. Chem. B **109**, 22067 (2005)
9. W. Gorzkowski, Phys. Stat. Sol. **3**, 599 (1963)
10. J.L. Verble, R.F. Wallis, Phys. Rev. **182**, 783 (1969)
11. R.W.G. Wyckoff, *Crystal Structures* (Interscience, New York, 1948)

[8] It should be borne in mind that the weighted representations of [13] (projective representations), which are often used to obtain irreducible representations of the group of the wavevector, depend on the choice of the origin (see [18]). This circumstance is aggravated by the fact that the position of the origin is not indicated in [13].

12. M.I. Aroyo (ed.), *International Tables for Crystallography, Volume A, Space-group symmetry*, 6th edn. (Wiley, New York, 2016)
13. O.V. Kovalev, *Irreducible Representations of Space Groups* [in Russian] (Acad. Sci. Ukrain. SSR, Kiev, 1961). Engl. transl.: Gordon and Breach, New York, 1965
14. L.P. Bouckaert, R. Smoluchowski, E. Wigner, Phys. Rev. **50**, 58 (1936)
15. J.L. Warren, Rev. Mod. Phys. **40**, 38 (1968)
16. L.D. Landau, E.M. Lifshitz, *Quantum Mechanics (Course on Theoretical Physics)*, 3rd edn., vol. 3, Chapter XII (Pergamon, Oxford, 1977)
17. G.F. Koster, J.O. Dimmock, R.G. Wheeler, H. Statz, *Properties of the Thirty-Two Point Groups* (M.I.T. Press, Cambridge, Mass., 1963)
18. O.V. Kovalev, Solid State Phys. **12**, 1529 (1970)

Chapter 12
Group-Theoretical Analysis of the Cr_3Si Structure

In my experience, there is no ideal way to teach the application of group to the quantum mechanics of atoms, molecules, and solids; the reason is the inherent nature of the subject. (V. Heine, Physics Today (2008) ***61****, 57)*

Compounds with the Cr_3Si structure have interesting physical properties (comparatively high temperatures of transition to the superconducting state, anomalies in electrical, magnetic, elastic and structural properties) and have been intensively studied (see reviews [1–3] and monograph [4]). Despite being bypassed in the quest for higher T_C by the cuprates in 1986, by MgB_2 in 2001, and by the iron-based superconductors in 2008 (see, e.g. [5, 6]), the intermetallic materials with the Cr_3Si structure are still among widely used superconductors in technological applications (see, e.g. [7]). The β-W thin films, that have been determined to have the Cr_3Si crystal structure, have been studied due to a giant spin Hall effect (see, e.g. [8, 9]).

All this motivated the study of symmetry properties of crystals with the Cr_3Si structure. The study of symmetry properties of this structure has been carried out in a number of papers, but by no means exhausts all the questions of the group-theoretical analysis of the crystal structure. The works by Gorzkowski [10, 11] contain representations of the small groups of the wavevectors (including the spinor representations) and some applications. The group-theoretical analysis of the Cr_3Si structure, performed by Mattheiss and used in his papers [12, 13], has not been published. Group-theoretical analysis for the phonon modes was carried out by [14]. The selection rules for electric dipole transitions were investigated by [11] and for the double space group—by [15]. Symmetry analysis was applied to study optical transitions by [16].

Another reason for the group-theoretical analysis of the Cr_3Si structure is the methodological one. By the example of a fairly complex and important crystal structure, we demonstrate how the group-theoretical analysis can be carried out in practice. Indeed, despite a large number of paper and monographs devoted to the use of the group theory in solid-state physics (see, e.g. [17–22]), there is no simple and universal method for analyzing a crystal structure.

N. B. Melnikov and B. I. Reser, *Space Group Representations*,
https://doi.org/10.1007/978-3-031-13991-8_12

Here we present a detailed study of the symmetry properties of this structure, following the work by [23]. Note that our consideration is limited to the case when spin is not taken into account. If necessary, the case with spin can be considered in complete analogy (see, e.g. [10, 11]).

12.1 Crystal Structure

12.1.1 Geometric Description

The Cr_3Si structure[1] has the form shown at Fig. 12.1. The atoms are in the following equivalent positions[2] [24]:

$$\text{Cr}: \quad (6\,\text{c}) \quad \frac{1}{4}, 0, \frac{1}{2}; \quad \frac{1}{2}, \frac{1}{4}, 0; \quad 0, \frac{1}{2}, \frac{1}{4}; \quad \frac{3}{4}, 0, \frac{1}{2}; \quad \frac{1}{2}, \frac{3}{4}, 0; \quad 0, \frac{1}{2}, \frac{3}{4}.$$

$$\text{Si}: \quad (2\,\text{a}) \quad 0, 0, 0; \quad \frac{1}{2}, \frac{1}{2}, \frac{1}{2}.$$

The compounds of the Cr_3Si structure have the chemical formula A_3B, where the A atoms are always transition elements of groups IV–VI and B can be atoms of either transition or nontransition metals (see [34] and the literature cited therein). As can be seen from Fig. 12.1, the B atoms are located at the vertices and in the center of the cube, the A atoms are on the faces. The nearest neighbors of an atom A are two A-atoms located at a distance of $1/2$, the second neighbors are four B-atoms at a distance of $\sqrt{5}/4$, the third are eight A atoms at a distance of $\sqrt{6}/4$.[3] The nearest neighbors of a B atom are twelve A atoms, located at a distance of $\sqrt{5}/4$, the second neighbors are eight B atoms at a distance of $\sqrt{12}/4$, and the third ones are twelve A atoms at a distance of $\sqrt{13}/4$. Table 12.1 contains information about all neighbors of the A and B atoms located within one unit cell ($\sqrt{3}$).

[1] This structure is found in the literature under three more names: β-tungsten (β-W), Cr_3O and A15. Whether the metastable β-W is a true allotrope of tungsten has been a long-debated question [24–27]; the second name is unjustified, since the Cr_3Si compound was discovered earlier than all other compounds of this type [28]; finally, the third title (A15) is out of use because the systematization of all types of crystal structures undertaken in the first volumes of the reference book [29, 30] has not been fulfilled and after 1939 has not been continued (in modern reference books—see, e.g. [31, 32]—the symbols introduced in [29] are absent at all).

[2] The number and the Latin letter in brackets after the name of the chemical element indicate the number of equivalent positions, also known as *Wyckoff positions* [33]; the coordinates of the atoms are given in units of the lattice constant a. (For details on the generally accepted way of describing crystal structures, see, e.g. [31, Vol. 4, Chap. XIII].)

[3] One can often come across the statement that the A-atoms form three mutually perpendicular chains. The meaning of this statement should not be exaggerated: the distance between neighboring atoms in a chain is only 1.2 times smaller than the distance between the nearest atoms of different chains.

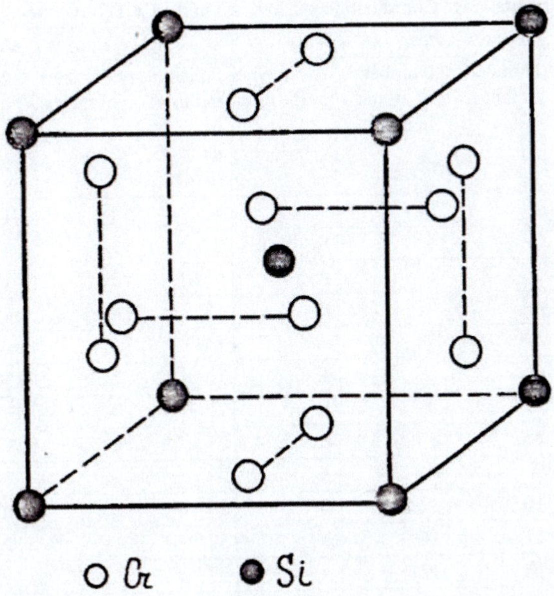

Fig. 12.1 Cr_3Si structure

12.1.2 Space Group O_h^3

The Cr_3Si structure has the O_h^3 space group [24]. The group O_h^3, like any space group, is completely determined by its translation subgroup, point group and nonprimitive translations. We give a complete description of these three components, using the crystallographic description of the space groups [35].

a. The translation subgroup of the space group O_h^3 consists of the transformations

$$\{\varepsilon|\mathbf{R}_n\}, \qquad \mathbf{R}_n = a(n_1, n_2, n_3), \tag{12.1}$$

where a is the lattice constant and n_i are integers. The set of vectors $\mathbf{R}_n$ forms the simple cubic (SC) Bravais lattice.

b. The point transformations of the elements of the space group O_h^3 form the point group (crystal class) $O_h = O \times C_i$. Elements of this group are shown in Table 12.2.
Here the numbering of the elements is the same as in monograph [36] and notation for the classes of conjugate elements is the same as in [37] (henceforth referred to as the BSW notation). As before, the point transformations α in Table 12.2 represent proper rotations α' or rotations with inversion $i\alpha'$. They are specified by indicating the rotation axis and rotation angle, selected in the range $-\pi < \theta \leq \pi$; the angles are measured counterclockwise as viewed from the point that defines the axis position to the origin. In addition, we indicate the action of each element α on the components x, y, z of an arbitrary vector.

Table 12.1 Coordination spheres (cs) in the Cr_3Si structure

For the A atom				For the B atom			
Number of the cs	Number of atoms in the cs	Atom type in the cs	Squared radius of the cs in units of $(a/4)^2$	Number of the cs	Number of atoms in the cs	Atom type in the cs	Squared radius of the cs in units of $(a/4)^2$
1	2	A	4	1	12	A	5
2	4	B	5	2	8	B	12
3	8	A	6	3	12	A	13
4	4	B	13	4	6	B	16
5	16	A	14	5	24	A	21
6	6	A	16	6	36	A	29
7	8	A	20	7	12	B	32
8	8	B	21	8	12	A	37
9	8	A	22	9	24	B	44
10	12	B	29	10	36	A	45
11	16	A	30	11	8	B	48
12	12	A	32				
13	10	A	36				
14	4	B	37				
15	24	A	38				
16	12	B	45				
17	16	A	46				
18	8	A	48				

c. We choose the origin according to [38], then the nonprimitive translations $\boldsymbol{\tau}_\alpha$ have the form shown in Table 12.2.

12.1.3 Wigner–Seitz Cell

In most of solid-state physics problems, instead of a unit cell whose sides are elementary translations, one uses the so-called *symmetric* unit cell, or the Wigner–Seitz cell. The advantage of the Wigner–Seitz cell is that it is invariant with respect to the corresponding holohedry.

The Wigner–Seitz cell of a simple cubic lattice is a cube of side a. In the considered Cr_3Si structure, it contains eight atoms: six A atoms and two B atoms (see Fig. 12.2).

The coordinates of these eight atoms, called the basis vectors, are given in Table 12.3. (The vector indicating the position of the t-th atom of the s-th sort in the Wigner–Seitz cell is denoted by $\boldsymbol{\tau}_t^s$.)

Table 12.2 Elements of the point group O_h (with the nonprimitive translations $\boldsymbol{\tau}_\alpha$ for the space group O_h^3)

No. i of the element α_i	Class of conj. elements	Rotation α_i'		$\mathbf{r}' = \alpha\mathbf{r}$	$\boldsymbol{\tau}_\alpha$
		Axis	Angle θ		
1	E		0	(x, y, z)	0
2		$(1, 0, 0)$	π	$(x, -y, -z)$	0
3	C_4^2	$(0, 1, 0)$	π	$(-x, y, -z)$	0
4		$(0, 0, 1)$	π	$(-x, -y, z)$	0
5	C_3	$(1, 1, 1)$	$-2\pi/3$	(y, z, x)	0
6		$(1, 1, -1)$	$2\pi/3$	$(y, -z, -x)$	0
7		$(-1, 1, 1)$	$2\pi/3$	$(-y, z, -x)$	0
8		$(1, -1, 1)$	$2\pi/3$	$(-y, -z, x)$	0
9		$(1, 1, 1)$	$2\pi/3$	(z, x, y)	0
10		$(1, -1, 1)$	$-2\pi/3$	$(z, -x, -y)$	0
11		$(1, 1, -1)$	$-2\pi/3$	$(-z, x, -y)$	0
12		$(-1, 1, 1)$	$-2\pi/3$	$(-z, -x, y)$	0
13	C_2	$(-1, 1, 0)$	π	$(-y, -x, -z)$	$\boldsymbol{\tau}$
14	C_4	$(0, 0, 1)$	$\pi/2$	$(-y, x, z)$	$\boldsymbol{\tau}$
15	C_4	$(0, 0, 1)$	$-\pi/2$	$(y, -x, z)$	$\boldsymbol{\tau}$
16	C_2	$(1, 1, 0)$	π	$(y, x, -z)$	$\boldsymbol{\tau}$
17	C_2	$(0, -1, 1)$	π	$(-x, -z, -y)$	$\boldsymbol{\tau}$
18	C_2	$(0, 1, 1)$	π	$(-x, z, y)$	$\boldsymbol{\tau}$
19	C_4	$(1, 0, 0)$	$\pi/2$	$(x, -z, y)$	$\boldsymbol{\tau}$
20	C_4	$(1, 0, 0)$	$-\pi/2$	$(x, z, -y)$	$\boldsymbol{\tau}$
21	C_2	$(-1, 0, 1)$	π	$(-z, -y, -x)$	$\boldsymbol{\tau}$
22	C_4	$(0, 1, 0)$	$-\pi/2$	$(-z, y, x)$	$\boldsymbol{\tau}$
23	C_2	$(1, 0, 1)$	π	$(z, -y, x)$	$\boldsymbol{\tau}$
24	C_4	$(0, 1, 0)$	$\pi/2$	$(z, y, -x)$	$\boldsymbol{\tau}$
25	I		0	$(-x, -y, -z)$	0
26	IC_4^2	$(1, 0, 0)$	π	$(-x, y, z)$	0
27		$(0, 1, 0)$	π	$(x, -y, z)$	0
28		$(0, 0, 1)$	π	$(x, y, -z)$	0
29	IC_3	$(1, 1, 1)$	$-2\pi/3$	$(-y, -z, -x)$	0
30		$(1, 1, -1)$	$2\pi/3$	$(-y, z, x)$	0
31		$(-1, 1, 1)$	$2\pi/3$	$(y, -z, x)$	0
32		$(1, -1, 1)$	$2\pi/3$	$(y, z, -x)$	0
33		$(1, 1, 1)$	$2\pi/3$	$(-z, -x, -y)$	0
34		$(1, -1, 1)$	$-2\pi/3$	$(-z, x, y)$	0
35		$(1, 1, -1)$	$-2\pi/3$	$(z, -x, y)$	0
36		$(-1, 1, 1)$	$-2\pi/3$	$(z, x, -y)$	0

(continued)

Table 12.2 (continued)

No. i of the element α_i	Class of conj. elements	Rotation α'_i		$\mathbf{r}' = \alpha\mathbf{r}$	$\boldsymbol{\tau}_\alpha$
		Axis	Angle θ		
37	IC_2	(−1, 1, 0)	π	(y, x, z)	$\boldsymbol{\tau}$
38	IC_4	(0, 0, 1)	$\pi/2$	$(y, -x, -z)$	$\boldsymbol{\tau}$
39	IC_4	(0, 0, 1)	$-\pi/2$	$(-y, x, -z)$	$\boldsymbol{\tau}$
40	IC_2	(1, 1, 0)	π	$(-y, -x, z)$	$\boldsymbol{\tau}$
41	IC_2	(0, −1, 1)	π	(x, z, y)	$\boldsymbol{\tau}$
42	IC_2	(0, 1, 1)	π	$(x, -z, -y)$	$\boldsymbol{\tau}$
43	IC_4	(1, 0, 0)	$\pi/2$	$(-x, z, -y)$	$\boldsymbol{\tau}$
44	IC_4	(1, 0, 0)	$-\pi/2$	$(-x, -z, y)$	$\boldsymbol{\tau}$
45	IC_2	(−1, 0, 1)	π	(z, y, x)	$\boldsymbol{\tau}$
46	IC_4	(0, 1, 0)	$-\pi/2$	$(z, -y, -x)$	$\boldsymbol{\tau}$
47	IC_2	(1, 0, 1)	π	$(-z, y, -x)$	$\boldsymbol{\tau}$
48	IC_4	(0, 1, 0)	$\pi/2$	$(-z, -y, x)$	$\boldsymbol{\tau}$

$\boldsymbol{\tau} = a(1/2, 1/2, 1/2)$

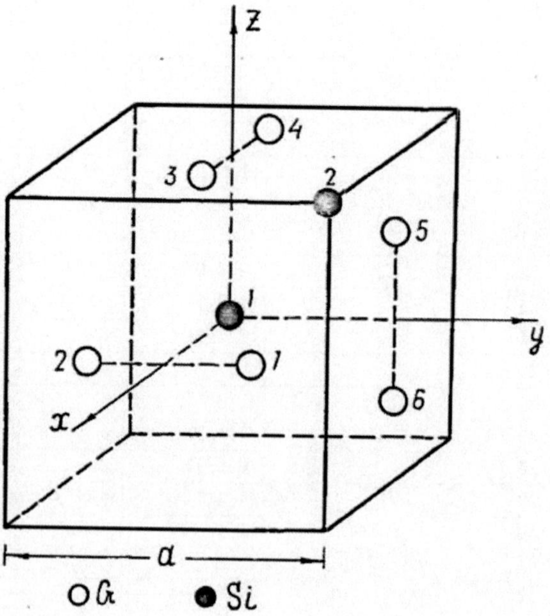

Fig. 12.2 Wigner–Seitz cell of the simple cubic lattice and basis atoms of the Cr_3Si structure

Table 12.3 Basis vectors $\boldsymbol{\tau}_t^s$ of the Cr_3Si structure

Sort of the atom s	Number of the atom t	Basis vector $\boldsymbol{\tau}_t^s$ in units of a
A	1	(1/2, 1/4, 0)
	2	(1/2, −1/4, 0)
	3	(1/4, 0, 1/2)
	4	(−1/4, 0, 1/2)
	5	(0, 1/2, 1/4)
	6	(0, 1/2, −1/4)
B	1	(0, 0, 0)
	2	(1/2, 1/2, 1/2)

12.2 Small Representations of the Group of the Wavevector

12.2.1 Brillouin Zone

The set of vectors $\mathbf{K}_\mu = (2\pi/a)(\mu_1, \mu_2, \mu_3)$, where μ_i are integers, forms the reciprocal lattice of the simple cubic lattice. The Brillouin zone (symmetric unit cell of the reciprocal lattice) is a cube of side $2\pi/a$.

To construct all irreducible representations of a space group, it is sufficient to use only one arm from each star of the wavevector (see Chap. 5). Usually, the arms are selected from a region called the irreducible Brillouin zone (IBZ), which consists of one $n(G^0)$-th part of the Brillouin zone ($n(G^0)$ is the order of the point group G^0). All internal points of the IBZ are general points, and hence the group of the wavevector $G_\mathbf{k}$ coincides with the translation subgroup T. A point at the boundary of the IBZ can be in a symmetrical position, i.e. the point is invariant under the action of certain elements of the point group G^0 that form the group $G^0_\mathbf{k}$.[4] It is these high symmetry points in the Brillouin zone that are of primary interest.

Figure 12.3 shows the irreducible Brillouin zone that we have chosen and all points of the **k**-space in a symmetrical position in the BSW notation.

Table 12.4 gives a complete list of wavevectors **k** in the Brillouin zone of the SC-lattice with an indication of the elements α of the crystal class O_h that transform the **k** to itself and thus form the point group $G^0_\mathbf{k}$.[5]

12.2.2 Structure of the Small Representations

As we have shown in Chap. 5, irreducible representations of a space group G are expressed in terms of the small representations $d^{\mathbf{k}j}$ of the groups of wavevector $G_\mathbf{k}$.

---

[4] By the invariance of the wavevector **k** under the action of α, we always mean its invariance up to a reciprocal lattice. vector, i.e. $\alpha\mathbf{k} = \mathbf{k} + \mathbf{K}_{\alpha,\mathbf{k}}$; which for brevity we sometimes write in the form $\alpha\mathbf{k} \approx \mathbf{k}$.

[5] In Table 4 of [23] the point B is missing.

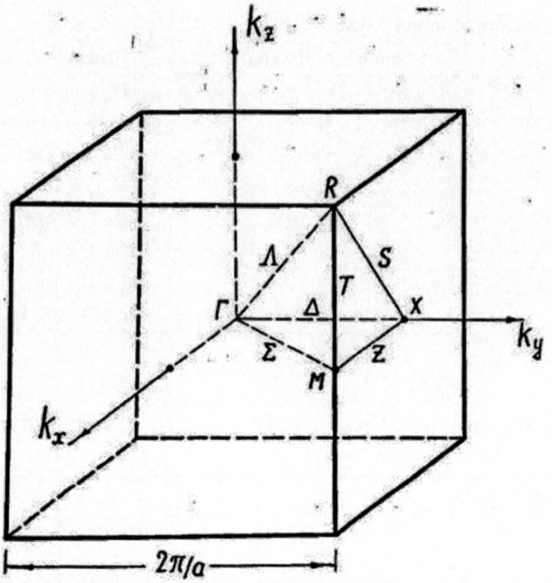

Fig. 12.3 Brillouin zone of the simple cubic lattice and irreducible Brillouin zone corresponding to the crystal class O_h

Table 12.4 High symmetry points in the Brillouin zone of the SC lattice

Point in the Brillouin zone	k-vector in units of $2\pi/a$	Elements of the crystal class O_h transforming **k** to itself	Point group $G_{\mathbf{k}}^0$
Γ	(0, 0, 0)	1–48	O_h
R	(1/2, 1/2, 1/2)	1–48	O_h
X	(0, 1/2, 0)	1–4, 21–24, 25–28, 45–48	D_{4h}
M	(1/2, 1/2, 0)	1–4, 13–16, 25–28, 37–40	D_{4h}
Δ	$(0, \eta, 0)$	1, 3, 22, 24, 26, 28, 45, 47	C_{4v}
T	$(1/2, 1/2, \zeta)$	1, 4, 14, 15, 26, 27, 37, 40	C_{4v}
Λ	(ξ, ξ, ξ)	1, 5, 9, 37, 41, 45	C_{3v}
Σ	$(\xi, \xi, 0)$	1, 16, 28, 37	C_{2v}
Z	$(\xi, 1/2, 0)$	1, 2, 27, 28	C_{2v}
S	$(\xi, 1/2, \xi)$	1, 23, 27, 45	C_{2v}
Θ	(ξ, ξ, ζ)	1, 37	C_s
Ξ	$(\xi, \eta, 0)$	1, 28	C_s
A	$(\xi, 1/2, \zeta)$	1, 27	C_s
B	(ξ, η, ξ)	1, 45	C_s

$0 < \xi, \eta, \zeta < 1/2$

The matrices of small representations $d^{\mathbf{k}j}$ are written in the form

$$d^{\mathbf{k}j}(\{\varepsilon|\mathbf{R}_n\}\{\alpha|\boldsymbol{\tau}_\alpha\}) = \mathrm{e}^{-\mathrm{i}\mathbf{k}\mathbf{R}_n}\,\Delta^{\mathbf{k}j}(\{\alpha|\boldsymbol{\tau}_\alpha\}), \tag{12.2}$$

where $\Delta^{\mathbf{k}j}$ are the irreducible projective representations of the factor group $G_{\mathbf{k}}/T$ isomorphic to the point group $G^0_{\mathbf{k}}$. The irreducible projective representations $\Delta^{\mathbf{k}j}$ correspond to the system of projective factors

$$\omega^{\mathbf{k}}_{\{\alpha|\boldsymbol{\tau}_\alpha\},\{\beta|\boldsymbol{\tau}_\beta\}} = \mathrm{e}^{-\mathrm{i}\mathbf{k}(\alpha\boldsymbol{\tau}_\beta+\boldsymbol{\tau}_\alpha-\boldsymbol{\tau}_{\alpha\beta})}, \tag{12.3}$$

i.e. the matrices $\Delta^{\mathbf{k}j}$ satisfy the relation

$$\Delta^{\mathbf{k}j}(\{\alpha|\boldsymbol{\tau}_\alpha\})\,\Delta^{\mathbf{k}j}(\{\beta|\boldsymbol{\tau}_\beta\}) = \mathrm{e}^{-\mathrm{i}\mathbf{k}(\alpha\boldsymbol{\tau}_\beta+\boldsymbol{\tau}_\alpha-\boldsymbol{\tau}_{\alpha\beta})}\Delta^{\mathbf{k}j}(\{\alpha\beta|\boldsymbol{\tau}_{\alpha\beta}\}). \tag{12.4}$$

Thus, the problem of constructing irreducible representations of the space group G is reduced to constructing the irreducible projective representations $\Delta^{\mathbf{k}j}$ of the corresponding point groups $G^0_{\mathbf{k}}$.

12.2.3 Small Representations

In the case of the space group O_h^3, according to Table 12.4, the problem of constructing small representations is reduced to constructing projective representations of the point groups C_s, C_{2v}, C_{3v}, C_{4v}, D_{4h} and O_h. We constructed these representations using the method described in Chaps. 3–5.

Tables 12.5, 12.6, 12.7, 12.8, 12.9, 12.10, 12.11, 12.12, 12.13, 12.14, 12.15, 12.16 and 12.17 show the matrices and characters of the small representations for all high symmetry points. The representations are numbered in full accord with the work of Gorzkowski [10], where matrices of the small representations for the space group under consideration (O_h^3) were obtained for the first time.[6]

For the representations at the point Γ, along with Gorzkowski's notation (G), we also give the BSW notation. For the remaining points of the Brillouin zone, the representations are not indexed using the BSW notation: it is either absent (points R, X, S), or is inconvenient due to the fact that the choice of the irreducible Brillouin zone in the BSW paper differs from the generally accepted one,[7] and the small representations for different arms of the star do not match.

Note that our representation matrices are in agreement with the matrices of [10].[8]

---

[6] In order to save space, for the points Γ, X, M and R, Gorzkowski gives only the matrices of the generators of the group, which creates certain difficulties for using his results. For instance, to obtain all matrices of the representation R_4, one has to multiply six-dimensional matrices many times.

[7] In the BSW paper, the irreducible Brillouin zone was chosen along the x axis, not y. In addition, BSW used the left-handed coordinate system.

[8] Only at the point M did Gorzkowski make an inaccuracy: the matrix corresponding to the generator 31 of the representation M_{10} must be $\begin{pmatrix}1 & 0\\ 0 & -1\end{pmatrix}$.

Table 12.5 Small representations for the point $\Gamma^\dagger$

G	BSW	1	2	3	4	5	6	7	8	9	10	11	12
Γ_1^+	Γ_1	1	1	1	1	1	1	1	1	1	1	1	1
Γ_2^+	Γ_2	1	1	1	1	1	1	1	1	1	1	1	1
Γ_3^+	Γ_{12}	1 0 0 1	1 0 0 1	1 0 0 1	1 0 0 1	ε^2 0 0 ε	ε^2 0 0 ε	ε^2 0 0 ε	ε^2 0 0 ε	ε 0 0 ε^2	ε 0 0 ε^2	ε 0 0 ε^2	ε 0 0 ε^2
		2	2	2	2	−1	−1	−1	−1	−1	−1	−1	−1
Γ_4^+	Γ'_{15}	1 0 0 0 1 0 0 0 1	−1 0 0 0 1 0 0 0 −1	1 0 0 0 −1 0 0 0 −1	−1 0 0 0 −1 0 0 0 1	0 0 1 1 0 0 0 1 0	0 0 −1 1 0 0 0 −1 0	0 0 1 −1 0 0 0 −1 0	0 0 −1 −1 0 0 0 1 0	0 1 0 0 0 1 1 0 0	0 −1 0 0 0 1 −1 0 0	0 1 0 0 0 −1 −1 0 0	0 −1 0 0 0 −1 1 0 0
		3	−1	−1	−1	0	0	0	0	0	0	0	0
Γ_5^+	Γ'_{25}	1 0 0 0 1 0 0 0 1	−1 0 0 0 1 0 0 0 −1	1 0 0 0 −1 0 0 0 −1	−1 0 0 0 −1 0 0 0 1	0 0 1 1 0 0 0 1 0	0 0 −1 1 0 0 0 −1 0	0 0 1 −1 0 0 0 −1 0	0 0 −1 −1 0 0 0 1 0	0 1 0 0 0 1 1 0 0	0 −1 0 0 0 1 −1 0 0	0 1 0 0 0 −1 −1 0 0	0 −1 0 0 0 −1 1 0 0
		3	−1	−1	−1	0	0	0	0	0	0	0	0

(continued)

Table 12.5 (continued)

G	BSW	13	14	15	16	17	18	19	20	21	22	23	24
Γ_1^+	Γ_1	1	1	1	1	1	1	1	1	1	1	1	1
Γ_2^+	Γ_2	−1	−1	−1	−1	−1	−1	−1	−1	−1	−1	−1	−1
Γ_3^+	Γ_{12}	0 1 1 0	0 1 1 0	0 1 1 0	0 1 1 0	$0\ \varepsilon^2$ $\varepsilon\ 0$	$0\ \varepsilon^2$ $\varepsilon\ 0$	$0\ \varepsilon^2$ $\varepsilon\ 0$	$0\ \varepsilon^2$ $\varepsilon\ 0$	$0\ \varepsilon$ $\varepsilon^2\ 0$	$0\ \varepsilon$ $\varepsilon^2\ 0$	$0\ \varepsilon$ $\varepsilon^2\ 0$	$0\ \varepsilon$ $\varepsilon^2\ 0$
		0	0	0	0	0	0	0	0	0	0	0	0
Γ_4^+	Γ'_{15}	0 −1 0 −1 0 0 0 0 −1	0 1 0 −1 0 0 0 0 1	0 −1 0 1 0 0 0 0 1	0 1 0 1 0 0 0 0 −1	0 0 −1 0 −1 0 −1 0 0	0 0 1 0 −1 0 1 0 0	0 0 −1 0 1 0 1 0 0	0 0 1 0 1 0 −1 0 0	−1 0 0 0 0 −1 0 −1 0	1 0 0 0 0 −1 0 1 0	−1 0 0 0 0 1 0 1 0	1 0 0 0 0 1 0 −1 0
		−1	1	1	−1	−1	−1	1	1	−1	1	−1	1
Γ_5^+	Γ'_{25}	0 1 0 1 0 0 0 0 1	0 −1 0 1 0 0 0 0 −1	0 1 0 −1 0 0 0 0 −1	0 −1 0 −1 0 0 0 0 1	0 0 1 0 1 0 1 0 0	0 0 −1 0 1 0 −1 0 0	0 0 1 0 −1 0 −1 0 0	0 0 −1 0 −1 0 1 0 0	1 0 0 0 0 1 0 1 0	−1 0 0 0 0 1 0 −1 0	1 0 0 0 0 −1 0 −1 0	−1 0 0 0 0 −1 0 1 0
		1	−1	−1	1	1	1	−1	−1	1	−1	1	−1

$\varepsilon = \exp(i2\pi/3)$

† In order to save space, we do not present the matrices of the representations Γ_1^+–Γ_5^+ for elements 25–48, because these matrices coincide with the corresponding matrices of elements 1–24, and we also do not present the matrices of the representations Γ_1^-–Γ_5^- (Γ'_1, Γ'_2, Γ'_{12}, Γ'_{15} and Γ'_{25} in the BSW notation), because these matrices coincide with the corresponding matrices of the representations Γ_1^+–Γ_5^+ for elements 1–24 and differ by the factor −1 from the corresponding matrices of the representations Γ_1^+–Γ_5^+ for elements 25–48

Table 12.6 Small representations for the point R

	1	2	3	4	5	6	7	8	9	10	11	12
R_1	$\begin{matrix}1&0\\0&1\end{matrix}$	$\begin{matrix}1&0\\0&1\end{matrix}$	$\begin{matrix}1&0\\0&1\end{matrix}$	$\begin{matrix}1&0\\0&1\end{matrix}$	$\begin{matrix}\varepsilon^2&0\\0&\varepsilon\end{matrix}$	$\begin{matrix}\varepsilon^2&0\\0&\varepsilon\end{matrix}$	$\begin{matrix}\varepsilon^2&0\\0&\varepsilon\end{matrix}$	$\begin{matrix}\varepsilon^2&0\\0&\varepsilon\end{matrix}$	$\begin{matrix}\varepsilon&0\\0&\varepsilon^2\end{matrix}$	$\begin{matrix}\varepsilon&0\\0&\varepsilon^2\end{matrix}$	$\begin{matrix}\varepsilon&0\\0&\varepsilon^2\end{matrix}$	$\begin{matrix}\varepsilon&0\\0&\varepsilon^2\end{matrix}$
	2	2	2	2	−1	−1	−1	−1	−1	−1	−1	−1
R_2	$\begin{matrix}1&0\\0&1\end{matrix}$	$\begin{matrix}1&0\\0&1\end{matrix}$	$\begin{matrix}1&0\\0&1\end{matrix}$	$\begin{matrix}1&0\\0&1\end{matrix}$	$\begin{matrix}\varepsilon^2&0\\0&\varepsilon\end{matrix}$	$\begin{matrix}\varepsilon^2&0\\0&\varepsilon\end{matrix}$	$\begin{matrix}\varepsilon^2&0\\0&\varepsilon\end{matrix}$	$\begin{matrix}\varepsilon^2&0\\0&\varepsilon\end{matrix}$	$\begin{matrix}\varepsilon&0\\0&\varepsilon^2\end{matrix}$	$\begin{matrix}\varepsilon&0\\0&\varepsilon^2\end{matrix}$	$\begin{matrix}\varepsilon&0\\0&\varepsilon^2\end{matrix}$	$\begin{matrix}\varepsilon&0\\0&\varepsilon^2\end{matrix}$
	2	2	2	2	−1	−1	−1	−1	−1	−1	−1	−1
R_3	$\begin{matrix}1&0\\0&1\end{matrix}$	$\begin{matrix}1&0\\0&1\end{matrix}$	$\begin{matrix}1&0\\0&1\end{matrix}$	$\begin{matrix}1&0\\0&1\end{matrix}$	$\begin{matrix}1&0\\0&1\end{matrix}$	$\begin{matrix}1&0\\0&1\end{matrix}$	$\begin{matrix}1&0\\0&1\end{matrix}$	$\begin{matrix}1&0\\0&1\end{matrix}$	$\begin{matrix}1&0\\0&1\end{matrix}$	$\begin{matrix}1&0\\0&1\end{matrix}$	$\begin{matrix}1&0\\0&1\end{matrix}$	$\begin{matrix}1&0\\0&1\end{matrix}$
	2	2	2	2	2	2	2	2	2	2	2	2
$R_4^\dagger$	$\begin{matrix}M_1&0\\0&M_1\end{matrix}$	$\begin{matrix}M_2&0\\0&M_2\end{matrix}$	$\begin{matrix}M_3&0\\0&M_3\end{matrix}$	$\begin{matrix}M_4&0\\0&M_4\end{matrix}$	$\begin{matrix}M_5&0\\0&M_5\end{matrix}$	$\begin{matrix}M_6&0\\0&M_6\end{matrix}$	$\begin{matrix}M_7&0\\0&M_7\end{matrix}$	$\begin{matrix}M_8&0\\0&M_8\end{matrix}$	$\begin{matrix}M_9&0\\0&M_9\end{matrix}$	$\begin{matrix}M_{10}&0\\0&M_{10}\end{matrix}$	$\begin{matrix}M_{11}&0\\0&M_{11}\end{matrix}$	$\begin{matrix}M_{11}&0\\0&M_{12}\end{matrix}$
	6	−2	−2	−2	0	0	0	0	0	0	0	0

	13	14	15	16	17	18	19	20	21	22	23	24
R_1	$\begin{matrix}0&1\\1&0\end{matrix}$	$\begin{matrix}0&1\\1&0\end{matrix}$	$\begin{matrix}0&1\\1&0\end{matrix}$	$\begin{matrix}0&1\\1&0\end{matrix}$	$\begin{matrix}0&\varepsilon^2\\\varepsilon&0\end{matrix}$	$\begin{matrix}0&\varepsilon^2\\\varepsilon&0\end{matrix}$	$\begin{matrix}0&\varepsilon^2\\\varepsilon&0\end{matrix}$	$\begin{matrix}0&\varepsilon^2\\\varepsilon&0\end{matrix}$	$\begin{matrix}0&\varepsilon\\\varepsilon^2&0\end{matrix}$	$\begin{matrix}0&\varepsilon\\\varepsilon^2&0\end{matrix}$	$\begin{matrix}0&\varepsilon\\\varepsilon^2&0\end{matrix}$	$\begin{matrix}0&\varepsilon\\\varepsilon^2&0\end{matrix}$
	0	0	0	0	0	0	0	0	0	0	0	0
R_2	$\begin{matrix}0&1\\1&0\end{matrix}$	$\begin{matrix}0&1\\1&0\end{matrix}$	$\begin{matrix}0&1\\1&0\end{matrix}$	$\begin{matrix}0&1\\1&0\end{matrix}$	$\begin{matrix}0&\varepsilon^2\\\varepsilon&0\end{matrix}$	$\begin{matrix}0&\varepsilon^2\\\varepsilon&0\end{matrix}$	$\begin{matrix}0&\varepsilon^2\\\varepsilon&0\end{matrix}$	$\begin{matrix}0&\varepsilon^2\\\varepsilon&0\end{matrix}$	$\begin{matrix}0&\varepsilon\\\varepsilon^2&0\end{matrix}$	$\begin{matrix}0&\varepsilon\\\varepsilon^2&0\end{matrix}$	$\begin{matrix}0&\varepsilon\\\varepsilon^2&0\end{matrix}$	$\begin{matrix}0&\varepsilon\\\varepsilon^2&0\end{matrix}$
	0	0	0	0	0	0	0	0	0	0	0	0
R_3	$\begin{matrix}1&0\\0&-1\end{matrix}$	$\begin{matrix}1&0\\0&-1\end{matrix}$	$\begin{matrix}1&0\\0&-1\end{matrix}$	$\begin{matrix}1&0\\0&-1\end{matrix}$	$\begin{matrix}1&0\\0&-1\end{matrix}$	$\begin{matrix}1&0\\0&-1\end{matrix}$	$\begin{matrix}1&0\\0&-1\end{matrix}$	$\begin{matrix}1&0\\0&-1\end{matrix}$	$\begin{matrix}1&0\\0&-1\end{matrix}$	$\begin{matrix}1&0\\0&-1\end{matrix}$	$\begin{matrix}1&0\\0&-1\end{matrix}$	$\begin{matrix}1&0\\0&-1\end{matrix}$
	0	0	0	0	0	0	0	0	0	0	0	0
$R_4^\dagger$	$\begin{matrix}M_{13}&0\\0&-M_{13}\end{matrix}$	$\begin{matrix}M_{14}&0\\0&-M_{14}\end{matrix}$	$\begin{matrix}M_{15}&0\\0&-M_{15}\end{matrix}$	$\begin{matrix}M_{16}&0\\0&-M_{16}\end{matrix}$	$\begin{matrix}M_{17}&0\\0&-M_{17}\end{matrix}$	$\begin{matrix}M_{18}&0\\0&-M_{18}\end{matrix}$	$\begin{matrix}M_{19}&0\\0&-M_{19}\end{matrix}$	$\begin{matrix}M_{20}&0\\0&-M_{20}\end{matrix}$	$\begin{matrix}M_{21}&0\\0&-M_{21}\end{matrix}$	$\begin{matrix}M_{22}&0\\0&-M_{22}\end{matrix}$	$\begin{matrix}M_{23}&0\\0&-M_{23}\end{matrix}$	$\begin{matrix}M_{24}&0\\0&-M_{24}\end{matrix}$
	0	0	0	0	0	0	0	0	0	0	0	0

(continued)

Table 12.6 (continued)

	25	26	27	28	29	30	31	32	33	34	35	36
R_1	$\begin{matrix}1&0\\0&-1\end{matrix}$	$\begin{matrix}1&0\\0&-1\end{matrix}$	$\begin{matrix}1&0\\0&-1\end{matrix}$	$\begin{matrix}1&0\\0&-1\end{matrix}$	$\begin{matrix}\varepsilon^2&0\\0&-\varepsilon\end{matrix}$	$\begin{matrix}\varepsilon^2&0\\0&-\varepsilon\end{matrix}$	$\begin{matrix}\varepsilon^2&0\\0&-\varepsilon\end{matrix}$	$\begin{matrix}\varepsilon^2&0\\0&-\varepsilon\end{matrix}$	$\begin{matrix}\varepsilon&0\\0&-\varepsilon^2\end{matrix}$	$\begin{matrix}\varepsilon&0\\0&-\varepsilon^2\end{matrix}$	$\begin{matrix}\varepsilon&0\\0&-\varepsilon^2\end{matrix}$	$\begin{matrix}\varepsilon&0\\0&-\varepsilon^2\end{matrix}$
	0	0	0	0	$-i\sqrt{3}$	$-i\sqrt{3}$	$-i\sqrt{3}$	$-i\sqrt{3}$	$i\sqrt{3}$	$i\sqrt{3}$	$i\sqrt{3}$	$i\sqrt{3}$
R_2	$\begin{matrix}-1&0\\0&1\end{matrix}$	$\begin{matrix}-1&0\\0&1\end{matrix}$	$\begin{matrix}-1&0\\0&1\end{matrix}$	$\begin{matrix}-1&0\\0&1\end{matrix}$	$\begin{matrix}-\varepsilon^2&0\\0&\varepsilon\end{matrix}$	$\begin{matrix}-\varepsilon^2&0\\0&\varepsilon\end{matrix}$	$\begin{matrix}-\varepsilon^2&0\\0&\varepsilon\end{matrix}$	$\begin{matrix}-\varepsilon^2&0\\0&\varepsilon\end{matrix}$	$\begin{matrix}-\varepsilon&0\\0&\varepsilon^2\end{matrix}$	$\begin{matrix}-\varepsilon&0\\0&\varepsilon^2\end{matrix}$	$\begin{matrix}-\varepsilon&0\\0&\varepsilon^2\end{matrix}$	$\begin{matrix}-\varepsilon&0\\0&\varepsilon^2\end{matrix}$
	0	0	0	0	$i\sqrt{3}$	$i\sqrt{3}$	$i\sqrt{3}$	$i\sqrt{3}$	$-i\sqrt{3}$	$-i\sqrt{3}$	$-i\sqrt{3}$	$-i\sqrt{3}$
R_3	$\begin{matrix}0&1\\1&0\end{matrix}$	$\begin{matrix}0&1\\1&0\end{matrix}$	$\begin{matrix}0&1\\1&0\end{matrix}$	$\begin{matrix}0&1\\1&0\end{matrix}$	$\begin{matrix}0&1\\1&0\end{matrix}$	$\begin{matrix}0&1\\1&0\end{matrix}$	$\begin{matrix}0&1\\1&0\end{matrix}$	$\begin{matrix}0&1\\1&0\end{matrix}$	$\begin{matrix}0&1\\1&0\end{matrix}$	$\begin{matrix}0&1\\1&0\end{matrix}$	$\begin{matrix}0&1\\1&0\end{matrix}$	$\begin{matrix}0&1\\1&0\end{matrix}$
	0	0	0	0	0	0	0	0	0	0	0	0
$R_4^{\dagger}$	$\begin{matrix}0&M_1\\M_1&0\end{matrix}$	$\begin{matrix}0&M_2\\M_2&0\end{matrix}$	$\begin{matrix}0&M_3\\M_3&0\end{matrix}$	$\begin{matrix}0&M_4\\M_4&0\end{matrix}$	$\begin{matrix}0&M_5\\M_5&0\end{matrix}$	$\begin{matrix}0&M_6\\M_6&0\end{matrix}$	$\begin{matrix}0&M_7\\M_7&0\end{matrix}$	$\begin{matrix}0&M_8\\M_8&0\end{matrix}$	$\begin{matrix}0&M_9\\M_9&0\end{matrix}$	$\begin{matrix}0&M_{10}\\M_{10}&0\end{matrix}$	$\begin{matrix}0&M_{11}\\M_{11}&0\end{matrix}$	$\begin{matrix}0&M_{12}\\M_{12}&0\end{matrix}$
	0	0	0	0	0	0	0	0	0	0	0	0

	37	38	39	40	41	42	43	44	45	46	47	48
R_1	$\begin{matrix}0&-1\\1&0\end{matrix}$	$\begin{matrix}0&-1\\1&0\end{matrix}$	$\begin{matrix}0&-1\\1&0\end{matrix}$	$\begin{matrix}0&-1\\1&0\end{matrix}$	$\begin{matrix}0&-\varepsilon^2\\\varepsilon&0\end{matrix}$	$\begin{matrix}0&-\varepsilon^2\\\varepsilon&0\end{matrix}$	$\begin{matrix}0&-\varepsilon^2\\\varepsilon&0\end{matrix}$	$\begin{matrix}0&-\varepsilon^2\\\varepsilon&0\end{matrix}$	$\begin{matrix}0&-\varepsilon\\\varepsilon^2&0\end{matrix}$	$\begin{matrix}0&-\varepsilon\\\varepsilon^2&0\end{matrix}$	$\begin{matrix}0&-\varepsilon\\\varepsilon^2&0\end{matrix}$	$\begin{matrix}0&-\varepsilon\\\varepsilon^2&0\end{matrix}$
	0	0	0	0	0	0	0	0	0	0	0	0
R_2	$\begin{matrix}0&1\\-1&0\end{matrix}$	$\begin{matrix}0&1\\-1&0\end{matrix}$	$\begin{matrix}0&1\\-1&0\end{matrix}$	$\begin{matrix}0&1\\-1&0\end{matrix}$	$\begin{matrix}0&\varepsilon^2\\-\varepsilon&0\end{matrix}$	$\begin{matrix}0&\varepsilon^2\\-\varepsilon&0\end{matrix}$	$\begin{matrix}0&\varepsilon^2\\-\varepsilon&0\end{matrix}$	$\begin{matrix}0&\varepsilon^2\\-\varepsilon&0\end{matrix}$	$\begin{matrix}0&\varepsilon\\-\varepsilon^2&0\end{matrix}$	$\begin{matrix}0&\varepsilon\\-\varepsilon^2&0\end{matrix}$	$\begin{matrix}0&\varepsilon\\-\varepsilon^2&0\end{matrix}$	$\begin{matrix}0&\varepsilon\\-\varepsilon^2&0\end{matrix}$
	0	0	0	0	0	0	0	0	0	0	0	0
R_3	$\begin{matrix}0&1\\-1&0\end{matrix}$	$\begin{matrix}0&1\\-1&0\end{matrix}$	$\begin{matrix}0&1\\-1&0\end{matrix}$	$\begin{matrix}0&1\\-1&0\end{matrix}$	$\begin{matrix}0&1\\-1&0\end{matrix}$	$\begin{matrix}0&1\\-1&0\end{matrix}$	$\begin{matrix}0&1\\-1&0\end{matrix}$	$\begin{matrix}0&1\\-1&0\end{matrix}$	$\begin{matrix}0&1\\-1&0\end{matrix}$	$\begin{matrix}0&1\\-1&0\end{matrix}$	$\begin{matrix}0&1\\-1&0\end{matrix}$	$\begin{matrix}0&1\\-1&0\end{matrix}$
	0	0	0	0	0	0	0	0	0	0	0	0
$R_4^{\dagger}$	$\begin{matrix}0&M_{13}\\-M_{13}&0\end{matrix}$	$\begin{matrix}0&M_{14}\\-M_{14}&0\end{matrix}$	$\begin{matrix}0&M_{15}\\-M_{15}&0\end{matrix}$	$\begin{matrix}0&M_{16}\\-M_{16}&0\end{matrix}$	$\begin{matrix}0&M_{17}\\-M_{17}&0\end{matrix}$	$\begin{matrix}0&M_{18}\\-M_{18}&0\end{matrix}$	$\begin{matrix}0&M_{19}\\-M_{19}&0\end{matrix}$	$\begin{matrix}0&M_{20}\\-M_{20}&0\end{matrix}$	$\begin{matrix}0&M_{21}\\-M_{21}&0\end{matrix}$	$\begin{matrix}0&M_{22}\\-M_{22}&0\end{matrix}$	$\begin{matrix}0&M_{23}\\-M_{23}&0\end{matrix}$	$\begin{matrix}0&M_{24}\\-M_{24}&0\end{matrix}$
	0	0	0	0	0	0	0	0	0	0	0	0

$\varepsilon = \exp(i2\pi/3)$

† This is a six-dimensional representation written in a block form. Each block is a 3×3-matrix, explained in Table 12.7

Table 12.7 The matrices M_1–M_{24} used in the six-dimensional representation R_4 of the point R (in the previous table)

M_1	M_2	M_3	M_4	M_5	M_6	M_7	M_8	M_9	M_{10}	M_{11}	M_{12}
1 0 0	−1 0 0	1 0 0	−1 0 0	0 0 1	0 0 −1	0 0 1	0 0 −1	0 1 0	0 −1 0	0 1 0	0 −1 0
0 1 0	0 1 0	0 −1 0	0 −1 0	1 0 0	1 0 0	−1 0 0	−1 0 0	0 0 1	0 0 1	0 0 −1	0 0 −1
0 0 1	0 0 −1	0 0 −1	0 0 1	0 1 0	0 −1 0	0 −1 0	0 1 0	1 0 0	−1 0 0	−1 0 0	1 0 0

M_{13}	M_{14}	M_{15}	M_{16}	M_{17}	M_{18}	M_{19}	M_{20}	M_{21}	M_{22}	M_{23}	M_{24}
0 −1 0	0 1 0	0 −1 0	0 1 0	0 0 −1	0 0 1	0 0 −1	0 0 1	−1 0 0	1 0 0	−1 0 0	1 0 0
−1 0 0	−1 0 0	1 0 0	1 0 0	0 −1 0	0 −1 0	0 1 0	0 1 0	0 0 −1	0 0 −1	0 0 1	0 0 1
0 0 −1	0 0 1	0 0 1	0 0 −1	−1 0 0	1 0 0	1 0 0	−1 0 0	0 −1 0	0 1 0	0 1 0	0 −1 0

Table 12.8 Small representations for the point X

	1	2	3	4	21	22	23	24	25	26	27	28	45	46	47	48
X_1	$\begin{matrix}1&0\\0&1\end{matrix}$	$\begin{matrix}0&-i\\i&0\end{matrix}$	$\begin{matrix}1&0\\0&1\end{matrix}$	$\begin{matrix}0&-i\\i&0\end{matrix}$	$\begin{matrix}0&1\\1&0\end{matrix}$	$\begin{matrix}i&0\\0&-i\end{matrix}$	$\begin{matrix}0&1\\1&0\end{matrix}$	$\begin{matrix}i&0\\0&-i\end{matrix}$	$\begin{matrix}0&-i\\i&0\end{matrix}$	$\begin{matrix}1&0\\0&1\end{matrix}$	$\begin{matrix}0&-i\\i&0\end{matrix}$	$\begin{matrix}1&0\\0&1\end{matrix}$	$\begin{matrix}i&0\\0&-i\end{matrix}$	$\begin{matrix}0&1\\1&0\end{matrix}$	$\begin{matrix}i&0\\0&-i\end{matrix}$	$\begin{matrix}0&1\\1&0\end{matrix}$
	2	0	2	0	0	0	0	0	0	2	0	2	0	0	0	0
X_2	$\begin{matrix}1&0\\0&1\end{matrix}$	$\begin{matrix}0&-i\\i&0\end{matrix}$	$\begin{matrix}1&0\\0&1\end{matrix}$	$\begin{matrix}0&-i\\i&0\end{matrix}$	$\begin{matrix}0&1\\1&0\end{matrix}$	$\begin{matrix}i&0\\0&-i\end{matrix}$	$\begin{matrix}0&1\\1&0\end{matrix}$	$\begin{matrix}i&0\\0&-i\end{matrix}$	$\begin{matrix}0&i\\-i&0\end{matrix}$	$\begin{matrix}-1&0\\0&-1\end{matrix}$	$\begin{matrix}0&i\\-i&0\end{matrix}$	$\begin{matrix}-1&0\\0&-1\end{matrix}$	$\begin{matrix}-i&0\\0&i\end{matrix}$	$\begin{matrix}0&-1\\-1&0\end{matrix}$	$\begin{matrix}-i&0\\0&i\end{matrix}$	$\begin{matrix}0&-1\\-1&0\end{matrix}$
	2	0	2	0	0	0	0	0	0	−2	0	−2	0	0	0	0
X_3	$\begin{matrix}1&0\\0&1\end{matrix}$	$\begin{matrix}-1&0\\0&-1\end{matrix}$	$\begin{matrix}-1&0\\0&-1\end{matrix}$	$\begin{matrix}1&0\\0&1\end{matrix}$	$\begin{matrix}-1&0\\0&1\end{matrix}$	$\begin{matrix}-1&0\\0&1\end{matrix}$	$\begin{matrix}1&0\\0&-1\end{matrix}$	$\begin{matrix}1&0\\0&-1\end{matrix}$	$\begin{matrix}0&1\\1&0\end{matrix}$	$\begin{matrix}0&-1\\-1&0\end{matrix}$	$\begin{matrix}0&-1\\-1&0\end{matrix}$	$\begin{matrix}0&1\\1&0\end{matrix}$	$\begin{matrix}0&-1\\1&0\end{matrix}$	$\begin{matrix}0&-1\\1&0\end{matrix}$	$\begin{matrix}0&1\\-1&0\end{matrix}$	$\begin{matrix}0&1\\-1&0\end{matrix}$
	2	−2	−2	2	0	0	0	0	0	0	0	0	0	0	0	0
X_4	$\begin{matrix}1&0\\0&1\end{matrix}$	$\begin{matrix}1&0\\0&1\end{matrix}$	$\begin{matrix}-1&0\\0&-1\end{matrix}$	$\begin{matrix}-1&0\\0&-1\end{matrix}$	$\begin{matrix}1&0\\0&-1\end{matrix}$	$\begin{matrix}-1&0\\0&1\end{matrix}$	$\begin{matrix}-1&0\\0&1\end{matrix}$	$\begin{matrix}1&0\\0&-1\end{matrix}$	$\begin{matrix}0&1\\1&0\end{matrix}$	$\begin{matrix}0&1\\1&0\end{matrix}$	$\begin{matrix}0&-1\\-1&0\end{matrix}$	$\begin{matrix}0&-1\\1&0\end{matrix}$	$\begin{matrix}0&1\\-1&0\end{matrix}$	$\begin{matrix}0&-1\\1&0\end{matrix}$	$\begin{matrix}0&-1\\1&0\end{matrix}$	$\begin{matrix}0&1\\-1&0\end{matrix}$
	2	2	−2	−2	0	0	0	0	0	0	0	0	0	0	0	0

Table 12.9 Small representations for the point M

	1	2	3	4	13	14	15	16	25	26	27	28	37	38	39	40
M_1	1	1	−1	−1	−1	1	−1	1	−1	−1	1	1	1	−1	1	−1
M_2	1	−1	1	−1	−1	−1	1	1	−1	1	−1	1	1	1	−1	−1
M_3	1	−1	1	−1	1	1	−1	−1	1	−1	1	−1	1	1	−1	−1
M_4	1	1	−1	−1	1	−1	1	−1	1	1	−1	−1	1	−1	1	−1
M_5	1	1	−1	−1	1	−1	1	−1	−1	−1	1	1	−1	1	−1	1
M_6	1	−1	1	−1	1	1	−1	−1	−1	1	−1	1	−1	−1	1	1
M_7	1	−1	1	−1	−1	−1	1	1	1	−1	1	−1	−1	−1	1	1
M_8	1	1	−1	−1	−1	1	−1	1	1	1	−1	−1	−1	1	−1	1
M_9	1 0 0 1	0 −i i 0	0 −i i 0	1 0 0 1	0 1 1 0	i 0 0 −i	i 0 0 −i	0 1 1 0	1 0 0 1	0 −i i 0	0 −i i 0	1 0 0 1	0 1 1 0	i 0 0 −i	i 0 0 −i	0 1 1 0
	2	0	0	2	0	0	0	0	2	0	0	2	0	0	0	0
M_{10}	1 0 0 1	0 −i i 0	0 −i i 0	1 0 0 1	0 1 1 0	i 0 0 −i	i 0 0 −i	0 1 1 0	−1 0 0 −1	0 i −i 0	0 i −i 0	−1 0 0 −1	0 −1 1 0	−i 0 0 i	−i 0 0 i	0 −1 −1 0
	2	0	0	2	0	0	0	0	−2	0	0	−2	0	0	0	0

Table 12.10 Small representations for the point Δ

	1	3	22	24	26	28	45	47
Δ_1	1	1	γ	γ	1	1	γ	γ
Δ_2	1	1	$-\gamma$	$-\gamma$	1	1	$-\gamma$	$-\gamma$
Δ_3	1	1	$-\gamma$	$-\gamma$	-1	-1	γ	γ
Δ_4	1	1	γ	γ	-1	-1	$-\gamma$	$-\gamma$
Δ_5	$\begin{matrix} 1 & 0 \\ 0 & 1 \end{matrix}$	$\begin{matrix} -1 & 0 \\ 0 & -1 \end{matrix}$	$\begin{pmatrix} -i & 0 \\ 0 & i \end{pmatrix}\gamma$	$\begin{pmatrix} i & 0 \\ 0 & -i \end{pmatrix}\gamma$	$\begin{matrix} 0 & i \\ -i & 0 \end{matrix}$	$\begin{matrix} 0 & -i \\ i & 0 \end{matrix}$	$\begin{pmatrix} 0 & -1 \\ -1 & 0 \end{pmatrix}\gamma$	$\begin{pmatrix} 0 & 1 \\ 1 & 0 \end{pmatrix}\gamma$
	2	-2	0	0	0	0	0	0

$\gamma = \exp(-i\pi\eta)$

Table 12.11 Small representations for the point T

	1	4	14	15	26	27	37	40
T_1	1	-1	$-\gamma$	γ	1	-1	γ	$-\gamma$
T_2	1	-1	γ	$-\gamma$	-1	1	γ	$-\gamma$
T_3	1	-1	γ	$-\gamma$	1	-1	$-\gamma$	γ
T_4	1	-1	$-\gamma$	γ	-1	1	$-\gamma$	γ
T_5	$\begin{matrix} 1 & 0 \\ 0 & 1 \end{matrix}$	$\begin{matrix} 1 & 0 \\ 0 & 1 \end{matrix}$	$\begin{pmatrix} i & 0 \\ 0 & -i \end{pmatrix}\gamma$	$\begin{pmatrix} i & 0 \\ 0 & -i \end{pmatrix}\gamma$	$\begin{matrix} 0 & -i \\ i & 0 \end{matrix}$	$\begin{matrix} 0 & -i \\ i & 0 \end{matrix}$	$\begin{pmatrix} 0 & 1 \\ 1 & 0 \end{pmatrix}\gamma$	$\begin{pmatrix} 0 & 1 \\ 1 & 0 \end{pmatrix}\gamma$
	2	2	0	0	0	0	0	0

$\gamma = \exp(-i\pi\zeta)$

Table 12.12 Small representations for the point Λ

	1	5	9	37	41	45
Λ_1	1	1	1	γ	γ	γ
Λ_2	1	1	1	$-\gamma$	$-\gamma$	$-\gamma$
Λ_3	$\begin{matrix} 1 & 0 \\ 0 & 1 \end{matrix}$	$\begin{matrix} \varepsilon & 0 \\ 0 & \varepsilon^2 \end{matrix}$	$\begin{matrix} \varepsilon^2 & 0 \\ 0 & \varepsilon \end{matrix}$	$\begin{pmatrix} 0 & 1 \\ 1 & 0 \end{pmatrix}\gamma$	$\begin{pmatrix} 0 & \varepsilon \\ \varepsilon^2 & 0 \end{pmatrix}\gamma$	$\begin{pmatrix} 0 & \varepsilon^2 \\ \varepsilon & 0 \end{pmatrix}\gamma$
	2	-1	-1	0	0	0

$\varepsilon = \exp(i2\pi/3)$, $\gamma = \exp(-i3\pi\xi)$

Table 12.13 Small representations for the point Σ

	1	16	28	37
Σ_1	1	γ	1	γ
Σ_2	1	$-\gamma$	1	$-\gamma$
Σ_3	1	γ	-1	$-\gamma$
Σ_4	1	$-\gamma$	-1	γ

$\gamma = \exp(-i2\pi\xi)$

Table 12.14 Small representations for the point Z

	1	2	27	28
Z_1	1	1	1	1
Z_2	1	−1	1	−1
Z_3	1	1	−1	−1
Z_4	1	−1	−1	1

Table 12.15 Small representations for the point S

	1	23	27	45
S_1	$\begin{matrix}1 & 0\\ 0 & 1\end{matrix}$	$\begin{pmatrix}1 & 0\\ 0 & -1\end{pmatrix}\gamma$	$\begin{matrix}0 & 1\\ 1 & 0\end{matrix}$	$\begin{pmatrix}0 & 1\\ -1 & 0\end{pmatrix}\gamma$
	2	0	0	0

$\gamma = \exp(-\mathrm{i}2\pi\xi)$

Table 12.16 Small representations for the point Θ

	1	37
Θ_1	1	γ
Θ_2	1	$-\gamma$

$\gamma = \exp(-\mathrm{i}\pi(2\xi + \zeta))$

Table 12.17 Small representations for the points A and Ξ

	1	27
	1	28
A_1 Ξ_1	1	1
A_2 Ξ_2	1	−1

Table 12.18 Small representations for the point B

	1	45
B_1	1	γ
B_2	1	$-\gamma$

$\gamma = \exp(-\mathrm{i}\pi(2\xi + \eta))$

12.3 Compatibility Relations

If we choose a high symmetry point $\mathbf{k}$ in the Brillouin zone and a neighboring to it point $\mathbf{k}+\boldsymbol{\kappa}$ and analyze the small representations $d^{\mathbf{k},i}$ and $d^{\mathbf{k}+\boldsymbol{\kappa},j}$, it turns out that these representations are not independent, but are related to each other in a certain way. The representations $d^{\mathbf{k},i}$ and $d^{\mathbf{k}+\boldsymbol{\kappa},j}$ must be compatible, or satisfy certain *compatibility relations* (Sect. 5.8).

The compatibility relations at the point $\mathbf{k}$ can be obtained by restricting the representation $d^{\mathbf{k},i}$ to the subgroup $G_{\mathbf{k}+\boldsymbol{\kappa}}$ and decomposing the restriction $d^{\mathbf{k}(\boldsymbol{\kappa}),i}$ (sometimes called subduction) into irreducible components. For this, it is enough to know only the characters of the small representations at the points $\mathbf{k}$ and $\mathbf{k}+\boldsymbol{\kappa}$. The multiplicity of the small representation $d^{\mathbf{k}+\boldsymbol{\kappa},j}$ in the restriction $d^{\mathbf{k}(\boldsymbol{\kappa}),i}$, i.e. the number of times $d^{\mathbf{k}+\boldsymbol{\kappa},j}$ appears in the expansion of $d^{\mathbf{k}(\boldsymbol{\kappa}),i}$, is calculated by formula (5.37)

$$a_{i,j}^{\mathbf{k},\mathbf{k}+\boldsymbol{\kappa}} = \frac{1}{n(G_{\mathbf{k}+\boldsymbol{\kappa}}^0)} \sum_{\{\alpha|\boldsymbol{\tau}_\alpha\}\in G_{\mathbf{k}+\boldsymbol{\kappa}}^0} \chi^{\mathbf{k},i}(\{\alpha|\boldsymbol{\tau}_\alpha\})\, \chi^{\mathbf{k}+\boldsymbol{\kappa},j*}(\{\alpha|\boldsymbol{\tau}_\alpha\}), \qquad (12.5)$$

where $n(G_{\mathbf{k}+\boldsymbol{\kappa}}^0)$ is the order of the group $G_{\mathbf{k}+\boldsymbol{\kappa}}^0$, and $\chi^{\mathbf{k},i}$ and $\chi^{\mathbf{k}+\boldsymbol{\kappa},j}$ are the characters of the matrices $d^{\mathbf{k},i}$ and $d^{\mathbf{k}+\boldsymbol{\kappa},j}$, respectively.

Let us consider all possible choices of the direction $\boldsymbol{\kappa}$.

1. The direction $\boldsymbol{\kappa}$ is chosen so that $\mathbf{k}+\boldsymbol{\kappa}$ is a general point. Then $G_{\mathbf{k}+\boldsymbol{\kappa}} = T$, and the representation $d^{\mathbf{k}+\boldsymbol{\kappa}}(\{\varepsilon|\mathbf{R}_n\}) = \mathrm{e}^{-\mathrm{i}(\mathbf{k}+\boldsymbol{\kappa})\mathbf{R}_n}$ is compatible with all representations $d^{\mathbf{k},i}$, because $d^{\mathbf{k},i}$ is a small representation.[9]
2. The direction $\boldsymbol{\kappa}$ is chosen so that $\mathbf{k}+\boldsymbol{\kappa}$ has the same symmetry as $\mathbf{k}$. Then $G_{\mathbf{k}+\boldsymbol{\kappa}} = G_{\mathbf{k}}$, and the restriction $d^{\mathbf{k}(\boldsymbol{\kappa}),i}$ coincides with the representation $d^{\mathbf{k},i}$, and therefore is irreducible.
3. Direction $\boldsymbol{\kappa}$ is chosen so that $\mathbf{k}+\boldsymbol{\kappa}$ is a symmetry point, but its symmetry is lower than the one of the point $\mathbf{k}$. In this case $G_{\mathbf{k}+\boldsymbol{\kappa}} \subset G_{\mathbf{k}}$, and the restriction $d^{\mathbf{k}(\boldsymbol{\kappa}),i}$ can be decomposed into irreducible components.

Thus, compatibility relations should be obtained for neighboring points of different symmetry.

In Tables 12.19, 12.20, 12.21, 12.22, 12.23, 12.24, 12.25, 12.26, 12.27 and 12.28, we present the compatibility relations for all symmetry points and lines, which are calculated using formula (12.5) and Tables 12.5, 12.6, 12.7, 12.8, 12.9, 12.10, 12.11, 12.12, 12.13, 12.14, 12.15, 12.16, 12.17 and 12.18.

---

[9] Recall that, for translations, the matrices of small representations have the form $d^{\mathbf{k},i}(\{\varepsilon|\mathbf{R}_n\}) = \mathrm{e}^{-\mathrm{i}\mathbf{k}\mathbf{R}_n} E$ (E is the identity matrix).

Table 12.19 Compatibility relations at the point Γ

Γ_1^+	Γ_2^+	Γ_3^+	Γ_4^+	Γ_5^+	Γ_1^-	Γ_2^-	Γ_3^-	Γ_4^-	Γ_5^-
Δ_1	Δ_2	$\Delta_1\Delta_2$	$\Delta_4\Delta_5$	$\Delta_3\Delta_5$	Δ_4	Δ_3	$\Delta_3\Delta_4$	$\Delta_1\Delta_5$	$\Delta_2\Delta_5$
Λ_1	Λ_2	Λ_3	$\Lambda_2\Lambda_3$	$\Lambda_1\Lambda_3$	Λ_2	Λ_1	Λ_3	$\Lambda_1\Lambda_3$	$\Lambda_2\Lambda_3$
Σ_1	Σ_2	$\Sigma_1\Sigma_2$	$\Sigma_2\Sigma_3\Sigma_4$	$\Sigma_1\Sigma_3\Sigma_4$	Σ_3	Σ_4	$\Sigma_3\Sigma_4$	$\Sigma_1\Sigma_2\Sigma_4$	$\Sigma_1\Sigma_2\Sigma_3$

Table 12.20 Compatibility relations at the point R

R_1	R_2	R_3	R_4
S_1	S_1	S_1	$3S_1$
T_5	T_5	T_5	$T_1T_2T_3T_4T_5$
Λ_3	Λ_3	$\Lambda_1\Lambda_2$	$\Lambda_1\Lambda_2 2\Lambda_3$

Table 12.21 Compatibility relations at the point X

X_1	X_2	X_3	X_4
S_1	S_1	S_1	S_1
Z_1Z_4	Z_2Z_3	Z_2Z_4	Z_1Z_3
$\Delta_1\Delta_2$	$\Delta_3\Delta_4$	Δ_5	Δ_5

Table 12.22 Compatibility relations at the point M

M_1	M_2	M_3	M_4	M_5	M_6	M_7	M_8	M_9	M_{10}
Σ_2	Σ_2	Σ_3	Σ_3	Σ_1	Σ_1	Σ_4	Σ_4	$\Sigma_1\Sigma_2$	$\Sigma_3\Sigma_4$
Z_1	Z_4	Z_2	Z_3	Z_1	Z_4	Z_2	Z_3	Z_1Z_4	Z_2Z_3
T_2	T_1	T_2	T_1	T_4	T_3	T_4	T_3	T_5	T_5

Table 12.23 Compatibility relations at the point Δ

Δ_1	Δ_2	Δ_3	Δ_4	Δ_5
Ξ_1	Ξ_1	Ξ_2	Ξ_2	$\Xi_1\Xi_2$

Table 12.24 Compatibility relations at the point T

T_1	T_2	T_3	T_4	T_5
Θ_2	Θ_2	Θ_1	Θ_1	$\Theta_1\Theta_2$
A_2	A_1	A_2	A_1	A_1A_2

Table 12.25 Compatibility relations at the point Λ

Λ_1	Λ_2	Λ_3
Θ_1	Θ_2	$\Theta_1\ \Theta_2$

Table 12.26 Compatibility relations at the point Σ

Σ_1	Σ_2	Σ_3	Σ_4
Θ_1	Θ_2	Θ_2	Θ_1
Ξ_1	Ξ_1	Ξ_2	Ξ_2

Table 12.27 Compatibility relations at the point Z

Z_1	Z_2	Z_3	Z_4
Ξ_1	Ξ_2	Ξ_2	Ξ_1
A_1	A_1	A_2	A_2

Table 12.28 Compatibility relations at the point S

S_1
$A_1\ A_2$

12.4 Time-Reversal Symmetry

12.4.1 Theory

Symmetry with respect to time reversal leads to additional degeneracy (on top of the degeneracy due to the spatial symmetry). This issue is discussed in detail in Sect. 8.4. Let us recall the main points.

It is possible to find out whether the additional degeneracy is present by investigating the reality properties of the irreducible representations. Wigner [39] showed that[10]: (a) if the representation $\mathcal{D}$ is potentially real ($\mathcal{D}$ is equivalent to $\mathcal{D}^*$, and $\mathcal{D}$ can be reduced to the real form), there is no additional degeneracy; (b) if the representation $\mathcal{D}$ is complex ($\mathcal{D}$ is not equivalent to $\mathcal{D}^*$), there is an additional degeneracy; (c) if the representation $\mathcal{D}$ is pseudoreal ($\mathcal{D}$ is equivalent to $\mathcal{D}^*$, but $\mathcal{D}$ cannot be reduced to the real form), there is an additional degeneracy.

To determine the type of representation we use the theorem of Frobenius and Schur [40]. When applied to representations of space groups, it can be greatly simplified and reduced to a criterion for small representations (Herring's criterion [41]):

[10] Recall that our consideration is limited to the case without spin.

$$\sum_{\beta\mathbf{k}\approx-\mathbf{k}} \chi^{\mathbf{k}\lambda}(\{\beta|\boldsymbol{\tau}_\beta\}^2) = \begin{cases} n(G^0_{\mathbf{k}}), & \text{case a,} \\ 0, & \text{case b,} \\ -n(G^0_{\mathbf{k}}), & \text{case c,} \end{cases} \tag{12.6}$$

where the summation is carried out only over the elements $\{\beta|\boldsymbol{\tau}_\beta\}$ of the space group G such that $\beta\mathbf{k} \approx -\mathbf{k}$.

Criterion (12.6), although expressed through the characters of the small representations $d^{\mathbf{k}\lambda}$, describes additional degeneracy of the representations of the space group $\mathcal{D}^{\mathbf{k}\lambda}$. We are interested here in additional degeneracy of the small representations of $d^{\mathbf{k}\lambda}$ (additional degeneracy of the bands). When investigating this degeneracy for a given vector **k**, it is convenient to distinguish between the three possible cases: (I) $\mathbf{k} \approx -\mathbf{k}$; (II) $\mathbf{k} \not\approx -\mathbf{k}$, $\{\mathbf{k}\} = \{-\mathbf{k}\}$; (III) $\mathbf{k} \not\approx -\mathbf{k}$, $\{\mathbf{k}\} \neq \{-\mathbf{k}\}$. A detailed analysis of all possible variants shows [42] that additional degeneracy of the small representations of $d^{\mathbf{k}\lambda}$ occurs only in four cases: Ib, Ic, IIb, IIc. Here, the Roman numeral denotes the category of the vector **k**, and the Latin letter denotes the reality class of the representation $\mathcal{D}^{\mathbf{k}\lambda}$. In cases Ic and IIc, the two identical representations $d^{\mathbf{k}\lambda}$ turn out to be degenerate. In cases Ib and IIb, the two different representations $d^{\mathbf{k}\lambda}$ and $d^{\mathbf{k}\mu}$ are degenerate. Moreover, in case Ib, we have

$$\chi^{\mathbf{k}\mu}(\{\alpha|\boldsymbol{\tau}_\alpha\}) = \chi^{\mathbf{k}\lambda *}(\{\alpha|\boldsymbol{\tau}_\alpha\}), \qquad \{\alpha|\boldsymbol{\tau}_\alpha\} \in G_{\mathbf{k}}, \tag{12.7}$$

and the value of μ is easily determined from the table of the character $\chi^{\mathbf{k}\lambda}$. In the case of IIb, we have

$$\begin{aligned}\chi^{\mathbf{k}\mu}(\{\alpha|\boldsymbol{\tau}_\alpha\}) &= \chi^{\mathbf{k}\lambda *}(\{\beta|\boldsymbol{\tau}_\beta\}^{-1}\{\alpha|\boldsymbol{\tau}_\alpha\}\{\beta|\boldsymbol{\tau}_\beta\}) \\ &= \mathrm{e}^{-i\mathbf{k}(\alpha\boldsymbol{\tau}_\beta+\boldsymbol{\tau}_\alpha-\boldsymbol{\tau}_\beta-\beta\boldsymbol{\tau}_{\beta^{-1}\alpha\beta})}\chi^{\mathbf{k}\lambda *}(\{\beta^{-1}\alpha\beta|\boldsymbol{\tau}_{\beta^{-1}\alpha\beta}\}),\end{aligned} \tag{12.8}$$

where $\{\beta|\boldsymbol{\tau}_\beta\}$ is such that $\beta\mathbf{k} \approx -\mathbf{k}$, and $\{\alpha|\boldsymbol{\tau}_\alpha\} \in G_{\mathbf{k}}$. Here it is slightly more difficult to determine the value of μ: besides the table of characters $\chi^{\mathbf{k}\lambda}$, it is necessary to use the multiplication table of the group G^0. In some special (but very common) cases, formula (12.8) is simplified. If $\{\beta|\boldsymbol{\tau}_\beta\}$ can be chosen so that $\beta\alpha = \alpha\beta$, then

$$\chi^{\mathbf{k}\mu}(\{\alpha|\boldsymbol{\tau}_\alpha\}) = \mathrm{e}^{-i\mathbf{k}(\alpha\boldsymbol{\tau}_\beta+\boldsymbol{\tau}_\alpha-\boldsymbol{\tau}_\beta-\beta\boldsymbol{\tau}_\alpha)}\chi^{\mathbf{k}\lambda *}(\{\alpha|\boldsymbol{\tau}_\alpha\}), \tag{12.9}$$

and the value of μ is easily determined from the character table.[11] Furthermore, if $\{i|0\} \in G$, then we can put $\{\beta|\boldsymbol{\tau}_\beta\} = \{i|0\}$, and from (12.9) it immediately follows that

$$\chi^{\mathbf{k}\mu}(\{\alpha|\boldsymbol{\tau}_\alpha\}) = \mathrm{e}^{-i2\mathbf{k}\boldsymbol{\tau}_\alpha}\chi^{\mathbf{k}\lambda *}(\{\alpha|\boldsymbol{\tau}_\alpha\}). \tag{12.10}$$

Thus, to study the effect of time-reversal symmetry on band degeneracy, it is convenient to proceed as follows. First, determine the category of the considered vector

[11] For example, this is how the case $\{i|\boldsymbol{\tau}_i\} \in G$ can be considered.

k. If **k** falls into case III (**k** $\in$ III), the answer is known without further investigation: there is no additional degeneration of the bands. If **k** $\in$ I or **k** $\in$ II, it is necessary to calculate the sum (12.6) for each small representation $d^{\mathbf{k}\lambda}$, thereby determining the reality class a, b or c of the representation $\mathcal{D}^{\mathbf{k}\lambda}$. Only case b requires further consideration, since in case a there is no additional degeneracy, and in case c two identical representations of $d^{\mathbf{k}\lambda}$ are degenerate. To find the value μ of the representation $d^{\mathbf{k}\mu}$, which is degenerate together with the representation $d^{\mathbf{k}\lambda}$, in case Ib it is necessary to use formula (12.7), and in case IIb—to use formula (12.8) or (12.9) or (12.10).

As for the general points, their vectors **k** can fall into any of the three categories. For the categories I and II, **k** satisfies case a. This means that at general points (without spin!) there is no additional degeneracy of the bands.

12.4.2 Results

Let us apply the above theory to the Cr_3Si structure. Using Tables 12.2 and 12.4, we find that the symmetry points Γ, R, X and M belong to category I, and all the others belong to category II (the O_h group contains an inversion, and there can be no symmetry points of category III). Next, calculating the sum (12.6), we find that all representations $\mathcal{D}^{\mathbf{k}\lambda}$, except for $\mathcal{D}^{R_1}$ and $\mathcal{D}^{R_2}$, belong to the reality class a and are not related by time reversal. The representations $\mathcal{D}^{R_1}$ and $\mathcal{D}^{R_2}$ belong to the reality class b. Taking the category of the vector R into account, we conclude that case Ib holds. Using Table 12.6 and formula (12.7), we find that the small representations d^{R_1} and d^{R_2} are degenerate.

As expected, for crystals with the Cr_3Si structure, having O_h^3 space group, the results are less informative than for crystals with low symmetry, where time reversal leads to a significant band degeneracy.

12.5 Correspondence Between Atomic and Crystal States

12.5.1 Method

In many problems of solid state physics, it is necessary to make a correspondence between the electronic states in the crystal and in free atoms that form the crystal. The general method for establishing such a correspondence is described in detail in Chap. 11. The summary of the method is as follows.

Let $\{\varphi^s_{nlm}(\mathbf{r}-\boldsymbol{\tau}^s_t)\}$ be the initial set of atomic functions (n, l, m are atomic quantum numbers), and

$$\Phi^{\mathbf{k}st}_{nlm}(\mathbf{r}) = \sum_p \mathrm{e}^{\mathrm{i}\mathbf{k}(\mathbf{R}_p+\boldsymbol{\tau}^s_t)}\varphi^s_{nlm}(\mathbf{r}-\mathbf{R}_p-\boldsymbol{\tau}^s_t) \tag{12.11}$$

are combinations of these functions having the translational symmetry of the crystal. Similarly to Sect. 11.1.2, it can proved [43] that the set of functions $\Phi^{\mathbf{k}st}_{nlm}(\mathbf{r})$ for fixed $\mathbf{k}, s, n, l$ forms a basis of a certain, generally reducible, projective representation $\mathcal{D}^{\mathbf{k}sl}$ of the factor group $G_{\mathbf{k}}/T$ corresponding to the system of projective factors (12.3). The matrix elements of this reducible representation are

$$D^{\mathbf{k}sl}_{m't',mt}(\{\alpha|\boldsymbol{\tau}_\alpha\}) = D^{l\pm}_{m'm}(\alpha)\, \mathrm{e}^{\mathrm{i}(\mathbf{K}_{\alpha,\mathbf{k}}\boldsymbol{\tau}^s_{t'} - (\mathbf{k}+\mathbf{K}_{\alpha,\mathbf{k}})\boldsymbol{\tau}_\alpha)}\, \delta_{t',v(\alpha,s,t)}\,, \tag{12.12}$$

where $\mathcal{D}^{l\pm}$ is an irreducible representation of the full rotation group (plus sign for even l, minus sign for the odd ones); $\mathbf{K}_{\alpha,\mathbf{k}}$ is the reciprocal lattice vector in the relation $\alpha\mathbf{k} = \mathbf{k} + \mathbf{K}_{\alpha,\mathbf{k}}$; and $\delta_{t',v(\alpha,s,t)}$ is the Kronecker delta-symbol with $v(\alpha, s, t)$ determined by the equality

$$\boldsymbol{\tau}^s_v = \alpha\boldsymbol{\tau}^s_t + \boldsymbol{\tau}_\alpha - \mathbf{R}^s_{\alpha,t} \tag{12.13}$$

($\mathbf{R}^s_{\alpha,t}$ is the translation vector chosen so that the vector on the right-hand side of equality (12.13) belongs to the Wigner–Seitz cell (basis vector)).

The representation $\mathcal{D}^{\mathbf{k}sl}$ generally decomposes into irreducible projective components. The multiplicity $a^{\mathbf{k}sl}_j$, the number of times the irreducible projective representation $\Delta^{\mathbf{k}j}$ occurs in the representation $\mathcal{D}^{\mathbf{k}sl}$ (the multiplicity of occurrence), is calculated by the formula

$$a^{\mathbf{k}sl}_j = \frac{1}{n(G^0_{\mathbf{k}})} \sum_i c_i\, \chi^{\mathbf{k}sl}(\{\alpha_i|\boldsymbol{\tau}_{\alpha_i}\})\, \chi^{\mathbf{k}j*}(\{\alpha_i|\boldsymbol{\tau}_{\alpha_i}\}), \tag{12.14}$$

where c_i is the number of elements α in the i-th class of conjugate elements of the group $G^0_{\mathbf{k}}$, and $\chi^{\mathbf{k}sl}$ and $\chi^{\mathbf{k}j}$ are the characters of the matrices $D^{\mathbf{k}sl}$ and $\Delta^{\mathbf{k}j}$, respectively.

It remains to write down an expression for the character of the reduced representation $\mathcal{D}^{\mathbf{k}sl}$. According to (12.12), we have

$$\chi^{\mathbf{k}sl}(\{\alpha|\boldsymbol{\tau}_\alpha\}) = \chi^{l\pm}(\alpha)\, \mathrm{e}^{-\mathrm{i}(\mathbf{k}+\mathbf{K}_{\alpha,\mathbf{k}})\boldsymbol{\tau}_\alpha} \sum_t \mathrm{e}^{\mathrm{i}\mathbf{K}_{\alpha,\mathbf{k}}\boldsymbol{\tau}^s_t}\, \delta_{t',v(\alpha,s,t)}\,. \tag{12.15}$$

Here $\chi^{l\pm}(\alpha)$ is the character of the irreducible representation of the full rotation group calculated by the formulas

$$\chi^{l\pm}(\alpha') = \chi^l(\alpha'), \tag{12.16}$$
$$\chi^{l\pm}(\mathrm{i}\alpha') = (-1)^l\, \chi^l(\alpha'), \tag{12.17}$$

where[12]

---

[12] The values of the characters $\chi^l(\alpha')$ of the irreducible representations of the rotation group for rotations through the angles θ corresponding to the crystallographic point groups are given in Table 11.1.

$$\chi^l(\alpha') = \sum_{m=-l}^{l} e^{im\theta} = \frac{\sin(l+\frac{1}{2})\theta}{\sin\frac{1}{2}\theta}. \tag{12.18}$$

So, in order to establish a correspondence between atomic and crystal states, it is convenient to proceed as follows.

1. Use equality (12.13) for all α, s and t to find the values of $v(\alpha, s, t)$ that coincide with t. Indeed, according to (12.15), the basis atoms that contribute to $\chi^{\mathbf{k}sl}(\{\alpha|\boldsymbol{\tau}_\alpha\})$ are only the ones that do not change their position under the action of $\{\alpha|\boldsymbol{\tau}_\alpha\}$ (up to the reciprocal lattice vector).
2. Calculate the characters of the reducible representation $\chi^{\mathbf{k}sl}$ for the elements $\{\alpha_i|\boldsymbol{\tau}_{\alpha_i}\}$ taken one from each set of "conjugate" elements.[13] Since the characters $\chi^{\mathbf{k}sl}$ are used in expression (12.14), it is necessary to find them only for the elements $\{\alpha_i|\boldsymbol{\tau}_{\alpha_i}\}$ that have at least one irreducible representation such that the character $\chi^{\mathbf{k}j}(\{\alpha_i|\boldsymbol{\tau}_{\alpha_i}\})$ is nonzero. Next, before calculating the characters of the reducible representation $\chi^{\mathbf{k}sl}$ by formula (12.15) for the remaining elements $\{\alpha_i|\boldsymbol{\tau}_{\alpha_i}\}$, it is necessary to make sure that at least one of the basis atoms does not change its position under the action of the given $\{\alpha_i|\boldsymbol{\tau}_{\alpha_i}\}$ and that the character of the irreducible representation of the rotation group χ^l for α_i' is not equal to zero.
3. Using formula (12.14), find the multiplicity of occurrence $a_j^{\mathbf{k}sl}$ of each irreducible representation of the group of the wavevector $\mathbf{k}$.

Finally, note that it is sufficient to establish the correspondence by formulas (12.13)–(12.18) only at the points of high symmetry of the Brillouin zone. For the neighboring points, the correspondence is then easily obtained with the help of the compatibility relations between irreducible representations at points of high symmetry and irreducible representations on lines and planes adjacent to these points.

12.5.2 Results

Let us apply the above method to establish correspondence between the atomic and crystal states for the Cr_3Si structure.

First, for each element $\{\alpha|\boldsymbol{\tau}_\alpha\}$ (Table 12.2), going sequentially through all basis vectors $\boldsymbol{\tau}_t^s$ (Table 12.3), we determine the corresponding values of $v(\alpha, s, t)$ using formula (12.13). For example, applying the element $\{\alpha_{13}|\boldsymbol{\tau}_{\alpha_{13}}\}$ to the basis vector $\boldsymbol{\tau}_2^A$, we obtain $\alpha_{13}\boldsymbol{\tau}_2^A + \boldsymbol{\tau}_{\alpha_{13}} = a(3/4, 0, 1/2)$, which is equal to $\boldsymbol{\tau}_4^A$ up to the lattice vector $\mathbf{R}^A_{\alpha_{13},2} = a(1, 0, 0)$, thus $v(\alpha_{13}, A, 2) = 4$. All values of $v(\alpha, s, t)$ that coincide with t are presented in Table 12.29.

[13] By the "conjugate" elements we mean the elements of the group $G_{\mathbf{k}}$, whose rotational parts are conjugate in $G^0_{\mathbf{k}}$.

Table 12.29 Numbers of the basis atoms that do not change their position under the action of $\{\alpha|\boldsymbol{\tau}_\alpha\}$ (values of $v(\alpha, s, t)$ coinciding with t)

Element α	A atoms	B atoms	Element α	A atoms	B atoms
α_1	1, 2, 3, 4, 5, 6	1, 2	α_{25}	–	1, 2
α_2	3, 4	1, 2	α_{26}	1, 2, 5, 6	1, 2
α_3	1, 2	1, 2	α_{27}	3, 4, 5, 6	1, 2
α_4	5, 6	1, 2	α_{28}	1, 2, 3, 4	1, 2
α_5	–	1, 2	α_{29}	–	1, 2
α_6	–	1, 2	α_{30}	–	1, 2
α_7	–	1, 2	α_{31}	–	1, 2
α_8	–	1, 2	α_{32}	–	1, 2
α_9	–	1, 2	α_{33}	–	1, 2
α_{10}	–	1, 2	α_{34}	–	1, 2
α_{11}	–	1, 2	α_{35}	–	1, 2
α_{12}	–	1, 2	α_{36}	–	1, 2
α_{13}	5, 6	–	α_{37}	–	–
α_{14}	–	–	α_{38}	5, 6	–
α_{15}	–	–	α_{39}	5, 6	–
α_{16}	5, 6	–	α_{40}	–	–
α_{17}	3, 4	–	α_{41}	–	–
α_{18}	3, 4	–	α_{42}	–	–
α_{19}	–	–	α_{43}	3, 4	–
α_{20}	–	–	α_{44}	3, 4	–
α_{21}	1, 2	–	α_{45}	–	–
α_{22}	–	–	α_{46}	1, 2	–
α_{23}	1, 2	–	α_{47}	–	–
α_{24}	–	–	α_{48}	1, 2	–

Then, in full accord with the described methodology, we calculate the characters of the reducible representations $\chi^{\mathbf{k}sl}(\{\alpha_i|\boldsymbol{\tau}_{\alpha_i}\})$. Let us explain how this is done by the example of the symmetry point R and s-states of A atoms.[14]

At the point R, according to Table 12.2, one can take elements $\{\alpha_i|\boldsymbol{\tau}_{\alpha_i}\}$, where $i = 1, 2, 5, 13, 14, 25, 26, 29, 37, 38$. There is no need to calculate the reducible characters χ^{Rsl} for the elements with $i = 13, 14, 25, 26, 37, 38$ because the corresponding characters of all irreducible representations $\chi^{R,j}$ $(j = 1, 2, 3, 4)$ are equal to zero (see Table 12.6). In two of the four remaining cases, namely for $i = 5, 29$, none of the basis A atoms remains invariant (see Table 12.29), so that one needs to calculate by formula (12.15) only the characters $\chi^{RAl}(\{\alpha_i|\boldsymbol{\tau}_{\alpha_i}\})$ for $i = 1, 2$. It follows directly from formula (12.15) that $\chi^{\mathbf{k}sl}(\{\alpha_1|\boldsymbol{\tau}_{\alpha_1}\}) = (2l+1)t_s$, where t_s is the number of the s-sort atoms in the Wigner–Seitz cell, and in practice it is necessary

[14] Do not confuse the s-state $(l = 0)$ with the s-sort of atoms.

to calculate χ^{RAl} only for the element $\{\alpha_2|\boldsymbol{\tau}_{\alpha_2}\}$. Applying formula (12.15) and using Tables 12.2, 12.3 and 12.29, for $l = 0$ we obtain $\chi^{RA0}(\{\alpha_2|\boldsymbol{\tau}_{\alpha_2}\}) = -2$.

The reducible characters for the remaining values of **k**, s and l are calculated in a similar way.

Finally, calculating the characters of the reducible representations $\chi^{\mathbf{k}sl}$ by formula (12.14) for all necessary elements $\{\alpha_i|\boldsymbol{\tau}_{\alpha_i}\}$, we find the multiplicities of occurrence $a_j^{\mathbf{k}sl}$.

In the case of the s-states of A atoms at point R, we obtain: $a_j^{RA0} = 0$ for $j = 1, 2, 3$ and $a_4^{RA0} = 1$. That means the s-states, which in crystals with one atom per cell are always transformed according to the identical (one-dimensional) representation, in crystals of the Cr_3Si type at the point R are transformed according to a six-dimensional representation. The result is rather unusual.

Table 12.30 shows the calculated multiplicities of occurrence for all irreducible representations of the points Γ, R, M, and X in the reducible representations obtained from the s-, p-, d- and f-states of A and B atoms of the Cr_3Si-type crystals.

12.6 Concluding Remarks

Among the many works devoted to the calculation of energy bands of crystals with the Cr_3Si structure, the works of Mattheiss [12, 13] deserve special attention. In our opinion, these works represent the most comprehensive theoretical studies of the energy bands of these crystals. Therefore, it seems appropriate to compare the results of our group-theoretical analysis with the ones of Mattheiss.

Comparison of our results with the results of works [12, 13] reveals a number of differences. These differences are due to the following circumstances. For the small representations at points R, X, M, T and S, Mattheiss uses Gorzkowski's notation [10]; for representations at points Γ, Δ, Λ, Σ and Z—BSW notation. However, this does not take into account the fact that the irreducible Brillouin zone of Gorzkowski and BSW (and Mattheiss) is chosen differently (compare [10, Fig. 1], [37, Fig. 2], [13, Fig. 1b]). This leads to a number of inaccuracies in the identification of representations. Indeed, the same arms (of the star) correspond to the points Γ, R, M, T, Λ and Σ of Gorzkowski and BSW, but different—to the points X, Δ, Z and S (see Table 12.31). Therefore, before comparing the Gorzkowski and BSW representations, they had to be reduced to the same arms.

The transition from a small representation of one arm of a star to corresponding representation of another arm is carried out by formulas (5.29) and (5.30):

$$d^{\beta\mathbf{k},j}(\{\beta\alpha\beta^{-1}|\boldsymbol{\tau}_{\beta\alpha\beta^{-1}}\}) = \mathrm{e}^{\mathrm{i}\beta\mathbf{k}\mathbf{R}_{\alpha\beta}}\, d^{\mathbf{k}j}(\{\alpha|\boldsymbol{\tau}_\alpha\})\,, \tag{12.19}$$

where

$$\mathbf{R}_{\alpha\beta} = -\boldsymbol{\tau}_{\beta\alpha\beta^{-1}} - \beta\alpha\beta^{-1}\boldsymbol{\tau}_\beta + \beta\boldsymbol{\tau}_\alpha + \boldsymbol{\tau}_\beta\,, \tag{12.20}$$

Table 12.30 Correspondence between atomic and crystal states of electrons (multiplicity of occurrence $a_j^{\mathbf{k}sl}$) for the Cr_3Si structure†

	A atom				B atom			
	s	p	d	f	s	p	d	f
Γ_1^+	1		1	1	1			
Γ_2^+		1	1	1	1			
Γ_3^+	1	1	2	2			2	
Γ_4^+		1	1	3			1	
Γ_5^+		1	2	2			1	
Γ_1^-			1					1
Γ_2^-				1				1
Γ_3^-			1	1				
Γ_4^-		2	2	3		1		2
Γ_5^-	1	1	2	3		1		2
R_1		1	1	2			1	
R_2		1	1	2			1	
R_3		1	1	2	1			1
R_4	1	2	4	5		1	1	2
M_1	1	1	2	3		1		2
M_2		2	2	3		1		2
M_3		1	2	2			1	
M_4	1		2	2			1	
M_5	1	1	3	2		1		2
M_6	1	1	2	3		1		2
M_7		1	1	3			1	
M_8		1	2	2			1	
M_9	1	3	4	6	1		3	
M_{10}		2	3	5		1		3
X_1	2	3	5	6	1	1	2	2
X_2		1	3	4			1	1
X_3		3	3	6		1	1	2
X_4	1	2	4	5		1	1	2

†Empty positions mean $a_j^{\mathbf{k}sl} = 0$

$\{\alpha|\boldsymbol{\tau}_\alpha\} \in G_{\mathbf{k}}$, and $\{\beta|\boldsymbol{\tau}_\beta\}$ is the transformation corresponding to the arm $\beta\mathbf{k}$. Using (12.19) and (12.20) at the points X, Δ, Z and S, we pass from the small representations of the arm $\mathbf{k}$ given in Tables 12.8, 12.10, 12.14 and 12.15, to the corresponding small representations of the arm $\beta\mathbf{k}$ of the BSW irreducible Brillouin zone (according to Table 12.31 $\beta = \alpha_{37}$). The characters of the obtained representations (for elements $\beta\alpha\beta^{-1}$) at the points Δ and S coincide, and at the points X and Z do not coincide with the characters of the original representations (for the elements α). From the

Table 12.31 Coordinates of the points X, Δ, Z and S (in units of $2\pi/a$) in Gorzkowski (G) and BSW (and Mattheiss (M))

Point	G	BSW (and M)
X	(0, 1/2, 0)	(1/2, 0, 0)
Δ	(0, μ, 0)	(μ, 0, 0)
Z	(μ, 1/2, 0)	(1/2, μ, 0)
S	(μ,1/2, μ)	(1/2, μ, μ)
$0 < \mu < 1/2$		

Table 12.32 Characters of the small representations for the point Z in the irreducible Brillouin zone of BSW

	1	3	26	28
Z_1	1	−1	−1	1
Z_2	1	1	−1	−1
Z_3	1	−1	1	−1
Z_4	1	1	1	1

Table 12.33 Correspondence between the representations of Gorzkowski (G) and BSW (and Mattheiss (M)) at the point Z

G	BSW (and M)
Z_1	Z_3
Z_2	Z_2
Z_3	Z_4
Z_4	Z_1

comparison of Table 12.32 with table VI of BSW, it can be seen that Mattheiss has an inaccuracy at the point Z. The correct correspondence between the Gorzkowski and BSW (and Mattheiss) representations at the point Z is shown in Table 12.33. Next, at the point X Mattheiss uses Gorzkowski's notation, but, judging, for example, by the compatibility relations ([12, Table III]), he does this again without taking into account the difference in the arms of the star between him and Gorzkowski (compare the characters of the representations in Tables 12.8 and 12.34). Hence, what Gorzkowski calls X_3, Mattheiss calls X_4, and *vice versa* (see Table 12.35).

Finally, Mattheiss's compatibility table ([12, Table III]) contains two inaccuracies: the representation R_3 is compatible with the representations Λ_1 and Λ_2, and the representation M_s—with Σ_1.

Concluding the group-theoretical analysis, we note that its significance goes beyond the Cr_3Si structure. Most of the results are about the O_h^3 space group and will be valid for any crystal structure that has this space group.

Table 12.34 Characters of the small representations for the point X in the irreducible Brillouin zone of BSW

	1	2	3	4	17	18	19	20	25	26	27	28	41	42	43	44
X_1	2	2	0	0	0	0	0	0	0	0	2	2	0	0	0	0
X_2	2	2	0	0	0	0	0	0	0	0	−2	−2	0	0	0	0
X_3	2	−2	2	−2	0	0	0	0	0	0	0	0	0	0	0	0
X_4	2	−2	−2	2	0	0	0	0	0	0	0	0	0	0	0	0

Table 12.35 Correspondence between the representations of Gorzkowski (G) and Mattheiss (M) at the point X

G	M
X_1	X_1
X_2	X_2
X_3	X_4
X_4	X_3

References

1. L.R. Testardi, in *Physical Acoustics*, eds. W.P. Mason, R.N. Thurston (Academic, New York, 1973), vol. 10, pp. 193–296
2. M. Weger, I.B. Goldberg, in *Solid State Physics*, vol. 28, eds. H. Ehrenreich, F. Seitz, D. Turnbull (Academic, New York, 1974), pp.1–177
3. Y.A. Izyumov, E.Z. Kurmaev, Sov. Phys. Usp. **17**, 356 (1974)
4. S.V. Vonsovsky, Y.A. Izyumov, E. Kurmaev, *Superconductivity of Transition Metals, Their Alloys and Compounds* (Springer, Berlin, 1982)
5. N. Plakida, *High-Temperature Cuprate Superconductors* (Springer, Berlin, 2010)
6. Y. Izyumov, E. Kurmaev, *High-Tc Superconductors Based on FeAs Compounds* (Springer, Berlin, 2010)
7. G.R. Stewart, Physica C **514**, 28 (2015)
8. C.F. Pai, L. Liu, Y. Li, H.W. Tseng, D.C. Ralph, R.A. Buhrman, Appl. Phys. Lett. **101**, 122404 (2012)
9. Q. Hao, W. Chen, G. Xiao, Appl. Phys. Lett. **106**, 182403 (2015)
10. W. Gorzkowski, Phys. Stat. Sol. **3**, 910 (1963)
11. W. Gorzkowski, Phys. Stat. Sol. **6**, 521 (1964)
12. L.F. Mattheiss, Phys. Rev. **138**, A 112 (1965)
13. L.F. Mattheiss, Phys. Rev. B **12**, 2161 (1975)
14. B.N.N. Achar, G.R. Barsch, Phys. Stat. Sol. (b) **76**, 677 (1976)
15. N. van Huong, P. Do Tien, H. Kunert, M. Suffczynski, J. Phys. (Paris) **38**, 51 (1977)
16. N.A. Mahammedi, M. Ferhat, T. Tsumuraya, T. Chikyo, J. Appl. Phys. **122**, 205103 (2017)
17. H. Jones, *Theory of Brillouin Zones and Electronic States in Crystals* (North-Holland, Amsterdam, 1960) Reprinted: LLC, Whitefish, MT, 2013
18. R.S. Knox, A. Gold, *Symmetry in the Solid State* (Benjamin, New York, 1964)
19. H.W. Streitwolf, *Group Theory in Solid State Physics* (MacDonald, London, 1971)
20. G.L. Bir, G.E. Pikus, *Symmetry and Strain-induced Effects in Semiconductors* [in Russian] (Nauka, Moscow, 1972) Engl. transl.: Wiley, New York, 1974

21. C.J. Bradley, A.P. Cracknell, *The Mathematical Theory of Symmetry in Solids. Representation Theory for Point Groups and Space Groups* (Clarendon, Oxford, 1972). Reprinted: Clarendon, Oxford, 2010
22. A.P. Cracknell, Adv. Phys. **23**, 673 (1974)
23. B.I. Reser, Y.S. Mitrokhin, *Group theoretical analysis of a structure of the type* Cr_3Si [in Russian] (1977). Dep. VINITI, No. 1599-77
24. G. Hägg, N. Schönberg, Acta Cryst. **7**, 351 (1954)
25. S. Basavaiah, S.R. Pollack, J. Appl. Phys. **39**, 5548 (1968)
26. Y.G. Shen, Y.W. Mai, Mater. Sci. Eng. A **284**, 176 (2000)
27. F.T.N. Vüllers, R. Spolenak, Thin Solid Films **577**, 26 (2015)
28. B. Borén, Ark. Kemi Min. Geol. **11 A**, 351 (1933)
29. *Strukturbericht, Bd. I–VII* (Academische Verlagsgesellschaft, Leipzig, 1928–1939)
30. *Structure Reports, Vol. VIII* (N. V. A. Oosthoek's Uitgevers MIJ, Utrecht, 1940)
31. R.W.G. Wyckoff, *Crystal Structures*, 2nd edn., vols. 1–6 (Interscience, New York, 1963–1968)
32. *Landolt-Börnstein, New Series*, vols. III/6, III/7 (Springer, Berlin, 1971, 1973 and later eds.)
33. R.W.G. Wyckoff, *The Analytical Expression of the Results of the Theory of Space Groups* (Carnegie Institution of Washington, Washington, 1922, 1930)
34. S. Geller, Acta Cryst. **9**, 885 (1956)
35. M.I. Aroyo (ed.), *International Tables for Crystallography, Volume A, Space-group symmetry*, 6th edn. (Wiley, New York, 2016)
36. O.V. Kovalev, *Irreducible representations of space groups* [in Russian] (Acad. Sci. Ukrain. SSR, Kiev, 1961). Engl. transl.: Gordon and Breach, New York, 1965
37. L.P. Bouckaert, R. Smoluchowski, E. Wigner, Phys. Rev. **50**, 58 (1936)
38. N.F.M. Henry, K. Lonsdale (eds.), *International Tables for X-Ray Crystallography*, vol. I (Kynoch Press, Birmingham, 1952)
39. E.P. Wigner, Göttingen Nachricht. (1932), pp. 546–559
40. F.G. Frobenius, I. Schur, Sitzber. Preuss. Akad. Wiss. (Berlin, 1906), pp. 186–208
41. C. Herring, Phys. Rev. **52**, 361 (1937)
42. V. Frei, Czech. J. Phys. B **16**, 207 (1966)
43. B.I. Reser, Y.S. Mitrokhin, *Correspondence between the classifications of electronic states according to symmetry in atoms and crystals* [in Russian] (1976). Dep. VINITI, No. 4142-76

Chapter 13
Symmetry in the Tight-Binding Method

...state everything briefly but in detail. (V. Voinovich, Pretender to the Throne: Further Adventures of Private Ivan Chonkin, transl. by R. Lourie, Northwestern Univ. Press, Evanston, IL, 1995)

The tight-binding method, also called the linear combination of atomic orbitals (LCAO) method, is one of the first methods proposed for electron energy band calculations of solids (see, e.g. [1–3]). The most serious objection to considering it as an accurate method of computation is that we cannot evaluate correctly overlapping of the atomic wave functions that is essential for real crystals. Slater and Koster [4] have suggested to treat the various integrals which appear in the theory as disposable parameters, which may be determined by accurate calculations of energy bands at symmetry points in the Brillouin zone. The LCAO method is then used for making calculations throughout the Brillouin zone. This procedure is known as the interpolation method of Slater and Koster.

For crystals with high symmetry the number of parameters is quite low and naturally depends on the chosen set of atomic functions and the number of neighbors taken into consideration. The genetic relation between the interpolation method and the LCAO method allows us to make reasonable assumptions on the effective overlapping radius of atomic wave functions, which make it possible to neglect more distant neighbors.

For a given crystal structure and a fixed set of atomic functions, the Hamiltonian matrix components may be obtained in general form, so it is not necessary to carry out new calculations for every particular material.

Tables for cubic structures including s-, p- and d-states have been given in [4]. In these tables the authors used an explicit form of atomic wave functions. Such a straightforward procedure requires, however, a lot of calculations. Therefore, for complicated crystal structures (in particular for those having nonprimitive translations) and for a large number of the atomic wave functions, the procedure becomes so extremely cumbersome that the results become unreliable. In addition, this approach is in principle inapplicable if we consider spin–orbit coupling.

N. B. Melnikov and B. I. Reser, *Space Group Representations*,
https://doi.org/10.1007/978-3-031-13991-8_13

In papers [5, 6] a general method for calculating the Hamiltonian matrix components between two Bloch sums is proposed which is based on the application of group theory and complete symmetry consideration of the problem. This method works for any crystal structure and any chosen set of atomic wave functions, both with and without spin. The calculation procedure may be broken up into simple stages.

We illustrate the application of group-theoretical techniques in the tight-binding method [5, 6] in the example of the diamond structure. An illustration in the example of a simple cubic structure with one atom per unit cell (space group O_h^9) is given in [7] (for cubic structures, see also [8]; for the hcp structure, see [5, 6]). In Chap. 15 the tight-binding method is applied to the quasi-2D dichalcogenides with the $2H$-$TaSe_2$ structure, which has six atoms per unit cell.

13.1 General Method

13.1.1 Properties of Symmetry Operators

In the Seitz notation [9], any space transformation may be written in the form $\{\alpha|\,\mathbf{t}\}$ and is acting upon an arbitrary vector $\mathbf{r}$ in the following manner:

$$\{\alpha|\,\mathbf{t}\}\mathbf{r} = \alpha\mathbf{r} + \mathbf{t}. \tag{13.1}$$

Here α is a point transformation corresponding to rotation, inversion or a combination of both, and $\mathbf{t}$ is a translation vector. It is easy to verify by direct calculations that

$$\{\alpha|\mathbf{t}\}^{-1} = \{\alpha^{-1}|-\alpha^{-1}\mathbf{t}\}, \qquad \{\alpha|\mathbf{t}\}\{\beta|\mathbf{t}'\} = \{\alpha\beta|\alpha\mathbf{t}' + \mathbf{t}\}. \tag{13.2}$$

Space transformations $\{\alpha|\,\mathbf{t}\}$ in the (x, y, z) space induce operators $P_{\{\alpha|\mathbf{t}\}}$ in the space of functions of these variables. The action of such operators on a function $f(\mathbf{r})$ is defined by the relation

$$P_{\{\alpha|\mathbf{t}\}}f(\mathbf{r}) = f(\{\alpha|\mathbf{t}\}^{-1}\mathbf{r}). \tag{13.3}$$

Based on this definition, it is easy to verify that the operators have properties which are similar to (13.2):

$$P_{\{\alpha|\mathbf{t}\}^{-1}} = P_{\{\alpha^{-1}|-\alpha^{-1}\mathbf{t}\}}, \qquad P_{\{\alpha|\mathbf{t}\}}P_{\{\beta|\mathbf{t}'\}} = P_{\{\alpha\beta|\alpha\mathbf{t}'+\mathbf{t}\}}. \tag{13.4}$$

The complete set of space transformations, which transform the crystal into itself, forms a space symmetry group. The set of operators corresponding to the symmetry transformations of the space group also forms a group which we call symmetry operators group. It may be shown that these groups are equivalent (isomorphic) in terms of group theory.

In addition, it should be noted that the symmetry operators commute with the Hamiltonian H:

$$P_{\{\alpha|\mathbf{t}\}}H = HP_{\{\alpha|\mathbf{t}\}} \tag{13.5}$$

and may always be chosen to be unitary (i.e. transforming an orthogonal basis to an orthogonal basis).

13.1.2 Problem Setup and Notation

We will label atomic functions $\varphi(\mathbf{r})$ by two indices (j, μ) which indicate that a given function belongs to the μ-th row of the irreducible representation Γ_j of the point subgroup G_0'[1]. Thus we write

$$\varphi(\mathbf{r}) = \varphi_\mu^j(\mathbf{r}). \tag{13.6}$$

Then the Bloch sum may be rewritten in the form

$$\Phi_\mu^{sj}(\mathbf{r}) = \frac{1}{\sqrt{N}}\sum_n \mathrm{e}^{\mathrm{i}\mathbf{k}(\mathbf{R}_n+\boldsymbol{\tau}_s)}\varphi_\mu^j(\mathbf{r}-\mathbf{R}_n-\boldsymbol{\tau}_s) = \frac{1}{\sqrt{N}}\sum_n \mathrm{e}^{\mathrm{i}\mathbf{k}(\mathbf{R}_n+\boldsymbol{\tau}_s)}P_{\{\varepsilon|\mathbf{R}_n+\boldsymbol{\tau}_s\}}\varphi_\mu^j(\mathbf{r}), \tag{13.7}$$

where N is the total number of unit cells in the crystal, $\mathbf{k}$ is the reduced wavevector, $\mathbf{R}_n$ is the translation vector of the appropriate Bravais lattice, $\boldsymbol{\tau}_s$ is a basis vector (characterizing the position of the atom in the unit cell), and $P_{\{\varepsilon|\mathbf{R}_n+\boldsymbol{\tau}_s\}}$ is the translation operator.

Now let us form the expression for the Hamiltonian matrix component between two Bloch sums[2]:

$$\begin{aligned}\int_{\Omega_{\mathrm{WS}}} \Phi_\mu^{sj*}(\mathbf{r})H\Phi_{\mu'}^{s'j'}(\mathbf{r})\mathrm{d}\mathbf{r} &\equiv \left(\Phi_\mu^{sj}, H\Phi_{\mu'}^{s'j'}\right)\\ &= \mathrm{e}^{i\mathbf{k}(\boldsymbol{\tau}_{s'}-\boldsymbol{\tau}_s)}\sum_n \mathrm{e}^{i\mathbf{k}\mathbf{R}_n}I_{\mu\mu'}^{jj'}(\boldsymbol{\tau}_s;\mathbf{R}_n+\boldsymbol{\tau}_{s'}),\end{aligned} \tag{13.8}$$

where

$$\begin{aligned}I_{\mu\mu'}^{jj'}(\boldsymbol{\tau}_s;\mathbf{R}_n+\boldsymbol{\tau}_{s'}) &= \int \varphi_\mu^{j*}(\mathbf{r}-\boldsymbol{\tau}_s)H\varphi_{\mu'}^{j'}(\mathbf{r}-\mathbf{R}_n-\boldsymbol{\tau}_{s'})\mathrm{d}\mathbf{r}\\ &= \int \left[P_{\{\varepsilon|\boldsymbol{\tau}_s\}}\varphi_\mu^j(\mathbf{r})\right]^* H\left[P_{\{\varepsilon|\mathbf{R}_n+\boldsymbol{\tau}_{s'}\}}\varphi_{\mu'}^{j'}(\mathbf{r})\right]\mathrm{d}\mathbf{r}.\end{aligned} \tag{13.9}$$

[1] Group G_0' is a subgroup of the point group G_0 (crystal class) consisting of the proper and improper rotations α that correspond to the space transformations $\{\alpha|0\}$.

[2] Since the summation is performed over the entire crystal, one of the summation indices can be set equal to zero and the factor $1/N$ can be omitted.

The energy integrals (13.9) are not all independent. There is a set of relations between them which follow from the Hermiticity and symmetry properties of the Hamiltonian. Our goal is to find these relations in order to express the matrix components (13.8) as functions of a minimum number of independent parameters.

13.1.3 Hermiticity of the Hamiltonian

Hermiticity of H means

$$\left(\Phi_{\mu}^{sj}, H\Phi_{\mu'}^{s'j'}\right) = \left(\Phi_{\mu'}^{s'j'}, H\Phi_{\mu}^{sj}\right)^{*}, \tag{13.10}$$

hence the energy integrals satisfy the following relation:

$$I_{\mu\mu'}^{jj'}(\boldsymbol{\tau}_s; \mathbf{R}_n + \boldsymbol{\tau}_{s'}) = I_{\mu'\mu}^{j'j*}(\boldsymbol{\tau}_{s'}; -\mathbf{R}_n + \boldsymbol{\tau}_s). \tag{13.11}$$

13.1.4 Inversion Symmetry

Atomic functions always have a definite parity. Therefore, if the set of symmetry transformations of the crystal contains a transformation of the form $\{i|\boldsymbol{\tau}\}$, it is convenient to consider it separately. The operator $P_{\{i|\boldsymbol{\tau}\}}$ is a unitary one and commutes with the Hamiltonian; using these properties we can write

$$\begin{aligned} I_{\mu\mu'}^{jj'}(\boldsymbol{\tau}_s; \mathbf{R}_n + \boldsymbol{\tau}_{s'}) &= \int \left[P_{\{\varepsilon|\boldsymbol{\tau}-\boldsymbol{\tau}_s\}} P_{\{i|0\}} \varphi_{\mu}^{j}(\mathbf{r})\right]^{*} H \left[P_{\{\varepsilon|-\mathbf{R}_n+\boldsymbol{\tau}-\boldsymbol{\tau}_{s'}\}} P_{\{i|0\}} \varphi_{\mu'}^{j'}(\mathbf{r})\right] \mathbf{dr} \\ &= \omega^{j}\omega^{j'} I_{\mu\mu'}^{jj'}(\boldsymbol{\tau}-\boldsymbol{\tau}_s; -\mathbf{R}_n + \boldsymbol{\tau} - \boldsymbol{\tau}_{s'}). \end{aligned} \tag{13.12}$$

The index of ω takes the values $+1$ or -1 for even and odd functions, respectively. As is known, parity of an atomic function is given by $\omega = (-1)^l$, where l is the azimuthal quantum number [10]. For instance, for s-orbital ($l = 0$) we have $\omega = 1$, for p-orbital ($l = 1$) we have $\omega = -1$, etc.

13.1.5 Time-Reversal Symmetry

The time-reversal operator θ is one of the symmetry operators, hence

$$\theta H = H\theta. \tag{13.13}$$

Contrary to the space symmetry operators discussed above, this operator is an antiunitary one. Of particularly importance is to understand to which of the three types the initial irreducible representations belong: (potentially) real, complex or pseudoreal (for details, see Sect. 8.4).

As the time-reversal operator commutes with all the space symmetry operators and is antiunitary, we can obtain

$$\begin{aligned} I^{jj'}_{\mu\mu'}(\boldsymbol{\tau}_s; \mathbf{R}_n + \boldsymbol{\tau}_{s'}) &= \left(\int \left[P_{\{\varepsilon|\boldsymbol{\tau}_s\}}\theta\varphi^j_\mu(\mathbf{r})\right]^* H \left[P_{\{\varepsilon|\mathbf{R}_n+\boldsymbol{\tau}_{s'}\}}\theta\varphi^{j'}_{\mu'}(\mathbf{r})\right] \mathbf{dr}\right)^* \\ &= \sum_{\nu\nu'} D^j_{\nu\mu}(\theta) D^{j'*}_{\nu'\mu'}(\theta) I^{jj'*}_{\nu\nu'}(\boldsymbol{\tau}_s; \mathbf{R}_n + \boldsymbol{\tau}_{s'}), \end{aligned} \tag{13.14}$$

where we have taken into account that the functions $\varphi(\mathbf{r})$ are transformed by the operator θ as

$$\theta\varphi^j_\mu(\mathbf{r}) = \sum_{\mu'} D^j_{\mu'\mu}(\theta)\varphi^j_{\mu'}(\mathbf{r}). \tag{13.15}$$

Here $D^j(\theta)$ means the matrix of the time-reversal operator in the j-th corepresentation. We choose the number of corepresentation corresponding to the number of the considered representation. To get concrete relations, the knowledge of concrete form of the matrix $D(\theta)$ is necessary.

We restrict ourselves to considering the case when the Hamiltonian is independent of spin (for the case with spin, see [5, 6]). In this case, the wave functions are also independent of spin and are transformed according to single-valued representations of the group G'_0. We consider potentially real representations, since all representations of the group G'_0 used below are of this type. As is known, potentially real representations can be reduced by a unitary transformation to the real form. Once this has been done, the matrix of the time-reversal operator may be chosen in form of the unit matrix:

$$D(\theta) = \mathbf{1}. \tag{13.16}$$

Note that this finally uniquely determines the choice of the basis functions, which were defined earlier only up to a common phase factor by specifying irreducible representations. Substituting (13.16) into (13.14), we obtain the following relation between the energy integrals:

$$I^{jj'}_{\mu\mu'}(\boldsymbol{\tau}_s; \mathbf{R}_n + \boldsymbol{\tau}_{s'}) = I^{jj'*}_{\mu\mu'}(\boldsymbol{\tau}_s; \mathbf{R}_n + \boldsymbol{\tau}_{s'}). \tag{13.17}$$

13.1.6 Relations Between the Matrix Components

From inversion symmetry (13.12) and time-reversal symmetry (13.17) it follows that

$$\begin{aligned}\left(\Phi_{\mu}^{\tau_s,j}, H\Phi_{\mu'}^{\tau_{s'},j'}\right) &= \omega^j \omega^{j'} e^{-i\mathbf{k}[(\tau-\tau_{s'})-(\tau-\tau_s)]} \\ &\quad \sum_n e^{-i\mathbf{k}\mathbf{R}_n} I_{\mu\mu'}^{jj'*}(\tau-\tau_s; \mathbf{R}_n+\tau-\tau_{s'}) \\ &= \omega^j \omega^{j'} \left(\Phi_{\mu}^{\tau-\tau_s,j}, H\Phi_{\mu'}^{\tau-\tau_{s'},j'}\right)^* .\end{aligned} \tag{13.18}$$

Combining this relation with (13.10), we have

$$\left(\Phi_{\mu}^{\tau_s,j}, H\Phi_{\mu'}^{\tau_{s'},j'}\right) = \omega^j \omega^{j'} \left(\Phi_{\mu'}^{\tau-\tau_{s'},j'}, H\Phi_{\mu}^{\tau-\tau_s,j}\right). \tag{13.19}$$

In this chapter and the next one, we will impose the following additional restrictions[3]:

(a) we consider only crystals with two atoms per (Wigner–Sietz) unit cell: $\boldsymbol{\tau}_s = \boldsymbol{\tau}_1,\ \boldsymbol{\tau}_2$;
(b) we assume that $\boldsymbol{\tau}_1$ is equal to zero (the origin is at the atom of the first type) and $\boldsymbol{\tau}_2$ is equal to $\boldsymbol{\tau}$.

In this case, from (13.19) it follows that it is sufficient to calculate only the matrix components of the form $\left(\Phi_{\mu}^{1j}, H\Phi_{\mu'}^{1j'}\right)$ and $\left(\Phi_{\mu}^{1j}, H\Phi_{\mu'}^{2j'}\right)$ for $j \le j'$, $\mu \le \mu'$. All other matrix components may be obtained from the matrix components of this form with the help of relation (13.19). Thus, the number of matrix components to be computed is considerably reduced.

13.1.7 Point Symmetry

So far we have obtained the relations between matrix elements (13.8) following from Hermiticity of the Hamiltonian and symmetry of the system. Hence we have reduced the number of the matrix elements that must be calculated. The next step is to simplify the expressions for the matrix components of the Hamiltonian.

To obtain the relations between the energy integrals (13.9), it is necessary to consider the effect of space symmetry of the system. It is sufficient to discuss only the point symmetry, because the inversion was considered above and the translation symmetry gives only a trivial permutation of the lattice site indices.

[3] The case without these restrictions is discussed in Chap. 15.

Let $P_{\{\alpha|0\}}$ be a symmetry operator, where α is an element of the subgroup G_0' of the point group G_0. Since the operator $P_{\{\alpha|0\}}$ is unitary and commutes with the Hamiltonian, we obtain

$$\begin{aligned} I_{\mu\mu'}^{jj'}(\boldsymbol{\tau}_s; \mathbf{R}_n + \boldsymbol{\tau}_{s'}) &= \int \left[P_{\{\varepsilon|\alpha\boldsymbol{\tau}_s\}} P_{\{\alpha|0\}} \varphi_\mu^j(\mathbf{r})\right]^* H \left[P_{\{\varepsilon|\alpha\mathbf{R}_n+\alpha\boldsymbol{\tau}_{s'}\}} P_{\{\alpha|0\}} \varphi_{\mu'}^{j'}(\mathbf{r})\right] \mathrm{d}\mathbf{r} \\ &= \sum_{\nu\nu'} D_{\nu\mu}^{j*}(\alpha) D_{\nu'\mu'}^{j'}(\alpha) I_{\nu\nu'}^{jj'}(\alpha\boldsymbol{\tau}_s; \alpha\mathbf{R}_n + \alpha\boldsymbol{\tau}_{s'}). \end{aligned} \tag{13.20}$$

Here $D^j(\alpha)$ is the matrix of the element $\{\alpha|0\}$ in the j-th irreducible representation of the subgroup G_0', i.e.

$$P_{\{\alpha|0\}} \varphi_\mu^j(\mathbf{r}) = \sum_\nu D_{\nu\mu}^j(\alpha) \varphi_\nu^j(\mathbf{r}). \tag{13.21}$$

13.1.8 Calculation of the Matrix Components

Let us rewrite expression (13.8) for the Hamiltonian matrix component between two Bloch sums as follows. We split the sum over $\mathbf{R}_n$ into two sums, one over the coordination spheres[4] ($p = 1, 2, \ldots$ is the number of the sphere) centered at the atom at the origin, and the other over the vectors of a given coordination sphere. For a given coordination sphere ($l = 1, 2, \ldots$ is the number of the vector), we can always choose a minimum set of vectors $(\mathbf{R} + \boldsymbol{\tau}_{s'})_p^l$ (though not uniquely) such that any vector of a coordination sphere is transformed into one of the vectors of this set by a transformation $\alpha \equiv \{\alpha|0\}$ from the subgroup G_0':

$$\alpha(\mathbf{R}_n + \boldsymbol{\tau}_{s'}) = (\mathbf{R} + \boldsymbol{\tau}_{s'})_p^l. \tag{13.22}$$

For brevity we will write $\mathbf{Q}_p^l$ instead of $(\mathbf{R} + \boldsymbol{\tau}_{s'})_p^l$. There can be several α from G_0' that transform a given vector $\mathbf{R}_n + \boldsymbol{\tau}_{s'}$ into $\mathbf{Q}_p^l$; it is sufficient to take one of them. As a result, (13.8) is written as

$$\left(\Phi_\mu^{sj}, H\Phi_{\mu'}^{s'j'}\right) = \mathrm{e}^{-\mathrm{i}\mathbf{k}\boldsymbol{\tau}_s} \sum_p \sum_l \sum_\alpha \mathrm{e}^{\mathrm{i}\mathbf{k}\alpha^{-1}\mathbf{Q}_p^l} I_{\mu\mu'}^{jj'}(\boldsymbol{\tau}_s; \alpha^{-1}\mathbf{Q}_p^l). \tag{13.23}$$

Next we take the point symmetry into account. Using (13.20), we have

$$\left(\Phi_\mu^{sj}, H\Phi_{\mu'}^{s'j'}\right) = \mathrm{e}^{-\mathrm{i}\mathbf{k}\boldsymbol{\tau}_s} \sum_p \sum_l \sum_{\nu\nu'} \left[\sum_\alpha \mathrm{e}^{\mathrm{i}\alpha\mathbf{k}\mathbf{Q}_p^l} D_{\nu\mu}^{j*}(\alpha) D_{\nu'\mu'}^{j'}(\alpha) I_{\nu\nu'}^{jj'}(\alpha\boldsymbol{\tau}_s; \mathbf{Q}_p^l)\right]. \tag{13.24}$$

[4] The coordination sphere is formed by the atoms having equal distance from the origin.

In the case of two atoms per unit cell, we have to calculate only the matrix elements for $\boldsymbol{\tau}_s = \boldsymbol{\tau}_1$, and we have set $\boldsymbol{\tau}_1 = 0$, so we write (13.24) as

$$\left(\Phi_{\mu}^{1j}, H\Phi_{\mu'}^{s'j'}\right) = \sum_{p}\sum_{l}\sum_{\nu\nu'} I_{\nu\nu'}^{jj'}(0; \mathbf{Q}_p^l)\left[\sum_{\alpha} \mathrm{e}^{\mathrm{i}\alpha\mathbf{k}\mathbf{Q}_p^l} D_{\nu\mu}^{j*}(\alpha) D_{\nu'\mu'}^{j'}(\alpha)\right]. \quad (13.25)$$

13.1.9 Determination of the Independent Parameters

Expression (13.25) involves only energy integrals with fixed vectors $\mathbf{Q}_p^l$. Now we need to determine the relations between these integrals. We do this in the case of two atoms per unit cell.

Let us consider transformations α of the point subgroup G_0' that do not change the vector $\mathbf{Q}_p^l$. It is easily established that they form a group; we call it a group of the vector $\mathbf{Q}_p^l$ and denote it by G_p^l. For α belonging to G_p^l the relation between the energy integrals with a fixed vector $\mathbf{Q}_p^l$ follows from (13.20):

$$I_{\mu\mu'}^{jj'}(\mathbf{Q}_p^l) = \sum_{\nu\nu'} D_{\nu\mu}^{j*}(\alpha) D_{\nu'\mu'}^{j'}(\alpha) I_{\nu\nu'}^{jj'}(\mathbf{Q}_p^l), \quad (13.26)$$

where

$$I_{\mu\mu'}^{jj'}(\mathbf{Q}_p^l) \equiv I_{\mu\mu'}^{jj'}(0; \mathbf{Q}_p^l). \quad (13.27)$$

It can be readily shown [5] that we may consider only the relations for the generators of the group G_p^l.

Generally, the system of equations (13.26) for the generators and relation (13.17), which follows from time-reversal symmetry, make up a complete set of relations between the energy integrals with fixed vectors $\mathbf{Q}_p^l$.

The exception is the case when we calculate energy integrals for functions which are transformed according to the same representation Γ_j of the group G_0'. If this representation can be reduced to irreducible representations of G_p^l in such a way that at least two of them are equal, then additional nontrivial relations appear. Namely, for $s' = 2$ from (13.11) and (13.12) it follows that

$$I_{\mu\mu'}^{jj}(0; (\mathbf{R} + \boldsymbol{\tau}_{s'})_p^l) = \omega\omega' I_{\mu'\mu}^{jj*}(0; (\mathbf{R} + \boldsymbol{\tau}_{s'})_p^l). \quad (13.28)$$

For $s' = 1$, when G_0' includes β which transforms $(\mathbf{R} + \boldsymbol{\tau}_{s'})_p^l$ into $-(\mathbf{R} + \boldsymbol{\tau}_{s'})_p^l$, from (13.11) and (13.20) for β we obtain

$$I_{\mu\mu'}^{jj}(0; (\mathbf{R} + \boldsymbol{\tau}_{s'})_p^l) = \sum_{\nu\nu'} D_{\nu\mu}^{j*}(\beta) D_{\nu'\mu'}^{j}(\beta) I_{\nu'\nu}^{jj*}(0; (\mathbf{R} + \boldsymbol{\tau}_{s'})_p^l). \quad (13.29)$$

Using these relations, we can express the energy integrals in terms of the minimum number of independent parameters. Repeating the described procedure for all $\mathbf{Q}_p^l$ and all representations Γ_j and $\Gamma_{j'}$ (under the condition $j \leq j'$) and using formula (13.25), we finally obtain the required expressions for the matrix elements of the Hamiltonian on the Bloch sums, which will include only independent parameters (energy integrals).

13.1.10 Calculation Algorithm

(a) Determine the point subgroup G_0' of the space group of the crystal. Find its irreducible representations and arrange them into a certain order. If the set of initial functions is fixed, it is sufficient to chose the representations according to which these functions transform. Define the transformation of an arbitrary vector under the operation of α from G_0'.
(b) Having classified the atoms into coordination spheres, choose a set of fixed vectors $\mathbf{Q}_p^l$ for every coordination sphere (see Sect. 13.1.8).
(c) For every $\mathbf{Q}_p^l$ find the group G_p^l of all transformations α that do not change this vector. Then find the generators for this group.
(d) For every $\mathbf{R}_n + \boldsymbol{\tau}_{s'} \neq \mathbf{Q}_p^l$ of a given coordination sphere choose one transformation α that transforms $\mathbf{R}_n + \boldsymbol{\tau}_{s'}$ to $\mathbf{Q}_p^l$ (see (13.22)).
(e) Write down equations (13.26) for the transformations defined in (c). By solving these equations express all the energy integrals in terms of independent integrals. Use (13.17) for the independent integrals.
(f) Compute the inner sums in (13.25):

$$\sum_{\alpha} \mathrm{e}^{\mathrm{i}\alpha\mathbf{k}\mathbf{Q}_p^l} D_{\nu\mu}^{j*}(\alpha) D_{\nu'\mu'}^{j'}(\alpha) \,, \tag{13.30}$$

where α runs over the set of transformations determined by (d).
(g) Substitute the results from (e) and (f) into (13.25) and obtain the final expressions for the matrix components.
(h) Determine all other matrix components with the help of formulas from Sect. 13.1.6.

Note: It is efficient to choose the irreducible representations of the group G_0' in such a way that for appropriate α they would be reduced to irreducible representations for a maximum number of groups of the vectors $\mathbf{Q}_p^l$. In this case, it is considerably easier to point out the independent parameters (see Sect. 13.1.9). Actually, in this case the functions φ_μ^j, which are classified according to the irreducible representations of the group G_0', will form the set of functions of the irreducible representations of the groups G_p^l. Then $\varphi_\mu^j = \varphi_\nu^i$, where ν is the number of the row of the i-th irreducible representation of G_p^l.

13.2 Application to the Diamond Structure

In this section the general method discussed above is applied to the diamond structure. We have chosen this structure because of two reasons. Firstly, the majority of semiconductors for which it is reasonable to calculate the electron energy bands in the tight-binding approximation have practically the diamond structure. Secondly, this structure represents a sufficiently complicated example for illustrating the idea of our method (large number of functions and complex symmetry elements).

In the case of the diamond structure the space group G is O_h^7 and the point subgroup G_0' is T_d, which consists of the elements α_1-α_{12}, α_{37}-α_{48} of the crystal class O_h (see Table 10.1). We will be interested in the valence bands and the lowest conduction bands formed by the outer s- and p-atomic states. The atomic s-function transforms according to the irreducible representation Γ_1, and p-function transforms according to the irreducible representation Γ_5 of the point group T_d.[5] The matrices of the representation Γ_5 are the matrices of elements α_1-α_{12} and α_{37}-α_{48} in representation Γ_{25}' (or Γ_{15}), given in Table F.2. The way the components of an arbitrary vector are transformed under the action of $\alpha \in T_d$ is shown in Table 10.1.

The space group O_h^7 contains the inversion transformation, which, in our choice of the origin, is written in the form $\{i|\tau\}$.

We will restrict ourselves to the first three coordination spheres. As the fixed vectors on the spheres, we choose $\mathbf{Q}_1 = a\left(\frac{1}{2}, \frac{1}{2}, \frac{1}{2}\right)$, $\mathbf{Q}_2 = a(110)$ and $\mathbf{Q}_3 = a\left(\frac{3}{2}\frac{\bar{1}}{2}\frac{1}{2}\right)$. (In our case, it suffices to take one fixed vector from each sphere, so the index l of the vectors Q is omitted.)

The transformations that leave the vector $\mathbf{Q}_p$ unchanged, and the transformations that map all the vectors of a given coordination sphere into $\mathbf{Q}_p$, are shown in Table 13.1. The same table contains decompositions of the representations Γ_j of the group G_0' into irreducible representations of the group G_p.

Table 13.1 Coordination spheres of the diamond structure

No. p of the coord. sphere	Fixed vector $\mathbf{Q}_p$ from coord. sphere	Tranform. α leaving the vector $\mathbf{Q}_p$ unchanged	Group G_p of the vector $\mathbf{Q}_p$	Decomp. of reps. of the subgroup G_0' into reps. of the group G_p		Tranform. α mapping vectors of the coord. sphere to $\mathbf{Q}_p$
				Γ_1	Γ_5	
1	$a\left(\frac{1}{2}\frac{1}{2}\frac{1}{2}\right)$	$\alpha_1, \alpha_5, \alpha_9,$ $\alpha_{37}, \alpha_{41}, \alpha_{45}$	C_{3v}	Γ_1	$\Gamma_1 + \Gamma_3$	α_1–α_4
2	$a(110)$	α_1, α_{37}	C_{1h}	Γ_1	$2\Gamma_1 + \Gamma_2$	α_1–α_{12}
3	$a\left(\frac{3}{2}\frac{\bar{1}}{2}\frac{1}{2}\right)$	α_1, α_{42}	C_{1h}	Γ_1	$2\Gamma_1 + \Gamma_2$	α_1–α_{12}

[5] Irreducible representations of the point groups are given in Heine's notation [11].

In the decomposition of the representation Γ_5 of the group T_d into irreducible representations of the group C_{1h} there are two identical representations. Therefore, to determine independent parameters, in addition to relations (13.26), we use relation (13.28) for $p = 1, 3$ and relation (13.29) for $p = 2$.

Let us illustrate how to determine independent parameters using the example of $p = 2$.

From (13.26) for α_{37} we obtain

$$I_{11}^{15}(110) = I_{12}^{15}(110), \quad I_{11}^{55}(110) = I_{22}^{55}(110), \quad I_{12}^{55}(110) = I_{21}^{55}(110),$$
$$I_{13}^{55}(110) = I_{23}^{55}(110), \quad I_{31}^{55}(110) = I_{32}^{55}(110).$$

Relation (13.29) for the transformation α_{40} mapping (110) to $(\bar{1}\bar{1}0)$ yields[6]

$$I_{13}^{55}(110) = -I_{32}^{55}(110), \quad I_{23}^{55}(110) = -I_{31}^{55}(110).$$

Comparison of the obtained equalities allows us to take the following ones as independent integrals: $I_{11}^{15}(110)$, $I_{11}^{55}(110)$, $I_{12}^{55}(110)$ and $I_{31}^{55}(110)$. These integrals together with $I_{11}^{11}(110)$, $I_{13}^{15}(110)$ and $I_{33}^{55}(110)$ form a complete set of independent parameters for the considered coordination sphere. In accord with (13.17), all the integrals are real.

The final expressions for the matrix elements of the Hamiltonian, calculated by formula (13.25), are presented in Table 13.2. For brevity, in Table 13.2, we present only a part of the matrix elements. The rest can be obtained from them using the relations[7]

$$\left.\begin{aligned} H_{13} &= H_{12}, \\ H_{17} &= H_{16}, \\ H_{33} &= H_{22}, \\ H_{34} &= H_{23}, \\ H_{37} &= H_{26}, \\ H_{38} &= H_{27}, \end{aligned}\right\} (\xi \to \eta \to \zeta \to \xi), \qquad \left.\begin{aligned} H_{14} &= H_{12}, \\ H_{18} &= H_{16}, \\ H_{24} &= H_{23}^{*}, \\ H_{28} &= H_{27}, \\ H_{44} &= H_{22}, \\ H_{48} &= H_{26}, \end{aligned}\right\} (\xi \to \zeta \to \eta \to \xi). \tag{13.31}$$

The Bloch sums Φ_{μ}^{sj} are numbered in the order shown in Table 13.3. Table 13.2 and relations (13.31) give expressions for the matrix elements schematically shown in Fig. 13.1 (indicated by hatching). Relations (13.18) and (13.19) allow to express all the remaining matrix elements in terms of the hatched ones.

[6] The transformation α_4 yields the identity: $I_{13}^{55}(110) = I_{13}^{55}(110)$.

[7] Here the notation $\xi \to \eta \to \zeta \to \xi$ means that, for example, H_{13} is obtained from H_{12} by replacing ξ by η, η by ζ and ζ by ξ.

Table 13.2 Matrix components of the Hamiltonian for the diamond structure

H_{11}	$I_{11}^{11}(000)+4I_{11}^{11}(110)(\cos\xi\cos\eta+\cos\eta\cos\zeta+\cos\xi\cos\zeta)$
H_{12}	$-4I_{13}^{15}(110)\sin\eta\sin\zeta+4iI_{11}^{15}(110)(\sin\xi\cos\eta+\sin\xi\cos\zeta)$
H_{15}	$4I_{11}^{11}\left(\frac{1}{2}\frac{1}{2}\frac{1}{2}\right)\left(\cos\frac{\xi}{2}\cos\frac{\eta}{2}\cos\frac{\zeta}{2}-i\sin\frac{\xi}{2}\sin\frac{\eta}{2}\sin\frac{\zeta}{2}\right)$
	$+4I_{11}^{11}\left(\frac{3}{2}\frac{\bar{1}}{2}\frac{1}{2}\right)\Big[\cos\frac{3}{2}\xi\cos\frac{\eta}{2}\cos\frac{\zeta}{2}+\cos\frac{\xi}{2}\cos\frac{3}{2}\eta\cos\frac{\zeta}{2}$
	$+\cos\frac{\xi}{2}\cos\frac{\eta}{2}\cos\frac{3}{2}\zeta+i\Big(\sin\frac{3}{2}\xi\sin\frac{\eta}{2}\sin\frac{\zeta}{2}$
	$+\sin\frac{\xi}{2}\sin\frac{3}{2}\eta\sin\frac{\zeta}{2}+\sin\frac{\xi}{2}\sin\frac{\eta}{2}\sin\frac{3}{2}\zeta\Big)\Big]$
H_{16}	$4I_{11}^{15}\left(\frac{1}{2}\frac{1}{2}\frac{1}{2}\right)\left(i\sin\frac{\xi}{2}\cos\frac{\eta}{2}\cos\frac{\zeta}{2}-\cos\frac{\xi}{2}\sin\frac{\eta}{2}\sin\frac{\zeta}{2}\right)$
	$+4I_{11}^{15}\left(\frac{3}{2}\frac{\bar{1}}{2}\frac{1}{2}\right)\left(i\sin\frac{3}{2}\xi\cos\frac{\eta}{2}\cos\frac{\zeta}{2}+\cos\frac{3}{2}\xi\sin\frac{\eta}{2}\sin\frac{\zeta}{2}\right)$
	$-4I_{12}^{15}\left(\frac{3}{2}\frac{\bar{1}}{2}\frac{1}{2}\right)\Big[i\left(\sin\frac{\xi}{2}\cos\frac{3}{2}\eta\cos\frac{\zeta}{2}+\sin\frac{\xi}{2}\cos\frac{\eta}{2}\cos\frac{3}{2}\zeta\right)$
	$+\cos\frac{\xi}{2}\sin\frac{3}{2}\eta\sin\frac{\zeta}{2}+\cos\frac{\xi}{2}\sin\frac{\eta}{2}\sin\frac{3}{2}\zeta\Big]$
H_{22}	$I_{11}^{55}(000)+4I_{11}^{55}(110)(\cos\xi\cos\eta+\cos\xi\cos\zeta)+4I_{33}^{55}(110)\cos\eta\cos\zeta$
H_{23}	$-4I_{12}^{55}(110)\sin\xi\sin\eta-4iI_{31}^{55}(110)(\cos\xi\sin\zeta-\cos\eta\sin\zeta)$
H_{26}	$4I_{11}^{55}\left(\frac{1}{2}\frac{1}{2}\frac{1}{2}\right)\left(\cos\frac{\xi}{2}\cos\frac{\eta}{2}\cos\frac{\zeta}{2}-i\sin\frac{\xi}{2}\sin\frac{\eta}{2}\sin\frac{\zeta}{2}\right)$
	$+4I_{11}^{55}\left(\frac{3}{2}\frac{\bar{1}}{2}\frac{1}{2}\right)\left(\cos\frac{3}{2}\xi\cos\frac{\eta}{2}\cos\frac{\zeta}{2}+i\sin\frac{3}{2}\xi\sin\frac{\eta}{2}\sin\frac{\zeta}{2}\right)$
	$+4I_{22}^{55}\left(\frac{3}{2}\frac{\bar{1}}{2}\frac{1}{2}\right)\Big[\cos\frac{\xi}{2}\cos\frac{3}{2}\eta\cos\frac{\zeta}{2}+\cos\frac{\xi}{2}\cos\frac{\eta}{2}\cos\frac{3}{2}\zeta$
	$+i\left(\sin\frac{\xi}{2}\sin\frac{3}{2}\eta\sin\frac{\zeta}{2}+\sin\frac{\xi}{2}\sin\frac{\eta}{2}\sin\frac{3}{2}\zeta\right)\Big]$
H_{27}	$4I_{12}^{55}\left(\frac{1}{2}\frac{1}{2}\frac{1}{2}\right)\left(i\cos\frac{\xi}{2}\cos\frac{\eta}{2}\sin\frac{\zeta}{2}-\sin\frac{\xi}{2}\sin\frac{\eta}{2}\cos\frac{\zeta}{2}\right)$
	$+4I_{12}^{55}\left(\frac{3}{2}\frac{\bar{1}}{2}\frac{1}{2}\right)\Big[i\left(\cos\frac{3}{2}\xi\cos\frac{\eta}{2}\sin\frac{\zeta}{2}+\cos\frac{\xi}{2}\cos\frac{3}{2}\eta\sin\frac{\zeta}{2}\right)$
	$+\sin\frac{3}{2}\xi\sin\frac{\eta}{2}\cos\frac{\zeta}{2}+\sin\frac{\xi}{2}\sin\frac{3}{2}\eta\cos\frac{\zeta}{2}\Big]$
	$+4I_{23}^{55}\left(\frac{3}{2}\frac{\bar{1}}{2}\frac{1}{2}\right)\left(i\cos\frac{\xi}{2}\cos\frac{\eta}{2}\sin\frac{3}{2}\zeta+\sin\frac{\xi}{2}\sin\frac{\eta}{2}\cos\frac{3}{2}\zeta\right)$

$\xi=ak_x,\quad \eta=ak_y,\quad \zeta=ak_z$

Table 13.3 Bloch sums numbering

Number	1	2	3	4	5	6	7	8
Bloch sum	Φ_1^{11}	Φ_1^{15}	Φ_2^{15}	Φ_3^{15}	Φ_1^{21}	Φ_1^{25}	Φ_2^{25}	Φ_3^{25}

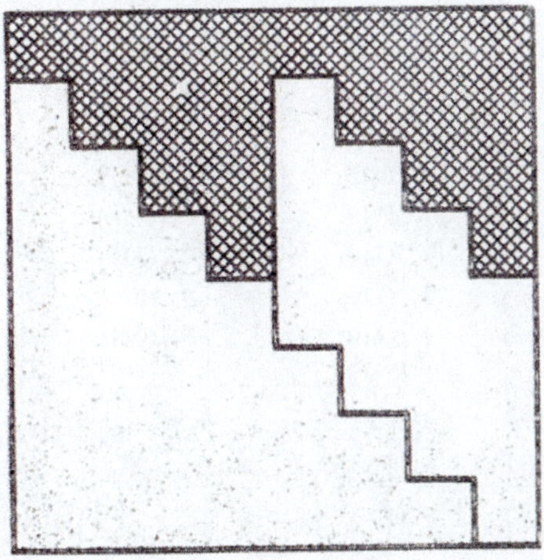

Fig. 13.1 Schematic form of the 8 × 8 Hamiltonian matrix. The matrix elements given by Table 13.2 and relations (13.31) are indicated by hatching

Matrix elements of the Hamiltonian on Bloch sums for the diamond structure, given in Table 13.2, were obtained using only the first three coordination spheres (in the third neighbor approximation). However, using the technique described in Sect. 13.1, it is easy to determine the matrix elements for any number of coordination spheres.

Note that the matrix elements for the diamond structure in the second neighbor approximation can be found in [4], and in the third neighbor approximation—in [12]. Unfortunately, mistakes were made in both works (see [13]). Comparison of our notation for energy parameters with the notation of [4, 12, 14] is given in Table 13.4.

The form of the matrix elements of the Hamiltonian for the diamond structure in the case with spin was obtained in [6]. In [15], a formalism was developed that allows one to express the relativistic energy integrals in terms of the nonrelativistic ones and hence to simplify the interpolation of electron bands in the relativistic case.

Table 13.4 Notation for the energy parameters

	E–R–S [1]	S–K [2]	D–D [3]	A–J [4]
x_1	$I_{11}^{11}(000)$	$E_{ss}(000)$	$\gamma_{01} - 4\gamma_{11} - 6\gamma_{21}$	h_{01}
x_2	$I_{11}^{55}(000)$	$E_{xx}(000)$	$\gamma_{02} - 4\gamma_{12} - 6\gamma_{22}$	h_{02}
x_3	$I_{11}^{11}\left(\frac{1}{2}\frac{1}{2}\frac{1}{2}\right)$	$E_{ss}\left(\frac{1}{2}\frac{1}{2}\frac{1}{2}\right)$	γ_{11}	h_{11}
x_4	$I_{11}^{15}\left(\frac{1}{2}\frac{1}{2}\frac{1}{2}\right)$	$E_{sx}\left(\frac{1}{2}\frac{1}{2}\frac{1}{2}\right)$	γ_{14}	h_{12}
x_5	$I_{11}^{55}\left(\frac{1}{2}\frac{1}{2}\frac{1}{2}\right)$	$E_{xx}\left(\frac{1}{2}\frac{1}{2}\frac{1}{2}\right)$	γ_{12}	h_{13}
x_6	$I_{12}^{55}\left(\frac{1}{2}\frac{1}{2}\frac{1}{2}\right)$	$E_{xy}\left(\frac{1}{2}\frac{1}{2}\frac{1}{2}\right)$	γ_{13}	h_{14}
x_7	$I_{11}^{11}(110)$	$E_{ss}(110)$	$\frac{1}{2}\gamma_{21}$	h_{21}
x_8	$I_{11}^{15}(110)$	$E_{sx}(110)$	$\frac{1}{2}\gamma_{24}$	h_{22}
x_9	$I_{13}^{15}(110)$	$E_{sx}(011)$	$\frac{1}{2}\gamma_{26}$	h'_{22}
x_{10}	$I_{11}^{55}(110)$	$E_{xx}(110)$	$\frac{1}{2}(\gamma_{22} - \gamma_{25})$	h_{23}
x_{11}	$I_{12}^{55}(110)$	$E_{xy}(110)$	$\frac{1}{2}\gamma_{23}$	h_{24}
x_{12}	$I_{31}^{55}(110)$	$E_{xy}(011)$	$\frac{1}{2}\gamma_{27}$ [5]	h'_{24}
x_{13}	$I_{33}^{55}(110)$	$E_{xx}(011)$	$\frac{1}{2}(\gamma_{22} + \gamma_{25})$	h'_{23}
x_{14}	$I_{11}^{11}\left(\frac{3}{2}\frac{\bar{1}}{2}\frac{1}{2}\right)$			h_{31}
x_{15}	$I_{11}^{15}\left(\frac{3}{2}\frac{\bar{1}}{2}\frac{1}{2}\right)$			$-h_{32}$
x_{16}	$I_{12}^{15}\left(\frac{3}{2}\frac{\bar{1}}{2}\frac{1}{2}\right)$			h'_{32}
x_{17}	$I_{11}^{55}\left(\frac{3}{2}\frac{\bar{1}}{2}\frac{1}{2}\right)$			h_{33}
x_{18}	$I_{12}^{55}\left(\frac{3}{2}\frac{\bar{1}}{2}\frac{1}{2}\right)$			h_{34}
x_{19}	$I_{22}^{55}\left(\frac{3}{2}\frac{\bar{1}}{2}\frac{1}{2}\right)$			h'_{33}
x_{20}	$I_{23}^{55}\left(\frac{3}{2}\frac{\bar{1}}{2}\frac{1}{2}\right)$			h'_{34}

[1] Ref. [5]
[2] Ref. [4]
[3] Ref. [14]. In [14], Appendix A, the parameters γ associated with the third and even fourth neighbors are introduced, but they are not used in the work itself
[4] Ref. [12]. In [12], nonorthogonalized atomic orbitals are used, and the parameters h have a different meaning compared to the corresponding values in the previous papers
[5] See [13], footnote at p. 4

References

1. W.A. Harrison, *Electronic Structure and the Properties of Solids*, 2nd edn. (Dover, New York, 1989)
2. J. Callaway, *Quantum Theory of the Solid State*, 2nd edn. (Academic, New York, 1991)
3. R.M. Martin, *Electronic structure: Basic Theory and Practical Methods*, 2nd edn. (Cambridge Univ. Press, Cambridge, 2020)
4. J.C. Slater, G.F. Koster, Phys. Rev. **94**, 1498 (1954)
5. R.F. Egorov, B.I. Reser, V.P. Shirokovskii, Phys. Stat. Sol. **26**, 391 (1968)
6. R.F. Egorov, B.I. Reser, V.P. Shirokovskii, Trudy Instituta Fiziki Metallov [in Russian] **27**, 98 (1968)

7. W. Hergert, R.M. Geilhufe, *Group Theory in Solid State Physics and Photonics: Problem Solving with Mathematica* (Wiley, New York, 2018)
8. R.F. Egorov, B.I. Reser, V.P. Shirokovskii, Fiz. Metallov i Metallovedenie [in Russian] **24**, 49 (1967)
9. F. Seitz, Ann. Math. **37**, 17 (1936)
10. L.D. Landau, E.M. Lifshitz, *Quantum Mechanics—Non-Relativistic Theory (Course of Theoretical Physics)*, vol. 3, 3rd edn. (Pergamon, Oxford, 1991)
11. V. Heine, *Group Theory in Quantum Mechanics* (Pergamon, Oxford, 1960). Reprinted: Dover, New York, 2009
12. I. Alstrup, K. Johansen, Phys. Stat. Sol. **28**, 555 (1968)
13. B.I. Reser, V.P. Shirokovskii, *Interpolation of the energy bands of semiconductors with the diamond structure on the basis of the tight binding method* [in Russian] (1971). Dep. VINITI, No. 2471–71
14. G. Dresselhaus, M. Dresselhaus, Phys. Rev. **160**, 649 (1967)
15. R.F. Egorov, V.K. Zvezdin, B.I. Reser, V.P. Shirokovskii, Phys. Stat. Sol. **39**, K93 (1970)

Chapter 14
Interpolation of Energy Bands

Throughout my life, I have found it best to seek physics problems whose solutions seem initially simple... Solving physics problems that are exacting from the first often becomes a hopeless undertaking. (The Recollections of Eugene P. Wigner as told to Andrew Szanton, Plenum, New York, 1992)

Slater and Koster [1] suggested using the tight-binding method as an interpolation scheme. It was suggested that the energy integrals considered as disposable parameters are calculated from the known eigenvalues at a few high symmetry **k**-points. By carrying out this analysis Slater and Koster were able to find expressions for energy integrals in the two-center approximation. Tables of matrix components for energy were presented for cubic and diamond structures by [1] and hexagonal close-packed (hcp) structure by [2]. The Slater-Koster parameters in the two- and three-center approximation for a large number of simple elements with the fcc, bcc, hcp and diamond crystal structures are tabulated in [3]. A way to combine the tight-binding interpolation with the density functional theory (DFT) was proposed in [4] (for a review, see, e.g. [5]).

Here we describe the interpolation scheme based on a general group-theoretical method that gives the matrix elements of the Hamiltonian in the tight-binding approximation using space symmetry of the problem, time-reversal symmetry and Hermiticity of the Hamiltonian (see Chap. 13). The interpolation of electron bands is illustrated in the example of the diamond structure [6].

14.1 Quasidiagonal Form of the Hamiltonian Matrix

Here we use the symmetry considerations in the tight-binding method to interpolate the energy bands $\varepsilon(\mathbf{k})$. In the Slater-Koster scheme [1], the interpolation process is divided into two steps. In the first step, we determine the disposable parameters

N. B. Melnikov and B. I. Reser, *Space Group Representations*,
https://doi.org/10.1007/978-3-031-13991-8_14

(energy integrals) that are present in the matrix elements $H_{ij}(\mathbf{k})$[1]. For this, we use the known values of $\varepsilon_n(\mathbf{k})$ at high symmetry points of the Brillouin zone, where the Hamiltonian matrix can be reduced to a quasidiagonal form. The block-diagonal form of the matrix greatly simplifies the process of determining the parameters. In the second step, we solve the characteristic equation

$$|H_{ij}(\mathbf{k}) - \varepsilon(\mathbf{k})\delta_{ij}| = 0 \tag{14.1}$$

with the obtained parameter values to calculate the eigenvalues $\varepsilon_n(\mathbf{k})$ in the entire Brillouin zone. Here $H_{ij}(\mathbf{k})$ is the matrix element of the Hamiltonian between two Bloch sums (constructed to be orthogonal to each other).

In order to make the matrix $H_{ij}(\mathbf{k})$ block-diagonal at a point $\mathbf{k}$, we need to combine the initial Bloch sums into linear combinations belonging to different rows of a small representation of the group of the wavevector $G_{\mathbf{k}}$, i.e. we wish to construct the *symmetrized* Bloch sums. Just as in the case of plane waves, this is done using the projection operator (8.20).

Let us write an expression for the symmetrized Bloch sum, which belongs to the ρ-th row of the J-th small representation of the group $G_{\mathbf{k}}$:

$$\begin{aligned}
\Phi^{sjJ}_{\mu\rho} &= \frac{l_J}{n(G^0_{\mathbf{k}})} \sum_{\{\alpha_m|\boldsymbol{\tau}_{\alpha m}\}} \Delta^{\mathbf{k}J*}_{\rho\rho}(\{\alpha_m|\boldsymbol{\tau}_{\alpha_m}\}) P_{\{\alpha_m|\boldsymbol{\tau}_{\alpha m}\}}\, \Phi^{sj}_{\mu}(\mathbf{r}) \\
&= \frac{l_J}{n(G^0_{\mathbf{k}})} \sum_{\{\alpha_m|\boldsymbol{\tau}_{\alpha m}\}} \Delta^{\mathbf{k}J*}_{\rho\rho}(\{\alpha_m|\boldsymbol{\tau}_{\alpha_m}\}) P_{\{\alpha_m|\boldsymbol{\tau}_{\alpha m}\}} \sum_n \mathrm{e}^{i\mathbf{k}(\mathbf{R}_n+\boldsymbol{\tau}_s)} P_{\{\varepsilon|\mathbf{R}_n+\boldsymbol{\tau}_s\}}\varphi^j_{\mu}(\mathbf{r}) \\
&= \frac{l_J}{n(G^0_{\mathbf{k}})} \sum_{\{\alpha_m|\boldsymbol{\tau}_{\alpha m}\}} \Delta^{\mathbf{k}J*}_{\rho\rho}(\{\alpha_m|\boldsymbol{\tau}_{\alpha_m}\}) \\
&\quad \times \sum_n \mathrm{e}^{i\mathbf{k}(\mathbf{R}_n+\boldsymbol{\tau}_s)} P_{\{\varepsilon|\alpha_m\mathbf{R}_n+\alpha_m\boldsymbol{\tau}_s+\boldsymbol{\tau}_{\alpha m}\}} P_{\{\alpha_m|0\}}\varphi^j_{\mu}(\mathbf{r}).
\end{aligned} \tag{14.2}$$

If we denote $\alpha_m\mathbf{R}_n + \alpha_m\boldsymbol{\tau}_s + \boldsymbol{\tau}_{\alpha_m}$ by $\mathbf{R}_{n'} + \boldsymbol{\tau}_{s'}$, then

$$\mathbf{R}_n + \boldsymbol{\tau}_s = \alpha_m^{-1}(\mathbf{R}_{n'} + \boldsymbol{\tau}_{s'} - \boldsymbol{\tau}_{\alpha_m}). \tag{14.3}$$

Taking (14.3) and (13.21) into account, we write (14.2) as

[1] The subscript j of $H_{ij}(\mathbf{k})$ should not be confused with the superscript of the representation j of the Bloch sum Φ^{sj}_{μ}.

$$\Phi^{sjJ}_{\mu\rho} = \frac{l_J}{n(G^0_{\mathbf{k}})} \sum_{\{\alpha_m|\tau_{\alpha m}\}} \Delta^{\mathbf{k}J*}_{\rho\rho}(\{\alpha_m|\boldsymbol{\tau}_{\alpha_m}\})\, \mathrm{e}^{-\mathrm{i}\mathbf{k}\boldsymbol{\tau}_{\alpha m}}\, \mathrm{e}^{\mathrm{i}\mathbf{K}_{\alpha m}(\boldsymbol{\tau}_{s'}-\boldsymbol{\tau}_{\alpha m})} \sum_{\nu} D^{j}_{\nu\mu}(\alpha_m)$$
$$\times \sum_{n'} \mathrm{e}^{\mathrm{i}\mathbf{k}(\mathbf{R}_{n'}+\boldsymbol{\tau}_{s'})} P_{\{\varepsilon|\mathbf{R}_{n'}+\boldsymbol{\tau}_{s'}\}}\varphi^{j}_{\mu}(\mathbf{r}) \tag{14.4}$$
$$= \frac{l_J}{n(G^0_{\mathbf{k}})} \sum_{\{\alpha_m|\tau_{\alpha m}\}} \Delta^{\mathbf{k}J*}_{\rho\rho}(\{\alpha_m|\boldsymbol{\tau}_{\alpha_m}\})\, \mathrm{e}^{-\mathrm{i}\mathbf{k}\boldsymbol{\tau}_{\alpha m}}\, \mathrm{e}^{\mathrm{i}\mathbf{K}_{\alpha m}(\boldsymbol{\tau}_{s'}-\boldsymbol{\tau}_{\alpha m})} \sum_{\nu} D^{j}_{\nu\mu}(\alpha_m)\, \Phi^{s'j}_{\nu},$$

where $\mathbf{K}_{\alpha_m} = \alpha_m\mathbf{k} - \mathbf{k}$. It is clear that at each point of the Brillouin zone, from the Bloch sums Φ^{sj}_{μ} with a fixed j we can construct symmetrized combinations only for those irreducible representations of the group of the wavevector that are present in the reducible representation generated by the functions Φ^{sj}_{μ}.

In the diamond structure, symmetrized combinations can be constructed from the Bloch sums corresponding to the outer s- and p-atomic functions for the representations indicated in Table 14.1.

Table 14.1 Distribution of the Bloch sums constructed from outer s- and p-atomic functions over irreducible representations of the groups of the wavevectors in the diamond structure

Symmetry point	Bloch sums	
	s	p
	$\Phi^{11}_{1}, \Phi^{21}_{1}$	$\Phi^{15}_{\mu}, \Phi^{25}_{\mu}, \quad \mu = 1, 2, 3$
Γ	$\Gamma_1 + \Gamma'_2$	$\Gamma'_{25} + \Gamma_{15}$
X	X_1	$X_1 + X_3 + X_4$
L	$L_1 + L'_2$	$L_1 + L'_2 + L_3 + L'_3$
W	W_1	$W_1 + 2W_2$
K	$K_1 + K_3$	$2K_1 + K_2 + 2K_3 + K_4$
U	$U_1 + U_3$	$2U_1 + U_2 + 2U_3 + U_4$
Δ	$\Delta_1 + \Delta'_2$	$\Delta_1 + \Delta'_2 + 2\Delta_5$
Λ	$2\Lambda_1$	$2\Lambda_1 + 2\Lambda_3$
Σ	$\Sigma_1 + \Sigma_3$	$2\Sigma_1 + \Sigma_2 + 2\Sigma_3 + \Sigma_4$
Z	Z_1	$3Z_1$
S	$S_1 + S_3$	$2S_1 + S_2 + 2S_3 + S_4$
Q	$Q_1 + Q_2$	$3Q_1 + 3Q_2$
LK	$2(LK)_1$	$4(LK)_1 + 2(LK)_2$
LU	$2(LU)_1$	$4(LU)_1 + 2(LU)_2$
KW	$(KW)_1 + (KW)_2$	$3(KW)_1 + 3(KW)_2$
WU	$(WU)_1 + (WU)_2$	$3(WU)_1 + 3(WU)_2$
$\Gamma\Lambda\Delta$	$2(\Gamma\Lambda\Delta)_1$	$4(\Gamma\Lambda\Delta)_1 + 2(\Gamma\Lambda\Delta)_2$
$\Gamma\Sigma\Lambda$	$2(\Gamma\Sigma\Lambda)_1$	$4(\Gamma\Sigma\Lambda)_1 + 2(\Gamma\Sigma\Lambda)_2$
$\Gamma\Delta\Sigma$	$(\Gamma\Delta\Sigma)_1 + (\Gamma\Delta\Sigma)_2$	$3(\Gamma\Delta\Sigma)_1 + 3(\Gamma\Delta\Sigma)_2$
WUX	$(WUX)_1 + (WUX)_2$	$3(WUX)_1 + 3(WUX)_2$

Explicit form of the blocks of the Hamiltonian matrix for the diamond structure at all points of high symmetry of the Brillouin zone is given in Appendix G. Since the blocks corresponding to different rows of the same representation are identical, in Appendix G we present only the blocks obtained from symmetrized Bloch sums that belong to the first row of irreducible representations. In calculations we used property (8.21), which makes it possible to take only one symmetrized function when calculating the matrix element.

14.2 Tight-Binding Method and kp-Method

The method of calculating the energy bands in the neighborhood of a point in the **k**-space using the perturbation theory is called the **kp**-method (see, e.g. [7, 8])[2]. Let us establish a connection between the energy integrals of the tight-binding method and the constants of the **kp**-method. Consider the point $\mathbf{k} = 0$ of the Brillouin zone of the diamond structure. (For any other structure and point **k**, the line of reasoning is the same.)

Let us construct symmetrized combinations of the initial Bloch sums corresponding to s- and p-atomic functions. In the previous section, when we calculated the matrix elements, it was sufficient to have functions that transform according to the first row of the J-th small representations of the group $G_{\mathbf{k}}$. Now with the help of the operator (8.20),

$$\tilde{O}^{\mathbf{k}J}_{\rho 1} = \frac{l_J}{n(G^0_{\mathbf{k}})} \sum_{\{\alpha_m|\boldsymbol{\tau}_{\alpha_m}\}} \Delta^{\mathbf{k}J*}_{\rho 1}(\{\alpha_m|\boldsymbol{\tau}_{\alpha_m}\}) P_{\{\alpha_m|\boldsymbol{\tau}_{\alpha_m}\}} \quad (\rho = 2, ..., l_J), \tag{14.5}$$

we find the partner functions. For the diamond structure, the functions f_i $(i = 1, 2, ..., 8)$ belonging to the same irreducible representations of the point Γ are as follows:

$$\begin{aligned}
&\Gamma_1\colon\ f_1 = \frac{1}{2}(\Phi^{11}_1 + \Phi^{21}_1);\\
&\Gamma'_2\colon\ f_2 = \frac{1}{2}(\Phi^{11}_1 - \Phi^{21}_1);\\
&\Gamma'_{25}\colon\ f_3 = \frac{1}{2}(\Phi^{15}_1 - \Phi^{25}_1),\quad f_4 = \frac{1}{2}(\Phi^{15}_2 - \Phi^{25}_2),\quad f_5 = \frac{1}{2}(\Phi^{15}_3 - \Phi^{25}_3);\\
&\Gamma_{15}\colon\ f_6 = \frac{1}{2}(\Phi^{15}_1 + \Phi^{25}_1),\quad f_7 = \frac{1}{2}(\Phi^{15}_2 + \Phi^{25}_2),\quad f_8 = \frac{1}{2}(\Phi^{15}_3 + \Phi^{25}_3).
\end{aligned} \tag{14.6}$$

We write down the Hamiltonian matrix between the functions (14.6), expand its elements h_{ij} $(i, j = 1, 2, ..., 8)$ in power series and restrict ourselves to the terms of order at most k^2. Thus, we obtain

[2] For a detailed exposition of the **kp**-method see, e.g. [9–11].

$$
\begin{aligned}
h_{11} &= H_{11}(0) + H_{15}(0) + \left(-\frac{1}{2}x_3 - 4x_7 - \frac{11}{2}x_{14}\right)(\xi^2 + \eta^2 + \zeta^2),\\
h_{12} &= h_{36} = h_{47} = h_{58} = 0,\\
h_{13} &= (x_4 - 4x_9 - x_{15} + 6x_{16})\eta\zeta,\\
h_{16} &= i(2x_4 + 8x_8 + 6x_{15} - 4x_{16})\xi,\\
h_{22} &= H_{11}(0) - H_{15}(0) + \left(\frac{1}{2}x_3 - 4x_7 + \frac{11}{2}x_{14}\right)(\xi^2 + \eta^2 + \zeta^2),\\
h_{23} &= i(-2x_4 + 8x_8 - 6x_{15} + 4x_{16})\xi,\\
h_{26} &= (-x_4 - 4x_9 + x_{15} - 6x_{16})\eta\zeta,\\
h_{33} &= H_{22}(0) - H_{26}(0) + \left(\frac{1}{2}x_5 - 4x_{10} + \frac{9}{2}x_{17} + x_{19}\right)\xi^2\\
&\quad + \left(\frac{1}{2}x_5 - 2x_{10} - 2x_{13} + \frac{1}{2}x_{17} + 5x_{19}\right)(\eta^2 + \zeta^2),\\
h_{34} &= (x_6 - 4x_{11} - 6x_{18} - x_{20})\,\xi\eta,\\
h_{37} &= h_{46} = i(2x_6 + 4x_{18} + 6x_{20})\zeta,\\
h_{66} &= H_{22}(0) + H_{26}(0) + \left(-\frac{1}{2}x_5 - 4x_{10} - \frac{9}{2}x_{17} - x_{19}\right)\xi^2\\
&\quad + \left(-\frac{1}{2}x_5 - 2x_{10} - 2x_{13} - \frac{1}{2}x_{17} - 5x_{19}\right)(\eta^2 + \zeta^2),\\
h_{67} &= (-x_6 - 4x_{11} + 6x_{18} + x_{20})\xi\eta,
\end{aligned}
\tag{14.7}
$$

$$
\begin{aligned}
&\left.\begin{array}{ll}
h_{14} = h_{13}, & h_{17} = h_{16}\\
h_{24} = h_{23}, & h_{27} = h_{26}\\
h_{44} = h_{33}, & h_{45} = h_{34}\\
h_{48} = h_{37}, & h_{57} = h_{37}\\
h_{77} = h_{66}, & h_{78} = h_{67}
\end{array}\right\}\ \xi \to \eta \to \zeta \to \xi,\\
&\left.\begin{array}{ll}
h_{15} = h_{13}, & h_{18} = h_{16}\\
h_{25} = h_{23}, & h_{28} = h_{26}\\
h_{35} = h_{34}, & h_{38} = h_{37}\\
h_{55} = h_{33}, & h_{56} = h_{37}\\
h_{68} = h_{67}, & h_{88} = h_{66}
\end{array}\right\}\ \xi \to \zeta \to \eta \to \xi.
\end{aligned}
\tag{14.8}
$$

Here $x_1, x_2, \ldots$ denote the energy integrals. The notation in (14.8) is explained in Table 13.4 and (13.31).

Now we will use the Löwdin perturbation theory [12]. We divide the states that define the Hamiltonian matrix (with the elements $h_{\mu\nu}$) into two classes, A and B, so that the class A includes the states corresponding to a representation Γ_j at the point Γ, and class B—all the others. Then, according to [12], the interactions between the states A and states B are eliminated if the matrix elements in A are recalculated by the formula

$$\bar{h}_{\alpha\alpha'} = h_{\alpha\alpha'} + \sum_{\beta} \frac{h_{\alpha\beta} h_{\beta\alpha'}}{\varepsilon_\alpha - h_{\beta\beta}} \qquad (\alpha, \alpha' \in A, \quad \beta \in B), \tag{14.9}$$

under the assumption that $|h_{\alpha\beta}| \ll |\varepsilon_\alpha - \varepsilon_\beta|$. Taking Γ_1, Γ'_2, Γ'_{25} and Γ_{15} one by one as the representation Γ_j and calculating the matrix elements $\bar{h}_{\alpha\alpha'}$ for each case, we obtain the Hamiltonian matrix in the quasidiagonal form.

The one-dimensional block, corresponding to representation Γ_1 or Γ'_2, immediately gives the dispersion law and hence an expression for the effective mass as the curvature of the band.

The three-dimensional block, corresponding to the representation Γ'_{25} or Γ_{15}, has the form

$$\begin{vmatrix} L\xi^2 + M(\eta^2 + \zeta^2) - \lambda & N\xi\eta & N\xi\zeta \\ N\xi\eta & L\eta^2 + M(\zeta^2 + \xi^2) - \lambda & N\eta\zeta \\ N\xi\zeta & N\eta\zeta & L\zeta^2 + M(\xi^2 + \eta^2) - \lambda \end{vmatrix} = 0 \tag{14.10}$$

($\lambda = \varepsilon(\mathbf{k}) - \varepsilon(\Gamma_j)$), i.e. it is identical to the determinant of the **kp** perturbation theory [13]. The quantities L, M, N in equation (14.10) and quantities L', M', N' in the paper [13] are related by

$$L' = a^2 L - 1, \qquad M' = a^2 M - 1, \qquad N' = a^2 N. \tag{14.11}$$

Knowing the quantities L', M', N', one can calculate the constants in the **kp**-approximation:

$$A = \frac{1}{3}(L' + 2M') + 1, \qquad B = \frac{1}{3}(L' - M'), \qquad C = \sqrt{\frac{1}{3}\left[(N')^2 - (L' - M')^2\right]}. \tag{14.12}$$

Let us give an explicit form of the quantities L, M, N corresponding to representation Γ'_{25} and the effective mass m^* corresponding to the representation Γ'_2:

$$\begin{aligned} L &= \frac{1}{2}x_5 - 4x_{10} + \frac{9}{2}x_{17} + x_{19} + \frac{4(-x_4 + 4x_8 - 3x_{15} + 2x_{16})^2}{\varepsilon(\Gamma'_{25}) - \varepsilon(\Gamma'_2)}, \\ M &= \frac{1}{2}x_5 - 2x_{10} - 2x_{13} + \frac{1}{2}x_{17} + 5x_{19} + \frac{4(x_6 + 2x_{18} + 3x_{20})^2}{\varepsilon(\Gamma'_{25}) - \varepsilon(\Gamma_{15})}, \\ N &= x_6 - 4x_{11} - 6x_{18} - x_{20} + \frac{4(-x_4 + 4x_8 - 3x_{15} + 2x_{16})^2}{\varepsilon(\Gamma'_{25}) - \varepsilon(\Gamma'_2)} \\ &\quad + \frac{4(x_6 + 2x_{18} + 3x_{20})^2}{\varepsilon(\Gamma'_{25}) - \varepsilon(\Gamma_{15})}, \end{aligned} \tag{14.13}$$

$$m^* = \left(\frac{\partial^2 \varepsilon(\mathbf{k})}{\partial \mathbf{k}^2}\right)^{-1} \tag{14.14}$$

$$= \left\{ a^2 \left[\frac{1}{2} x_3 - 4x_7 + \frac{11}{2} x_{14} + \frac{4(-x_4 + 4x_8 - 3x_{15} + 2x_{16})^2}{\varepsilon(\Gamma_2') - \varepsilon(\Gamma_{25}')} \right] \right\}^{-1} .$$

In the second neighbor approximation, formulas similar to (14.13) and (14.14) were obtained using a different formalism in the paper [14].

14.3 An Example: The Diamond Structure

At each high symmetry point (see Appendix G), we obtain a system of equations for calculating the unknown energy parameters $x_1, x_2, \ldots$ for the determinant equation (14.1) by using the following relationships between the matrix elements H_{ij} of the Hamiltonian H and its eigenvalues ε_i (see, e.g. [15, Sect. II-1-7] or [16, Sect. 1.2]):

$$\sum_{i=1}^{z} \varepsilon_i = \mathrm{Sp}\, H, \tag{14.15}$$

$$\sum_{1 \le i < j \le z} \varepsilon_i \varepsilon_j = \sum_{l=1}^{z-1} \begin{vmatrix} H_{l,l} & H_{l,l+1} \\ H_{l+1,l} & H_{l+1,l+1} \end{vmatrix} \tag{14.16}$$

In order to interpolate all eight energy bands simultaneously, first we write down all linear and quadratic equations at the chosen high symmetry points (reference points). Then we use the least squares method and replace the overdetermined linear subsystems with normal subsystems (with square matrices). We think that this approach is justified, because it makes the calculation scheme universal and does not require any modification when we pass from one specific material to another. At the same time, we tried to use as few reference points as possible.

The interpolation method has been implemented in two forms.

- **Second neighbor approximation**. In the second neighbor approximation, the matrix elements of the Hamiltonian $H_{ij}(\mathbf{k})$ depend on 13 parameters $x_1, x_2, \ldots, x_{13}$. Taking the points Γ, X and L as the reference points, we determine the parameters $x_1, \ldots, x_7, x_9, x_{10}, x_{11}, x_{13}$. To determine x_8, we use the point $\Delta(0\frac{1}{2}0)$, and for x_{12} we use $\Sigma\left(\frac{1}{2}\frac{1}{2}0\right)$. First we obtain the values of x_8 and x_{12} from each quadratic equation (given by the determinants Δ_1, Δ_2' and Σ_1, Σ_3, respectively), then we calculate their arithmetic mean.
- **Third neighbor approximation**. In this case, the number of the unknown parameters becomes equal to 20. As the reference points, we choose Γ, X, L, $\Delta(0\frac{1}{2}0)$ and $\Sigma\left(\frac{1}{2}\frac{1}{2}0\right)$. The resulting system consists of 18 linear and 8 quadratic equations. After easy algebraic manipulations and application of the least squares method for linear equations, the system is split into three subsystems containing all unknown parameters. Subsystem I consists of 7 linear and one quadratic equations for the unknowns $x_3, x_5, x_6, x_{14}, x_{17}, \ldots, x_{20}$; subsystem II—of 5 linear

equations for $x_1, x_2, x_7, x_{10}, x_{11}, x_{13}$; and subsystem III—of 7 quadratic equations. First, we solve subsystem I, then we solve simultaneously subsystems II and III by the steepest descent method. As the initial approximation of the unknowns, we take the corresponding values obtained in the second neighbor approximation.

For crystals with the diamond structure, interpolation of electron bands in the tight-binding method was carried out in [17] (see also [6, 18]). The energy values at the reference points were taken from [19], which are, in our opinion, one of the best band calculations for diamond, silicon and germanium.

First, the interpolation is performed in the second neighbor approximation. The calculation results show that, despite good qualitative agreement with [19], there are noticeable quantitative discrepancies, and, what is especially important, they occur near the edge of the forbidden band.

To study how the possible inaccuracy of far-apart energy values affects the region near the forbidden band, i.e. how sensitive the obtained picture is to these values, we have alternately perturbed the energy of the terms $L_{2'}^{(1)}$, $L_{2'}^{(2)}$, X_3 and $X_1^{(1)}$ within their admissible limits. Note that x_8 is determined only by the value of $\Delta_1^{(2)}$, and x_{12}—only by $\Sigma_3^{(2)}$, i.e. by the energy values in the studied area. Calculations have shown that the existing quantitative discrepancies in the energy bands cannot be explained by the possible inaccuracy of the indicated energy values. For all perturbations, the overall picture of the energy bands does not change significantly.

At the next stage of the calculation, we take into account the interaction with the third neighbors. The point is that 12 neighbors of this type are located at a distance of 1.66 a, which is quite close to the distance to the second neighbors - 1.414 a. The residuals of the order of 5×10^{-6} achieved in solving subsystems II and III guarantee us at least three correct decimals in the values of the energy parameters. The obtained values of the parameters have been used to calculate the electron bands of diamond, silicon and germanium. In the third neighbor approximation, the overall picture of the energy bands is closer to the results of [19] than in the second neighbor approximation. For illustration, we present the energy bands of silicon (Figs. 14.1 and 14.2). In diamond interpolation is better, and in germanium it is slightly worse than in silicon.

Since there is a small number of correct digits in the energy values used for interpolation, we study the influence of the inaccuracy of the initial data on the results. For this, a random error was added to each energy value within the range from -0.1 to $+0.1$. Calculations have shown that the energy parameters change quite significantly, and the electron bands change within the limits of random errors as expected.

The results of the interpolation of the electron bands of diamond, silicon and germanium have been compared with well-established experimental facts. Namely, using formulas (14.13) and (14.14) the constants of the **kp**-approximation near the Γ_{25}' state and the effective mass near Γ_2' were calculated. In addition, the longitudinal effective masses were calculated at the points L and Δ_1^m. To find the position of the minimum of the conduction band Δ_1^m, we solved the equation obtained by equating

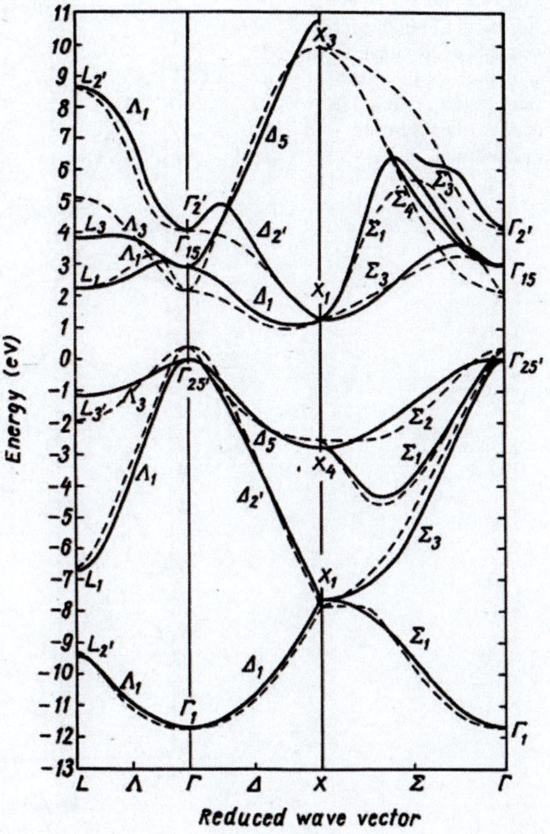

Fig. 14.1 Energy band structure of Si in the second neighbor approximation. The solid lines are the data of Herman [19], the dashed ones are the results of interpolation

to zero the derivative with respect to **k** in the direction Δ of the dispersion law of the lowest conduction band.

Comparison of the calculation results with experimental values is given in Table 14.2. It can be seen that for most of the considered characteristics of the energy bands there is good agreement with experiment.

The study carried out in [17] showed that interpolation based on the tight-binding method for crystals with the diamond structure gives a generally correct picture of the energy bands. In this regard, it seems appropriate to consider the work [14] in more detail. First of all, we doubt the underlying assumption of this work that it is sufficient to take into account the second neighbors to describe the energy bands of Si and Ge. It seems to us that the energy bands of these materials obtained by the authors are in agreement with experiment only to the extent that this experiment is taken into account in the interpolation. It is therefore not surprising that there are large discrepancies between the energy bands patterns of [14] and [19, 23–25] far from the forbidden band. The authors' assertion that, with an increase in the experimental material, a more accurate description of the spectrum is possible

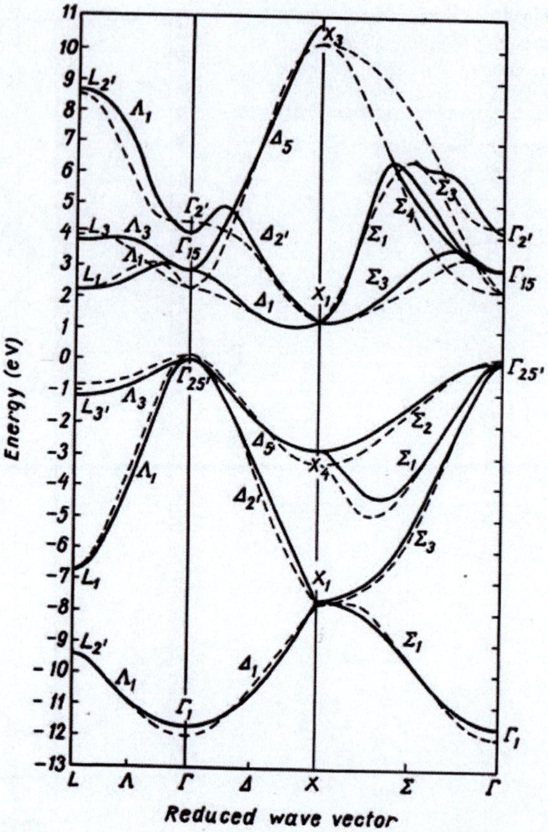

Fig. 14.2 Energy band structure of Si in the third neighbor approximation. The solid lines are the data of Herman [19], the dashed ones are the results of interpolation

by taking into account third neighbors, raises two objections. First, as shown in [17], taking third neighbors into account does not fundamentally change the overall picture of the energy bands. Second, since the values obtained from the experiment have a relatively low accuracy, the numerical solution of equations with such initial data excludes the possibility of accurately determining the parameters that are small in absolute value. The last remark allows us to think that, despite relatively good general picture of the energy bands in [26], the values of small interaction integrals and overlap integrals (of the order of 10^{-3} eV) do not make sense, since their reliable calculation requires 5–6 correct digits after the decimal point in the initial data.

Since interpolation in the tight-binding method in the described form gives an overall correct picture of energy bands, but does not give completely reliable results for the differential characteristics of the energy bands, we used the interpolation to calculate the density of states, which is not very sensitive to the details of the energy band picture.

Table 14.2 Comparison of some calculated values with experiment

	Diamond		Silicon		Germanium	
	Theory[a]	Experiment	Theory	Experiment[b]	Theory	Experiment[b]
Direct energy gap					1.08	0.90
Indirect energy gap	4.89	5.47[c]	0.96	1.17		
A			−2.63	−4.28	−10.30	−13.27
B			−0.76	−0.75	−7.60	−8.63
C			2.02	5.25	8.06	12.4
Δ_1^m	0.76	0.77[d]	0.85	0.85		
$m^*(\Gamma_2')$					0.02	0.04
$m_l^*(\Delta_1^m)$			0.91	0.90		
$m_l^*(L_1)$					0.91	1.59

[a] Calculation results in the third neighbor approximation
[b] Ref. [20]. The model is not accurate enough to distinguish between experimental data presented in [20] and in later works. Moreover, neither for germanium nor for silicon, there is complete agreement on the numerical values of the parameters of the valence band A, B and C (see [20], the footnote in the section on Germanium)
[c] Ref. [21]
[d] Ref. [22]

14.4 Electron Density of States

For the interpretation of X-ray and optical spectra of solids, it is of considerable interest to study the quantity characterizing the number of electronic states per given energy interval. This quantity is expressed in terms of the density of states function $n(\varepsilon)$, defined so that $n(\varepsilon)\,\mathrm{d}\varepsilon$ is equal to the number of electronic states in the crystal per atom, with energies ranging from ε to $\varepsilon + \mathrm{d}\varepsilon$. For crystals with two atoms per unit cell (in particular, for crystals with the diamond structure), the following equality follows directly from the definition of the function $n(\varepsilon)$ and the fact that two electrons with different spins can be in one state:

$$n(\varepsilon) = \frac{1}{\Omega_{\mathrm{BZ}}} \frac{\partial}{\partial \varepsilon} \int\limits_{\varepsilon(\mathbf{k})<\varepsilon} \mathbf{dk}\,, \tag{14.17}$$

where the integration is performed over the volume of the $\mathbf{k}$-space, inside the surface of constant energy $\varepsilon(\mathbf{k}) = \varepsilon$.

Replacing the derivative in (14.17) by the ratio of the function increment to the argument increment, we obtain

$$n(\varepsilon) = \frac{1}{\Omega_{BZ}\,\Delta\varepsilon} \int\limits_{\varepsilon \le \varepsilon(\mathbf{k}) < \varepsilon + \Delta\varepsilon} d\mathbf{k}. \tag{14.18}$$

Here $\int_{\varepsilon \le \varepsilon(\mathbf{k}) < \varepsilon + \Delta\varepsilon} d\mathbf{k} \,/\, \Omega_{BZ}$ is the ratio of the volume enclosed between the surfaces of constant energy $\varepsilon(\mathbf{k}) = \varepsilon$ and $\varepsilon(\mathbf{k}) = \varepsilon + \Delta\varepsilon$ to the volume of the Brillouin zone. This ratio, in turn, can be approximately calculated as the ratio of the number of random points at which the energy values fall within the interval $[\varepsilon, \varepsilon + \Delta\varepsilon)$ to the total number of random points uniformly distributed in the Brillouin zone. Thus, the approximate value of the density of states in a crystal is determined by the formula

$$n(\varepsilon) = \frac{1}{N\,\Delta\varepsilon} \sum_{\varepsilon \le \varepsilon_j < \varepsilon + \Delta\varepsilon} N(\varepsilon_j) \left(\frac{\text{electrons}}{\text{atom} \cdot \text{eV}}\right), \tag{14.19}$$

where $N(\varepsilon_j)$ the number of random points $\mathbf{k}$ with the energy ε_j and N is the total number of random points in the Brillouin zone.

The random vectors $\mathbf{k}$ for calculating $n(\varepsilon)$ were obtained by the Korobov method [27], which gives a uniform distribution of pseudorandom points in a unit cube.

It is clear that formula (14.19) remains valid for the irreducible Brillouin zone. Therefore, the choice of random points and calculation of the energy eigenvalues were carried out in the irreducible Brillouin zone, which, like the entire Brillouin zone, is contained in the unit cube of $\mathbf{k}$-space.

Given the coordinates of the vertices of the irreducible Brillouin zone (see Fig. 10.2), it is easy to write down the equations of the bounding planes of the polyhedron and inequalities for the coordinates of the points lying inside it:

$$k_1 < k_2\,, \qquad k_1 + k_2 + k_3 < \frac{3}{2}\,, \qquad k_1 > k_3\,, \tag{14.20}$$

moreover $k_2 < 1$ and $k_3 > 0$.[3]

The volume of the irreducible Brillouin zone is equal to 1/12 of the volume of a unit cube. Therefore, in order to obtain M points in the irreducible Brillouin zone, it is necessary to construct $12M$ pseudorandom vectors $\mathbf{k}$ in the unit cube.

The interpolation procedure described in the previous section is used here to calculate the density of states $n(\varepsilon)$ in diamond [28].

The initial values of the energy ε at the reference points (see Table 14.3) were calculated using the OPW method (see Chap. 9).

The energy parameters calculated from these data in the second neighbor approximation are given in Table 14.4

[3] The coordinates k_1, k_2, k_3 are defined by the relations $k_x = \frac{\pi}{a}k_1, k_y = \frac{\pi}{a}k_2, k_z = \frac{\pi}{a}k_3$.

Table 14.3 Energy values ε at the reference points for diamond (in eV)[a]

Term	ε	Term	ε	Term	ε	Term	ε
Γ_1	−20.41848	$L_{2'}^{(1)}$	−14.54260	$\Delta_1^{(1)}$	−18.26013	$\Sigma_1^{(1)}$	−16.01329
Γ'_{25}	00.00000	$L_1^{(1)}$	−13.05363	$\Delta_{2'}^{(1)}$	−3.82924	$\Sigma_3^{(1)}$	−7.93146
Γ_{15}	6.75141	L'_3	−2.34090	$\Delta_5^{(1)}$	−3.71973	$\Sigma_1^{(2)}$	−7.28855
Γ'_2	16.31172	$L_1^{(2)}$	9.32469	$\Delta_1^{(2)}$	5.22416	Σ_2	−2.87871
$X_1^{(1)}$	−11.97127	L_3	11.08756	$\Delta_{2'}^{(2)}$	11.09613	$\Sigma_3^{(2)}$	9.17781
X_4	−6.22683	$L_{2'}^{(2)}$	16.81020	$\Delta_5^{(2)}$	13.06971	Σ_4	11.49380
$X_1^{(2)}$	5.58640					$\Sigma_1^{(3)}$	14.20733
X_3	18.69206					$\Sigma_3^{(3)}$	15.93129

[a] The energy value of the term Γ'_{25} is taken to be zero

Table 14.4 Energy parameters for diamond (in eV)

x_1	−5.382
x_2	4.069
x_3	−4.547
x_4	2.024
x_5	0.618
x_6	3.023
x_7	0.279
x_8	−0.038
x_9	0.600
x_{10}	0.201
x_{11}	−0.175
x_{12}	−0.247
x_{13}	−0.564

Since the values of the function were calculated at a finite number of points, a histogram was plotted instead of the density of states. The calculation was made according to formula (14.19). When the number of random points in the Brillouin zone became $\approx 10^6$ and energy step (histogram step) reached $\Delta\varepsilon = 0.2$ eV, the histogram stabilized.

There are no experimental methods for measuring the density of states, so the calculated density of states is usually compared with the X-ray emission and absorption spectra.

In Fig. 14.3 the density of states for the valence bands of diamond is compared with the data on X-ray emission K-spectrum [29].

Fig. 14.3 Density of states for the valence bands of diamond: calculation (blue solid line) and X-ray emission K-spectrum [29] (red dashed line)

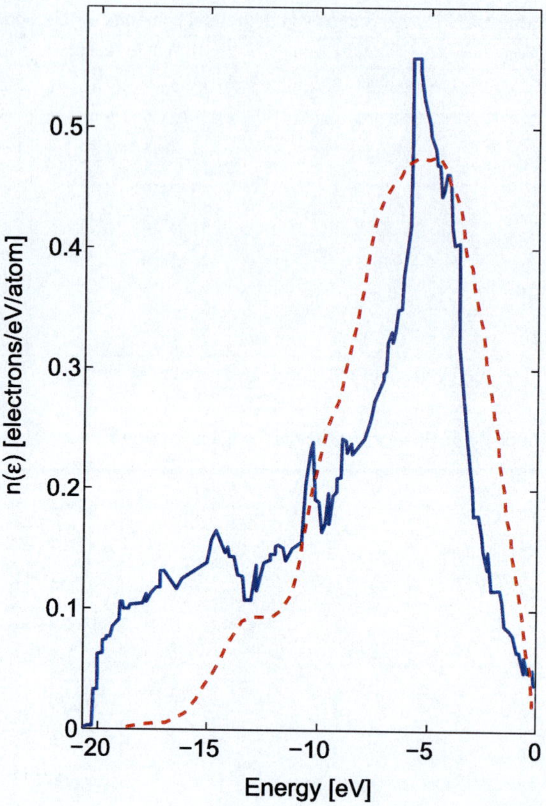

It can be seen that there is quite satisfactory agreement between the histogram of the density of states and the intensity curve. Two circumstances should be taken into account here. First, the resolution in the X-ray spectra does not exceed 0.5 eV, and therefore the details in the histogram of the density of states cannot be interpreted. Second, due to the dependence of the transition probability on the state, the intensity curve gives a distorted representation of the density of states. The discrepancy between the theoretically calculated histogram and the X-ray emission K-spectrum curve on the left side of Fig. 14.3 can be explained exactly by the fact that in diamond the lower valence band is of the s-type.

A similar comparison for the lowest conduction bands of diamond can be made from Fig. 14.4. The density of states for the conduction bands of diamond is compared with the data on X-ray absorption spectrum [30]. The calculated density of states is in qualitative agreement with experiment.

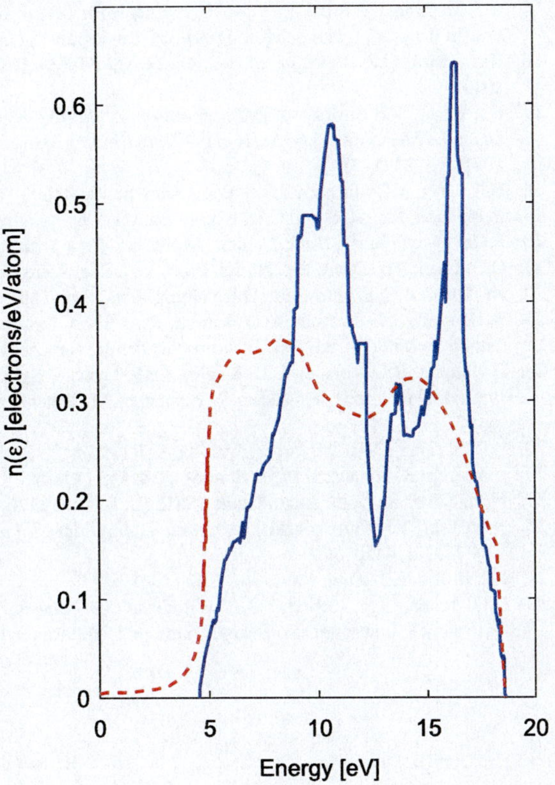

Fig. 14.4 Density of states for the conduction bands of diamond: calculation (blue solid line) and X-ray absorption spectrum [30] (red dashed line).

References

1. J.C. Slater, G.F. Koster, Phys. Rev. **94**, 1498 (1954)
2. M. Miasek, Phys. Rev. **107**, 92 (1957)
3. D.A. Papaconstantopoulos, *Handbook of the Band Structure of Elemental Solids*, 2nd edn. (Springer, New York, 2015)
4. R.E. Cohen, M.J. Mehl, D.A. Papaconstantopoulos, Phys. Rev. B **50**, 14694 (1994)
5. D.A. Papaconstantopoulos, M.J. Mehl, J. Phys.: Condens. Matter **15**, R413 (2003)
6. B.I. Reser, V.P. Shirokovskii, Phys. Stat. Sol. **38**, K135 (1970)
7. J. Callaway, *Quantum Theory of the Solid State*, 2nd edn. (Academic, New York, 1991)
8. M. Dresselhaus, G. Dresselhaus, A. Jorio, *Group Theory: Application to the Physics of Condensed Matter* (Springer, Berlin, 2008)
9. E.O. Kane, in *Semiconductors and Semimetals*, vol. 1, eds. R.K. Willardson, A.C. Beer (Academic, New York, 1966), pp. 75–100
10. G.L. Bir, G.E. Pikus, *Symmetry and Strain-induced Effects in Semiconductors* [in Russian] (Nauka, Moscow, 1972). Engl. transl.: Wiley, New York, 1974
11. L.C. Lew Yan Voon, M. Willatzen, *The* **k·p** *Method: Electronic Properties of Semiconductors* (Springer, Berlin, 2009)
12. P.O. Löwdin, J. Chem. Phys. **19**, 1396 (1951)
13. G. Dresselhaus, A.F. Kip, C. Kittel, Phys. Rev. **98**, 368 (1955)
14. G. Dresselhaus, M. Dresselhaus, Phys. Rev. **160**, 649 (1967)

15. A.P. Mishina, I.V. Proskuryakov, *Higher Algebra: Linear Algebra, Polynomials, General Algebra* [in Russian] (Nauka, Moscow, 1965). Engl. transl.: Pergamon, Oxford, 1965
16. R.A. Horn, C.R. Johnson, *Matrix Analysis*, 2nd edn. (Cambridge Univ. Press, Cambridge, 2013)
17. B.I. Reser, V.P. Shirokovskii, *Interpolation of the Energy Bands of Semiconductors with the Diamond Structure on the Basis of the Tight Binding Method* [in Russian] (1971). Dep. VINITI, 1971, No. 2471–71
18. B.I. Reser, V.P. Shirokovskii, Phys. Techn. Semicond. **5**, 1673 (1971)
19. F. Herman, R.L. Kortum, C.D. Kuglin, Internat. J. Quantum Chem. **1S**, 533 (1967)
20. J. Callaway, *Energy Band Theory* (Academic, New York, 1964)
21. C.D. Clark, P.J. Dean, P.V. Harris, Proc. Roy. Soc. **A277**, 312 (1964)
22. P.J. Dean, E.C. Lightowlers, D.R. Wight, Phys. Rev. **140**, A352 (1965)
23. F. Herman, R.L. Kortum, C.D. Kuglin, R.A. Short, in *Quantum Theory of Atoms, Molecules and the Solid State*. ed. P.O. Löwdin (Academic, New York, 1966), p. 381
24. F. Herman, R.L. Kortum, C.D. Kuglin, J.P.V. Dyke, S. Skillman, in *Methods in Computational Physics*, vol. 8, eds. B. Alder, S. Fernbach, M. Rotenberg (Academic, New York, 1968), p. 193
25. L.R. Saravia, D. Brust, Phys. Rev. **171**, 916 (1968)
26. I. Alstrup, K. Johansen, Phys. Stat. Sol. **28**, 555 (1968)
27. N.M. Korobov, Dokl. Akad. Nauk SSSR, [in Russian] **115**, 1062 (1957)
28. B.I. Reser, V.P. Shirokovskii, *Electronic energy bands of diamond* [in Russian] (1972). Dep. VINITI, No. 4450–72
29. J.E. Holliday, J. Appl. Phys. **38**, 4720 (1967)
30. G.O. Jones, O.R. Roberts, G.T. Williams, D.P. Langstaff, N.R.J. Poolton, P. Martineau, D. Fisher, S. Lawson, D.A. Evans, Diamond Light Source Proc. **1**, E131 (2011)

Chapter 15
Quasi-2D Dichalcogenides with the $2H$-$TaSe_2$ Structure

A complex problem always has a solution that is simple, intuitive and wrong. (Murphy's law)

In this chapter, we obtain an analytical form of the low-dimensional effective Hamiltonian describing the electronic structure of the conduction band of a trigonal prismatic tantalum diselenide $2H$-$TaSe_2$ having quasi-2D hexagonal crystal structure.

The notation $2H$-$TaSe_2$ means the $2H$ polytype of $TaSe_2$; there is also the $1T$ polytype and others [1]. In the $2H$-$TaSe_2$ polytype there are two Se-Ta-Se complexes per unit cell, located along the hexagonal axis, and the coordination of the metal atoms is trigonal prismatic [2].

Other transition-metal dichalcogenides (including $NbSe_2$, TaS_2 and NbS_2) have the same crystal structure. An analytical form of the matrix elements of the Hamiltonian is obtained by consistent treatment of symmetry in the tight-binding method, which is described in Chap. 13.

15.1 Introduction

Quasi-2D diselenides with the $2H$-$TaSe_2$ structure ($2H$-$TaSe_2$ and $2H$-$NbSe_2$) have long attracted attention [1, 3]. This interest has been promoted, firstly, by early attempts to find high-temperature superconductivity in layered structures, and secondly, by the charge density waves (CDW) formation in these systems (for a review, see, e.g. [4]). In particular, $2H$-$TaSe_2$ undergoes a second-order phase transition to an incommensurate CDW phase at 122.3 K and another (*first-order*) phase transition into the commensurate CDW phase near 90.0 K [3]. In $2H$-$NbSe_2$, the transition to an incommensurate CDW phase occurs at a much lower temperature, 33.5 K, and no transition to the commensurate CDW phase is observed at all. At higher temperatures in $2H$-$TaSe_2$ and $2H$-$NbSe_2$, a transition to the paramagnetic phase occurs. In

N. B. Melnikov and B. I. Reser, *Space Group Representations*,
https://doi.org/10.1007/978-3-031-13991-8_15

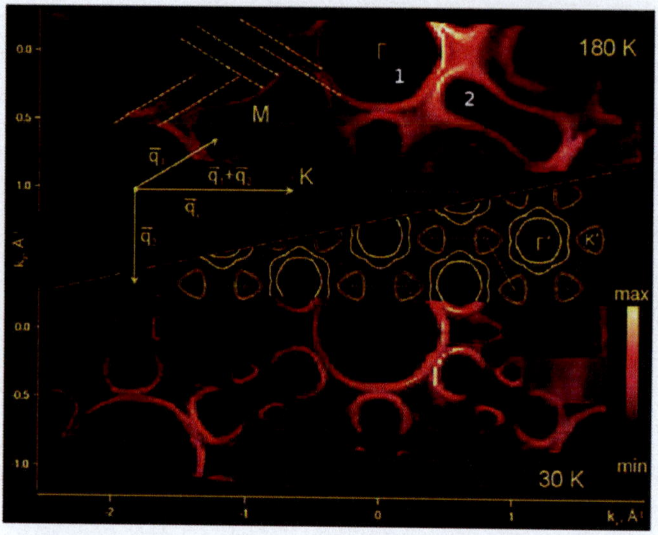

Fig. 15.1 Fermi surface of $2H$-$TaSe_2$ in the $k_z = 0$ plane of the Brillouin zone, obtained in the ARPES experiment [5]. Here the number 1 denotes the part of the Fermi surface corresponding to the barrel and number 2 denotes the part corresponding to the "bone"

addition, a superconducting phase is observed in $2H$-$NbSe_2$ below 7.3 K [1]. Thus, the quasi-2D diselenides $2H$-$TaSe_2$ and $2H$-$NbSe_2$ have a rich phase diagram.

Recently, interest in quasi-2D transition-metal dichalcogenides has been boosted by the progress in experimental studies using angle-resolved photoemission spectroscopy (ARPES). This progress is due to the fact that the layered compounds give a very good surface when the sample is prepared by the cleavage method.

In the study [5], the section of the $2H$-$TaSe_2$ Fermi surface by the $k_z = 0$ plane in the Brillouin zone was measured using the ARPES method (see Fig. 15.1). In the experiment for the normal-state $2H$-$TaSe_2$ (at a temperature of 180 K), two sheets of the Fermi surface were found: a barrel (number 1 in Fig. 15.1) and a "bone" (number 2 in Fig. 15.1). To describe the shape of the Fermi surface, an analytical dispersion law for $2H$-$TaSe_2$ in the $k_z = 0$ plane of the Brillouin zone was used (see [6] and refs. therein):

$$\begin{aligned}\varepsilon(\mathbf{k}) = t_0 &+ t_1\left(2\cos\frac{k_x}{2}\cos\frac{\sqrt{3}k_y}{2} + \cos k_x\right)\\ &+ t_2\left(2\cos\frac{3k_x}{2}\cos\frac{\sqrt{3}k_y}{2} + \cos\sqrt{3}k_y\right)\\ &+ t_3\left(2\cos k_x\cos\sqrt{3}k_y + \cos 2k_x\right)\\ &+ t_4\left(2\cos 3k_x\cos\sqrt{3}k_y + \cos 2\sqrt{3}k_y\right),\end{aligned} \tag{15.1}$$

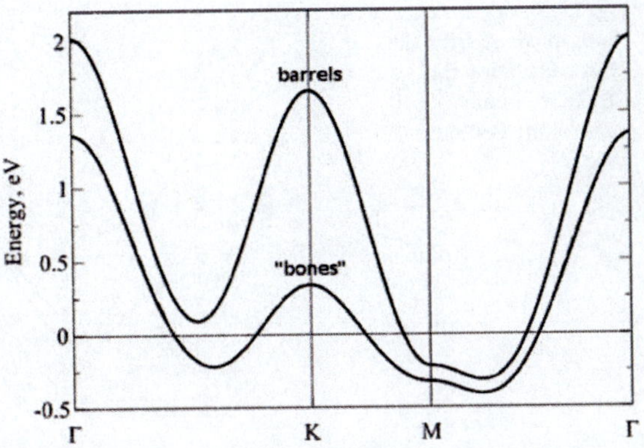

Fig. 15.2 "Experimental" energy bands for $2H$-$TaSe_2$ in the $k_z = 0$ plane of the Brillouin zone [6]. Zero corresponds to the Fermi level

where the parameters were obtained from the fit to "experimental" Fermi surface [7]. The form of the dispersion law (15.1) is the same for both sheets, but has different sets of values of the hopping integrals t_i (in eV): $t_0 = -0.027$, $t_1 = 0.199$, $t_2 = 0.221$, $t_3 = 0.028$, $t_4 = 0.013$ for the dispersion branch reproducing the barrel, and $t_0 = 0.407$, $t_1 = 0.114$, $t_2 = 0.444$, $t_3 = -0.033$, $t_4 = 0.011$ for the dispersion branch reproducing the "bone". The corresponding dispersion branches are shown in Fig. 15.2.

However, calculations of the electronic structure in the local density approximation (LDA) [2, 6, 8, 9] give dispersion curves that qualitatively differ from those obtained from experiment.[1] In particular, the Fermi level must be decreased by about 0.1 eV [6] to obtain the Fermi surface sheets of the "bone" type (Fig. 15.3). Furthermore, the dispersion curves obtained in the LDA calculation degenerate at the K point in the Brillouin zone, in contrast to the curves shown in Fig. 15.2, which correspond to the "experimental" dispersion law (15.1). But from the physical considerations discussed below, it is clear that this degeneracy at the point K is accidental and is not due to the symmetry of the problem. Spin-polarized calculation [10] gives a noticeable splitting of the dispersion curves, which happens at the point K as well.

To study the reason for such significant differences in the "experimental" and theoretical energy bands, it is useful to obtain an *analytical* form of the dispersion relations for the $2H$-$TaSe_2$ conduction bands in the tight-binding method taking symmetry into account.

In this chapter, we obtain an *analytical* form of the low-dimensional effective Hamiltonian describing the behavior of the energy bands of $2H$-$TaSe_2$ that cross

[1] Note that the band calculations correspond to $T = 0$, and the experiment was carried out at a temperature of 180 K. In this regard, two questions appear:

- How do the chemical potential and the entire band structure change with temperature?
- How do CDW affect the electronic structure?

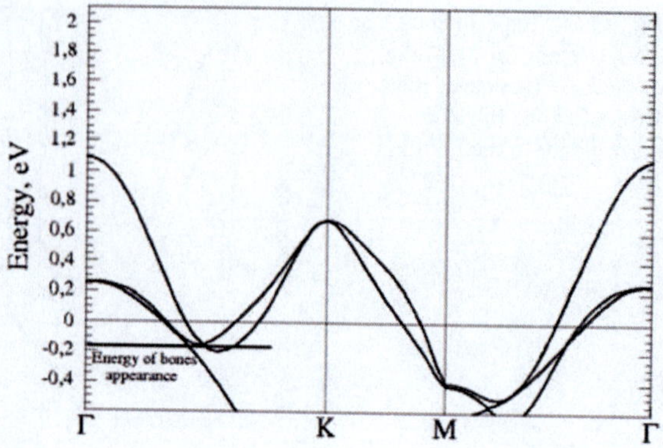

Fig. 15.3 Valence energy bands obtained from the LDA calculation for $2H$-TaSe$_2$ in the $k_z = 0$ plane of the Brillouin zone [6]

the Fermi level. We will use the method for consistent treatment of symmetry in the tight-binding method described in Chap. 13. Specifically, we do the following:

- we determine the types of orbitals of the effective low-dimensional Hamiltonian for $2H$-TaSe$_2$ based on the symmetry analysis of the dispersion curves from the LDA calculation;
- we apply the technique of Chap. 13 with this set of orbitals to the $2H$-TaSe$_2$ structure;
- we obtain analytical expressions for the matrix elements of the Hamiltonian for the $2H$-TaSe$_2$ structure.

15.2 Crystal Structure of $2H$-TaSe$_2$

The diselenides we study have a hexagonal crystal structure with the space group D_{6h}^4 [3]. Most of the authors (see, e.g. [2, 3]) choose the Cartesian coordinate system (x, y, z) following Slater [11], such that the lattice vectors have the form

$$\mathbf{t}_1 = \left(\frac{\sqrt{3}}{2}a, -\frac{1}{2}a, 0\right), \qquad \mathbf{t}_2 = (0, a, 0), \qquad \mathbf{t}_3 = (0, 0, c), \tag{15.2}$$

where a and c are the parameters of the hexagonal lattice (Fig. 15.4). For the $2H$-TaSe$_2$ compound, they are $a = 3.436$ Å and $c = 12.70$ Å.

The origin is chosen in the midpoint between the two metal atoms. The atomic positions in the Wigner–Seitz unit cell are as follows:[2]

[2] In contrast to our earlier notation of basis vectors $\boldsymbol{\tau}_t^s$ (s is the sort, t is the ordinal number of an atom of the sth sort), in this chapter, as well as in Chap. 13, we use a simpler notation $\boldsymbol{\tau}_s$ (s is the ordinal number of an atom in a unit cell).

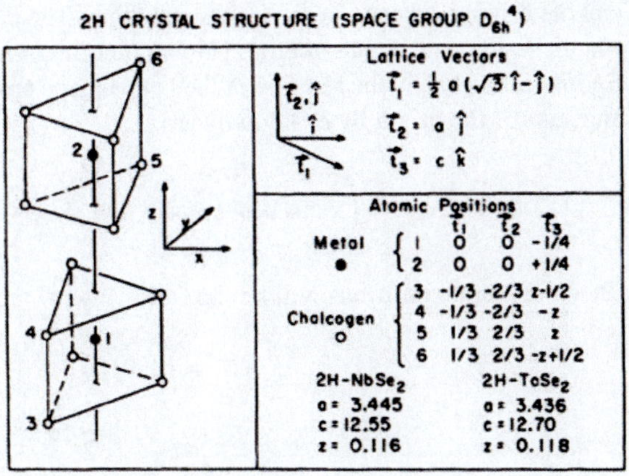

Fig. 15.4 $2H$ structure and crystallographic parameters for $2H$-$TaSe_2$ and $2H$-$NbSe_2$ [3]. The lattice vectors are of the form (15.2)

$$\boldsymbol{\tau}_1 = -\frac{1}{4}\mathbf{t}_3 = \left(0, 0, -\frac{1}{4}c\right),$$

$$\boldsymbol{\tau}_2 = \frac{1}{4}\mathbf{t}_3 = \left(0, 0, \frac{1}{4}c\right)$$

(metal) and

$$\boldsymbol{\tau}_3 = -\frac{1}{3}\mathbf{t}_1 - \frac{2}{3}\mathbf{t}_2 + \left(z_0 - \frac{1}{2}\right)\mathbf{t}_3 = \left(-\frac{\sqrt{3}}{6}a, -\frac{1}{2}a, \left(z_0 - \frac{1}{2}\right)c\right),$$

$$\boldsymbol{\tau}_4 = -\frac{1}{3}\mathbf{t}_1 - \frac{2}{3}\mathbf{t}_2 - z_0\,\mathbf{t}_3 = \left(-\frac{\sqrt{3}}{6}a, -\frac{1}{2}a, -z_0 c\right),$$

$$\boldsymbol{\tau}_5 = \frac{1}{3}\mathbf{t}_1 + \frac{2}{3}\mathbf{t}_2 + z_0\,\mathbf{t}_3 = \left(\frac{\sqrt{3}}{6}a, \frac{1}{2}a, z_0 c\right),$$

$$\boldsymbol{\tau}_6 = \frac{1}{3}\mathbf{t}_1 + \frac{2}{3}\mathbf{t}_2 + \left(-z_0 + \frac{1}{2}\right)\mathbf{t}_3 = \left(\frac{\sqrt{3}}{6}a, \frac{1}{2}a, \left(-z_0 + \frac{1}{2}\right)c\right)$$

(chalcogen), where $2z_0c$ is the distance between two nearest layers of Se atoms ($z_0 = 0.118$ Å for $2H$-$TaSe_2$). All the six atoms described above lie on the boundary of the Wigner–Seitz unit cell.

Egorov et al. [12] choose the Cartesian coordinate system (x', y', z') differently, such that the hexagonal lattice vectors have the form

$$\mathbf{a}_1 = \left(\frac{1}{2}a, \frac{\sqrt{3}}{2}a, 0\right), \quad \mathbf{a}_2 = \left(-\frac{1}{2}a, \frac{\sqrt{3}}{2}a, 0\right), \quad \mathbf{a}_3 = (0, 0, c), \qquad (15.3)$$

and the origin is chosen in the first metal atom. The crystal structure (15.2) used by Mattheiss [2] and Moncton et al. [3] can be transformed to the one (15.3) used by Egorov et al. [12] by the rotation in the (x, y) plane around the z-axis by $\varphi = -\pi/6$[3] and shift of the origin by $c/4$ downward:

$$(x', y', z') = \left(x \cos\varphi + y \sin\varphi,\ -x \sin\varphi + y \cos\varphi,\ z + \frac{1}{4}c\right).$$

Then the atomic positions will be the following:

$$\boldsymbol{\tau}'_1 = (0, 0, 0)\,, \tag{15.4}$$

$$\boldsymbol{\tau}'_2 = \left(0, 0, \frac{1}{2}c\right) \tag{15.5}$$

(metal) and

$$\boldsymbol{\tau}'_3 = \left(0, -\frac{\sqrt{3}}{3}a, \left(z_0 - \frac{1}{4}\right)c\right), \tag{15.6}$$

$$\boldsymbol{\tau}'_4 = \left(0, -\frac{\sqrt{3}}{3}a, -\left(z_0 - \frac{1}{4}\right)c\right), \tag{15.7}$$

$$\boldsymbol{\tau}'_5 = \left(0, \frac{\sqrt{3}}{3}a, \left(z_0 + \frac{1}{4}\right)c\right), \tag{15.8}$$

$$\boldsymbol{\tau}'_6 = \left(0, \frac{\sqrt{3}}{3}a, \left(-z_0 + \frac{3}{4}\right)c\right)$$

(chalcogen). To make the vectors $\boldsymbol{\tau}'_5$ and $\boldsymbol{\tau}'_6$ symmetric with respect to the horizontal plane $z = 0$ (just as $\boldsymbol{\tau}'_3$ and $\boldsymbol{\tau}'_4$), we subtract the lattice vector $(0, 0, c)$ from $\boldsymbol{\tau}'_6$:

$$\boldsymbol{\tau}'_6 = \left(0, \frac{\sqrt{3}}{3}a, -\left(z_0 + \frac{1}{4}\right)c\right). \tag{15.9}$$

Thus obtained atomic positions $\boldsymbol{\tau}'_1, \ldots, \boldsymbol{\tau}'_6$ belong to the Wigner–Seitz unit cell. Henceforth we omit the prime.

The Wigner–Seitz cell of the $2H$-TaSe$_2$ structure, as well as the atomic positions, is shown in Fig. 15.5. The top view is shown in Fig. 15.6 together with the basis vectors of the direct lattice.

The crystal class of the space group D^4_{6h} is the group D_{6h}. The symmetry elements of the crystal class D_{6h} and their action on an arbitrary vector $\mathbf{r}$ are listed in Table 6.4. The nonprimitive translations $\boldsymbol{\tau}_{\alpha_i}$ for the space group D^4_{6h} with our choice of the origin

[3] The angle φ is considered positive if the rotation is counterclockwise.

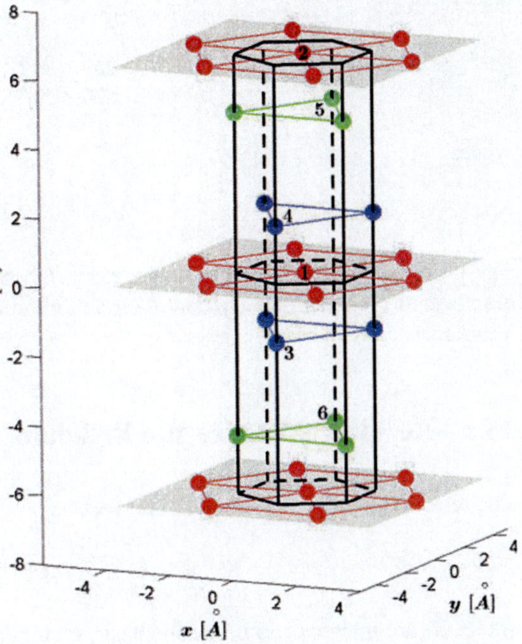

Fig. 15.5 $2H$-TaSe$_2$ crystal structure: metal atoms (red) and chalcogen atoms—the ones close to the $z = 0$ plane (blue) and the ones close to the plane $z = \pm c/2$ (green). The Wigner–Seitz cell is bounded by a hexagonal prism (black). The atomic positions in the Wigner–Seitz cell $\boldsymbol{\tau}_s$ given by (15.4)–(15.9) are indicated by their numbers

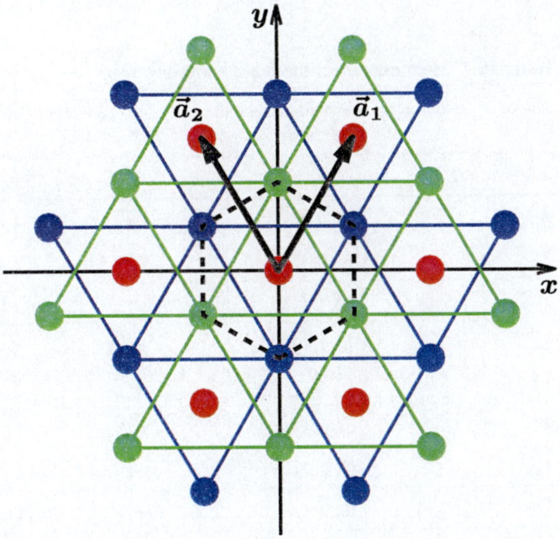

Fig. 15.6 2D projection along the z-axis of the $2H$-TaSe$_2$ crystal structure: metal atoms (red) and chalcogen atoms—the ones close to the $z = 0$ plane (blue) and the ones close to the plane $z = \pm c/2$ (green). The vectors $\mathbf{a}_1$ and $\mathbf{a}_2$ are the basis vectors (15.3), and projection of the boundary of the Wigner–Seitz cell is shown by the black dashed line

have the form

$$\boldsymbol{\tau}_{\alpha_i} = \begin{cases} 0 & \text{for i} = 1, 4, 5, 10\text{–}12, 14, 15, 18\text{–}21, \\ \boldsymbol{\tau} & \text{for i} = 2, 3, 6\text{–}9, 13, 16, 17, 22\text{–}24, \end{cases} \tag{15.10}$$

where

$$\boldsymbol{\tau} = \left(0, 0, \frac{1}{2}c\right). \tag{15.11}$$

The point subgroup G_0' of the space group D_{6h}^4 is the group D_{3h}. The symmetry elements of the subgroup D_{3h} and their action on an arbitrary vector **r** are listed in Table 15.1.

15.3 Reciprocal Lattice and Brillouin Zone

The reciprocal lattice is formed by the vectors

$$\mathbf{K}_\mu = \mu_1 \mathbf{b}_1 + \mu_2 \mathbf{b}_2 + \mu_3 \mathbf{b}_3, \tag{15.12}$$

where μ_i are integers and $\mathbf{b}_i$ are the basic vectors of the reciprocal lattice given by the formulas

Table 15.1 Elements α_i of the point subgroup D_{3h}

No. of α_i	Conj. class	Rotation axis unit vector **n**	Rot. angle θ	$\mathbf{r}' = \alpha\mathbf{r}$
1	E		0	(x, y, z)
4	C_6^2	(0, 0, 1)	$2\pi/3$	$\left(-\frac{1}{2}x - \frac{\sqrt{3}}{2}y, \frac{\sqrt{3}}{2}x - \frac{1}{2}y, z\right)$
5		(0, 0, 1)	$-2\pi/3$	$\left(-\frac{1}{2}x + \frac{\sqrt{3}}{2}y, -\frac{\sqrt{3}}{2}x - \frac{1}{2}y, z\right)$
10		$(\frac{\sqrt{3}}{2}, \frac{1}{2}, 0)$	π	$\left(\frac{1}{2}x + \frac{\sqrt{3}}{2}y, \frac{\sqrt{3}}{2}x - \frac{1}{2}y, -z\right)$
11	C_2'	(0, 1, 0)	π	$(-x, y, -z)$
12		$(-\frac{\sqrt{3}}{2}, \frac{1}{2}, 0)$	π	$\left(\frac{1}{2}x - \frac{\sqrt{3}}{2}y, -\frac{\sqrt{3}}{2}x - \frac{1}{2}y, -z\right)$
14	IC_6	(0, 0, 1)	$\pi/3$	$\left(-\frac{1}{2}x + \frac{\sqrt{3}}{2}y, -\frac{\sqrt{3}}{2}x - \frac{1}{2}y, -z\right)$
15		(0, 0, 1)	$-\pi/3$	$\left(-\frac{1}{2}x - \frac{\sqrt{3}}{2}y, \frac{\sqrt{3}}{2}x - \frac{1}{2}y, -z\right)$
18	IC_6^3	(0, 0, 1)	π	$(x, y, -z)$
19		(1, 0, 0)	π	$(-x, y, z)$
20	IC_2	$(\frac{1}{2}, \frac{\sqrt{3}}{2}, 0)$	π	$\left(\frac{1}{2}x - \frac{\sqrt{3}}{2}y, -\frac{\sqrt{3}}{2}x - \frac{1}{2}y, z\right)$
21		$(-\frac{1}{2}, \frac{\sqrt{3}}{2}, 0)$	π	$\left(\frac{1}{2}x + \frac{\sqrt{3}}{2}y, \frac{\sqrt{3}}{2}x - \frac{1}{2}y, z\right)$

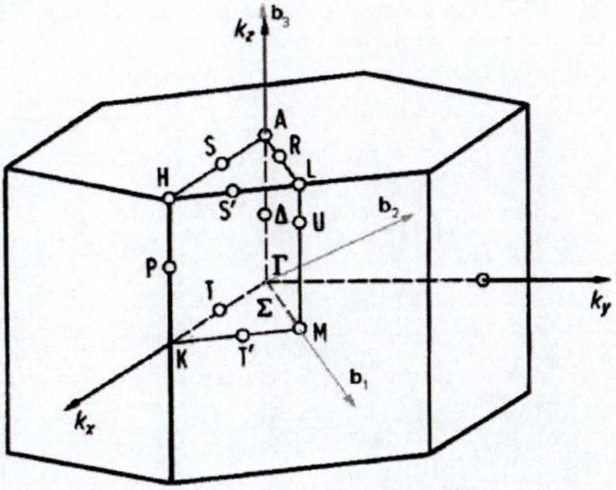

Fig. 15.7 Brillouin zone for a hexagonal crystal lattice, its irreducible part (irreducible Brillouin zone) and high symmetry points. The notation of the high symmetry points follows [13]

$$\mathbf{b}_1 = \frac{2\pi(\mathbf{a}_2 \times \mathbf{a}_3)}{\mathbf{a}_1 \cdot (\mathbf{a}_2 \times \mathbf{a}_3)}, \qquad \mathbf{b}_2 = \frac{2\pi(\mathbf{a}_3 \times \mathbf{a}_1)}{\mathbf{a}_1 \cdot (\mathbf{a}_2 \times \mathbf{a}_3)}, \qquad \mathbf{b}_3 = \frac{2\pi(\mathbf{a}_1 \times \mathbf{a}_2)}{\mathbf{a}_1 \cdot (\mathbf{a}_2 \times \mathbf{a}_3)}. \tag{15.13}$$

The volume of the Brillouin zone is calculated by the formula

$$\Omega_{\mathrm{BZ}} = \mathbf{b}_1 \cdot (\mathbf{b}_2 \times \mathbf{b}_3). \tag{15.14}$$

For the hexagonal crystal structure with the lattice vectors (15.3), we obtain

$$\mathbf{b}_1 = \frac{2\pi}{a}\left(1, \frac{\sqrt{3}}{3}, 0\right), \qquad \mathbf{b}_2 = \frac{2\pi}{a}\left(-1, \frac{\sqrt{3}}{3}, 0\right), \qquad \mathbf{b}_3 = \frac{2\pi}{c}(0, 0, 1). \tag{15.15}$$

The Brillouin zone of this lattice is a hexagonal prism (see Fig. 15.7). Taking (15.14) and (15.15) into account, we calculate the volume of the Brillouin zone for the hexagonal lattice

$$\Omega_{\mathrm{BZ}} = \frac{16\pi^3}{a^2 c\sqrt{3}}. \tag{15.16}$$

If the Hamiltonian H is invariant under transformations of a space group G, then its eigenvalues $\varepsilon(\mathbf{k})$ (energy dispersion) are invariant under transformations of the *point* group G_0 (crystal class):

$$\varepsilon_j(\alpha_i \mathbf{k}) = \varepsilon_j(\mathbf{k}), \qquad \alpha_i \in G_0.$$

This allows us to confine ourselves to the **k** vectors lying in the irreducible Brillouin zone. In our case, the irreducible Brillouin zone is a triangular prism $\Gamma KMLHA$ (see Fig. 15.7), which is 1/24 of the Brillouin zone. Complete description of the symmetry points of the Brillouin zone is given in Table 15.2.

Table 15.2 High symmetry points of the hexagonal lattice in Koster notation [13] and their symmetry elements from the *point* group G_0

Point	Coordinates of the equivalent points (in units of 2π)	$G^0_{\mathbf{k}}$	Nos. of elems. α_m from $G^0_{\mathbf{k}}$
Γ	$(0, 0, 0)$	D_{6h}	**1–24**
K	$K = \frac{1}{3}(\mathbf{b}_1 - \mathbf{b}_2) = \left(\frac{2}{3a}, 0, 0\right)$, $K + \mathbf{b}_2 = \left(-\frac{1}{3a}, \frac{\sqrt{3}}{3a}, 0\right)$, $K - \mathbf{b}_1 = \left(-\frac{1}{3a}, -\frac{\sqrt{3}}{3a}, 0\right)$	D_{3h}	**1**, 4, 5, **7**, 8, 9, 14, 15, **18**, 22, **23**, 24
M	$M = \frac{1}{2}\mathbf{b}_1 = \left(\frac{1}{2a}, \frac{\sqrt{3}}{6a}, 0\right)$, $M - \mathbf{b}_1 = \left(-\frac{1}{2a}, -\frac{\sqrt{3}}{6a}, 0\right)$	D_{2h}	**1**, 6, 9, **10**, 13, **18**, **21**, 22
L	$L = \frac{1}{2}\mathbf{b}_1 + \frac{1}{2}\mathbf{b}_3 = \left(\frac{1}{2a}, \frac{\sqrt{3}}{6a}, \frac{1}{2c}\right)$, $L - \mathbf{b}_1 = \left(-\frac{1}{2a}, -\frac{\sqrt{3}}{6a}, \frac{1}{2c}\right)$, $L - \mathbf{b}_3 = \left(\frac{1}{2a}, \frac{\sqrt{3}}{6a}, -\frac{1}{2c}\right)$, $L - (\mathbf{b}_1 + \mathbf{b}_3) = \left(-\frac{1}{2a}, -\frac{\sqrt{3}}{6a}, -\frac{1}{2c}\right)$	D_{2h}	**1**, 6, 9, 10, 13, 18, **21**, 22
H	$H = \frac{1}{3}(\mathbf{b}_1 - \mathbf{b}_2) + \frac{1}{2}\mathbf{b}_3 = \left(\frac{2}{3a}, 0, \frac{1}{2c}\right)$, $H + \mathbf{b}_2 = \left(-\frac{1}{3a}, \frac{\sqrt{3}}{3a}, \frac{1}{2c}\right)$, $H - \mathbf{b}_1 = \left(-\frac{1}{3a}, -\frac{\sqrt{3}}{3a}, \frac{1}{2c}\right)$, $H - \mathbf{b}_3 = \left(\frac{2}{3a}, 0, -\frac{1}{2c}\right)$, $H + \mathbf{b}_2 - \mathbf{b}_3 = \left(-\frac{1}{3a}, \frac{\sqrt{3}}{3a}, -\frac{1}{2c}\right)$, $H - \mathbf{b}_1 - \mathbf{b}_3 = \left(-\frac{1}{3a}, -\frac{\sqrt{3}}{3a}, -\frac{1}{2c}\right)$	D_{3h}	**1**, 4, 5, 7–9, 14, 15, 18, 22, **23**, 24
A	$A = \frac{1}{2}\mathbf{b}_3 = \left(0, 0, \frac{1}{2c}\right)$, $A - \mathbf{b}_3 = \left(0, 0, -\frac{1}{2c}\right)$	D_{6h}	**1–6**, 7–12, 13–18, **19–24**
T	$\left(\frac{2\mu}{3a}, 0, 0\right)$	C_{2v}	**1**, **7**, **18**, **23**
T′	$T' = \left(\frac{4-\mu}{6a}, \frac{\sqrt{3}\mu}{6a}, 0\right)$, $T' - \mathbf{b}_1 = \left(-\frac{2+\mu}{6a}, \frac{\sqrt{3}\mu}{6a}, 0\right)$	C_{2v}	**1**, **9**, **18**, 22
Σ	$\left(\frac{\mu}{2a}, \frac{\sqrt{3}(\mu-2)}{6a}, 0\right)$	C_{2v}	**1**, **10**, **18**, **21**

$0 < \mu < 1$

Bold numbers denote the elements that transform the symmetry point to itself, the rest denote the elements that transform a symmetry point to the equivalent points

15.4 Matrix Elements of the Hamiltonian

Let us turn to the electronic structure of $2H$-$TaSe_2$. Band calculations [14] in the LDA approximation (which use the TB-LMTO-ASA [15] method with the von Barth and Hedin potential [16]) give the electronic structure shown in Fig. 15.8. From the dispersion in the $k_z = 0$ plane of the Brillouin zone ($\Gamma K M \Gamma$ in Fig. 15.7), we see that the Fermi level is crossed by two bands. From the symmetry analysis of the band structure it follows that these bands have the Ta-$5d_{3z^2-r^2}$ character. It also follows that these two bands are not degenerate in the plane $k_z = 0$ of the Brillouin zone everywhere except for the point K, where an accidental degeneracy occurs (just as in earlier LDA calculations [8, 9]).

From further symmetry analysis of the $2H$-TaSe$_2$ electronic structure, we can conclude that of primary interest are the Se $4p_z$-orbitals, whose levels are closest to the Ta $5d_{3z^2-r^2}$-levels (see the LDA calculation in Fig. 15.8, where the $4p_z$-orbitals are highlighted[4]). The paper by [17] also discusses that Se $4p_z$-orbitals (in NbSe$_2$) are located near the Fermi level. Based on the fact that the energy bands corresponding to Se $4p_z$-orbitals intersect the bands of Nb d-states (transversally), it is stated that the $4p_z$-orbitals do not hybridize (do not interact) with the Nb d-states located near the Fermi level. Thus, the effective Hamiltonian of $2H$-TaSe$_2$ must include not only two Ta $5d_{3z^2-r^2}$-orbitals, but also four Se $4p_z$-orbitals corresponding to four Se atoms in the unit cell (see Fig. 15.5). These six wave functions will be considered as a set of initial functions for constructing the matrix elements of the Hamiltonian. Hence our effective Hamiltonian will be a 6×6 matrix.

In paper [14] it was suggested that the *direct* overlap of the Ta $5d_{3z^2-r^2}$-orbitals is too small and cannot lead to such a strong splitting of the energy bands. Therefore, it was proposed to consider *indirect* overlap of the Ta $5d_{3z^2-r^2}$-orbitals from different layers. Based on the LDA calculation presented in Fig. 15.8, it was conjectured that the overlap occurs through Se $4p$-orbitals, whose levels lie in the vicinity of the Fermi level. However, as we show further, this conjecture is incorrect. We will show that the Se $4p_z$-orbitals do not hybridize (do not interact) with Ta $5d_{3z^2-r^2}$-orbitals (just as with Nb $5d_{3z^2-r^2}$-orbitals in $2H$-NbSe$_2$ [17]).

The Ta $5d_{3z^2-r^2}$-wave function is transformed according to the representation Γ_1, and Se $4p_z$-wave function is transformed according to the representation Γ_4 [18]. Matrices of the corresponding irreducible representations and basis functions are given in Table 15.3. Since the representation matrices are one dimensional, we omit the row index μ in (13.7). We enumerate 6 initial functions (see Table 15.4) and use the abbreviation $H_{ii'}$ for the matrix elements of the Hamiltonian. Note that the wave functions Φ^{s1} ($s = 1, 2$) are localized on Ta atoms, and the wave functions Φ^{s4} ($s = 3, 4, 5, 6$)—on Se atoms.

Table 15.3 Irreducible representations Γ_1 and Γ_4 of the subgroup $G'_0 = D_{3h}$ [18]

Irreducible representation	Symmetry transformations						Basis functions
	E	C_6^2	C'_2	IC_6	IC_6^3	IC_2	
Γ_1	1	1	1	1	1	1	$3z^2 - r^2$
Γ_4	1	1	−1	−1	−1	1	z

Table 15.4 Numbering of $2H$-TaSe$_2$ wave functions

i	1	2	3	4	5	6
Φ^{sj}	Φ^{11}	Φ^{21}	Φ^{34}	Φ^{44}	Φ^{54}	Φ^{64}

[4] A similar figure is presented in [9], but without a discussion of the $4p_z$-orbitals' contribution.

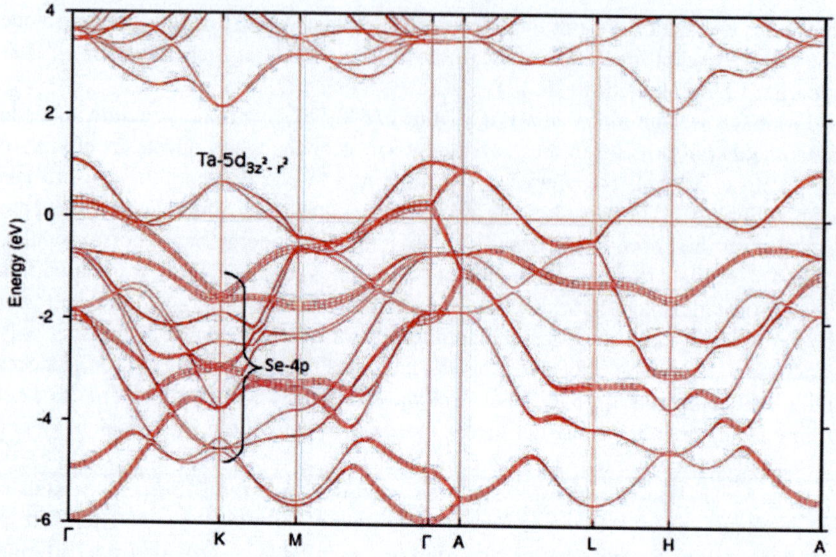

Fig. 15.8 Energy bands of $2H$-TaSe$_2$ obtained from the LDA calculation. The line thickness shows the contribution of the $4p_z$ orbital to the corresponding band

First we establish relations between the matrix elements of the Hamiltonian. To do this, we use relations (13.18) and (13.19), which were obtained from the symmetry of the Hamiltonian with respect to spatial and temporal inversion, as well as from its Hermiticity. The atomic function $3z^2 - r^2$ is even with respect to inversion. Hence $\omega^{(1)} = 1$ for the atomic function in the Bloch sums Φ^{11} and Φ^{21} of the metal atoms. The atomic function $4p_z$ is odd with respect to inversion (for details, see Sect. 13.1.4), and we have $\omega^{(4)} = -1$ for the Bloch sums $\Phi^{s4}, s = 3, 4, 5, 6$ of the chalcogen atoms.

Let us consider the matrix element H_{13} as an example. Taking into account

$$\boldsymbol{\tau}_2 - \boldsymbol{\tau}_3 = \left(0, 0, \frac{1}{2}c\right) - \left(0, -\frac{\sqrt{3}}{3}a, \left(z_0 - \frac{1}{4}\right)c\right)$$
$$= \left(0, \frac{\sqrt{3}}{3}a, \left(-z_0 + \frac{3}{4}\right)c \mod(c)\right) = \left(0, \frac{\sqrt{3}}{3}a, -\left(z_0 + \frac{1}{4}\right)c\right) = \boldsymbol{\tau}_6$$

and using Hermiticity (13.18), we have

$$H_{13} = \left(\Phi^{\tau_1,1}, H\Phi^{\tau_3,4}\right) = -\left(\Phi^{\tau_2-\tau_1,1}, H\Phi^{\tau_2-\tau_3,4}\right)^*$$
$$= -\left(\Phi^{\tau_2,1}, H\Phi^{\tau_6,4}\right)^* = -H_{26}^*.$$

Similarly, using (13.19), we obtain

$$H_{13} = \left(\Phi^{\tau_1,1}, H\Phi^{\tau_3,4}\right) = -\left(\Phi^{\tau_2-\tau_3,4}, H\Phi^{\tau_2-\tau_1,1}\right) = -\left(\Phi^{\tau_6,4}, H\Phi^{\tau_2,1}\right) = -H_{62}.$$

For the matrix element H_{31}, using Hermiticity (13.18), we have

$$\begin{aligned} H_{31} &= \left(\Phi^{\tau_3,4}, H\Phi^{\tau_1,1}\right) = -\left(\Phi^{\tau_2-\tau_3,4}, H\Phi^{\tau_2-\tau_1,1}\right)^* \\ &= -\left(\Phi^{\tau_6,4}, H\Phi^{\tau_2,1}\right)^* = -H_{62}^* \end{aligned}$$

and from (13.19) we obtain

$$H_{31} = \left(\Phi^{\tau_3,4}, H\Phi^{\tau_1,1}\right) = -\left(\Phi^{\tau_2-\tau_1,1}, H\Phi^{\tau_2-\tau_3,4}\right) = -\left(\Phi^{\tau_2,1}, H\Phi^{\tau_6,4}\right) = -H_{26}.$$

Summing up, we have

$$H_{13} = -H_{26}^* = -H_{62} = H_{31}^*. \tag{15.17}$$

Similarly, applying relations (13.18) and (13.19) to the rest of the matrix elements and taking into account

$$\begin{aligned} \boldsymbol{\tau}_2 - \boldsymbol{\tau}_3 &= \boldsymbol{\tau}_6, & \boldsymbol{\tau}_2 - \boldsymbol{\tau}_4 &= \boldsymbol{\tau}_5, \\ \boldsymbol{\tau}_2 - \boldsymbol{\tau}_5 &= \boldsymbol{\tau}_4, & \boldsymbol{\tau}_2 - \boldsymbol{\tau}_6 &= \boldsymbol{\tau}_3, \end{aligned}$$

we write the Hamiltonian matrix as

$$\begin{pmatrix} H_{11} & H_{12} & H_{13} & H_{14} & H_{15} & H_{16} \\ H_{12}^* & H_{11} & -H_{16}^* & -H_{15}^* & -H_{14}^* & -H_{13}^* \\ H_{13}^* & -H_{16} & H_{33} & H_{34} & H_{35} & H_{36} \\ H_{14}^* & -H_{15} & H_{34}^* & H_{44} & H_{45} & H_{35} \\ H_{15}^* & -H_{14} & H_{35}^* & H_{45}^* & H_{44} & H_{34} \\ H_{16}^* & -H_{13} & H_{36}^* & H_{35}^* & H_{34}^* & H_{33} \end{pmatrix}, \tag{15.18}$$

where the diagonal matrix elements H_{11}, H_{33}, H_{44} are real.

Now we obtain the *explicit* form of the matrix elements of the Hamiltonian. In calculating them, for simplicity, we restrict ourselves to one coordination sphere. The corresponding set of fixed vectors $\mathbf{Q}_p \equiv (\mathbf{R} + \boldsymbol{\tau}_{s'})_p$ is given in Table 15.5. It also lists the transformations of the group G_p, as well as the transformations α that map all vectors of the coordination sphere to a fixed $\mathbf{Q}_p$. In the case under consideration, it suffices to choose one fixed vector for each sphere, so the index l of $\mathbf{Q}_p$ is omitted. The vector $\mathbf{a}_1 = \left(\frac{1}{2}a, \frac{\sqrt{3}}{2}a, 0\right)$ gives fixed vectors $\mathbf{Q}_p$ for $s' = 1, 2$. The vectors $\mathbf{a}_1 = \left(\frac{1}{2}a, \frac{\sqrt{3}}{2}a, 0\right)$ and $-\mathbf{a}_2 = \left(\frac{1}{2}a, -\frac{\sqrt{3}}{2}a, 0\right)$ give fixed vectors $\mathbf{Q}_p$ for $s' = 3, 4, 5, 6$.

Since all representations are one dimensional (Table 15.3), relations (13.26), (13.28) and (13.29) do not provide any new information about the independent energy

Table 15.5 Fixed vectors $\mathbf{Q}_p \equiv (\mathbf{R}+\boldsymbol{\tau}_{s'})_p$, elements of the group G_p and transformations α that map all vectors of the coordination sphere to a fixed $\mathbf{Q}_p$ for the $2H$-TaSe$_2$ structure

Matrix elem.	Fixed vector $\mathbf{Q}_p \equiv (\mathbf{R}+\boldsymbol{\tau}_{s'})_p$	Nos. of $\alpha_i \in G_p$	Nos. of α_i that map vectors of the coord. sphere to $\mathbf{Q}_p$
H_{11}	$\mathbf{a}_1 = \left(\frac{1}{2}a, \frac{\sqrt{3}}{2}a, 0\right)$	1, 18	1, 4, 5, 10, 11, 12
H_{12}	$\boldsymbol{\tau}_2 = \left(0, 0, \frac{1}{2}c\right)$	1, 4, 5, 19, 20, 21	1, 10
H_{13}	$\boldsymbol{\tau}_3 = \left(0, -\frac{\sqrt{3}}{3}a, (z_0-\frac{1}{4})c\right)$	1, 19	1, 4, 5, 10, 11, 12
H_{14}	$\boldsymbol{\tau}_4 = \left(0, -\frac{\sqrt{3}}{3}a, -(z_0-\frac{1}{4})c\right)$	1, 19	1, 4, 5, 10, 11, 12
H_{15}	$\boldsymbol{\tau}_5 = \left(0, \frac{\sqrt{3}}{3}a, (z_0+\frac{1}{4})c\right)$	1, 19	1, 4, 5, 10, 11, 12
H_{16}	$\boldsymbol{\tau}_6 = \left(0, \frac{\sqrt{3}}{3}a, -(z_0+\frac{1}{4})c\right)$	1, 19	1, 4, 5, 10, 11, 12
H_{33}	$\mathbf{a}_1+\boldsymbol{\tau}_3 = \left(\frac{1}{2}a, \frac{\sqrt{3}}{6}a, (z_0-\frac{1}{4})c\right)$	1	1, 19
H_{34}	$\mathbf{a}_1+\boldsymbol{\tau}_4 = \left(\frac{1}{2}a, \frac{\sqrt{3}}{6}a, -(z_0-\frac{1}{4})c\right)$	1	1, 19
H_{35}	$-\mathbf{a}_2+\boldsymbol{\tau}_5 = \left(-\frac{1}{2}a, -\frac{\sqrt{3}}{6}a, (z_0+\frac{1}{4})c\right)$	1	1, 19
H_{36}	$-\mathbf{a}_2+\boldsymbol{\tau}_6 = \left(-\frac{1}{2}a, -\frac{\sqrt{3}}{6}a, -(z_0+\frac{1}{4})c\right)$	1	1, 19
H_{44}	$\mathbf{a}_1+\boldsymbol{\tau}_4 = \left(\frac{1}{2}a, \frac{\sqrt{3}}{6}a, -(z_0-\frac{1}{4})c\right)$	1	1, 19
H_{45}	$\mathbf{a}_1+\boldsymbol{\tau}_5 = \left(\frac{1}{2}a, \frac{\sqrt{3}}{6}a, (z_0+\frac{1}{4})c\right)$	1	1, 19

integrals. From equation (13.24) we obtain the following expressions for the matrix elements of the Hamiltonian[5]:

$$H_{11} = I^{11}(0;0) + 2I^{11}(0;\mathbf{a}_1)\left(\cos\xi + 2\cos\left(\frac{1}{2}\xi\right)\cos\left(\frac{\sqrt{3}}{2}\eta\right)\right),$$

$$H_{12} = 2I^{11}(0;\tau_2)\cos\left(\frac{1}{2}\zeta\right),$$

$$H_{13} = I^{14}(0;\tau_3)2i\sin\left(\left(z_0-\frac{1}{4}\right)\zeta\right)$$
$$\times\left[\exp\left(-i\frac{\sqrt{3}}{3}\eta\right) + \exp\left(i\frac{\sqrt{3}}{6}\eta\right)2\cos\left(\frac{1}{2}\xi\right)\right],$$

$$H_{14} = I^{14}(0;\tau_4)\left(-2i\sin\left(\left(z_0-\frac{1}{4}\right)\zeta\right)\right)$$

[5] The matrix elements contain energy integrals of the form $I^{jj'}(\boldsymbol{\tau}_s;\mathbf{Q}_p)$, where $\boldsymbol{\tau}_s$ is nonzero. This kind of integrals did not appear in Chap. 13.

$$\times \left[\exp\left(-i\frac{\sqrt{3}}{3}\eta\right) + \exp\left(i\frac{\sqrt{3}}{6}\eta\right) 2\cos\left(\frac{1}{2}\xi\right) \right],$$

$$H_{15} = I^{14}(0;\tau_5)2i\sin\left(\left(z_0 + \frac{1}{4}\right)\zeta\right)$$

$$\times \left[\exp\left(i\frac{\sqrt{3}}{3}\eta\right) + \exp\left(-i\frac{\sqrt{3}}{6}\eta\right) 2\cos\left(\frac{1}{2}\xi\right) \right],$$

$$H_{16} = I^{14}(0;\tau_6)\left(-2i\sin\left(\left(z_0 + \frac{1}{4}\right)\zeta\right)\right)$$

$$\times \left[\exp\left(i\frac{\sqrt{3}}{3}\eta\right) + \exp\left(-i\frac{\sqrt{3}}{6}\eta\right) 2\cos\left(\frac{1}{2}\xi\right) \right],$$

$$H_{33} = I^{44}(\tau_3;\tau_3) + I^{44}(\tau_3;\mathbf{a}_1 + \tau_3)\exp\left(i\frac{\sqrt{3}}{2}\eta\right) 2\cos\left(\frac{1}{2}\xi\right),$$

$$H_{34} = \exp\left(-i2\left(z_0 - \frac{1}{4}\right)\zeta\right)$$

$$\times \left[I^{44}(\tau_3;\tau_4) + I^{44}(\tau_3;\mathbf{a}_1 + \tau_4)\exp\left(i\frac{\sqrt{3}}{2}\eta\right) 2\cos\left(\frac{1}{2}\xi\right) \right],$$

$$H_{35} = \exp\left(i\left(2\frac{\sqrt{3}}{3}\eta + \frac{1}{2}\zeta\right)\right)$$

$$\times \left[I^{44}(\tau_3;\tau_5) + I^{44}(\tau_3;-\mathbf{a}_2 + \tau_5)\exp\left(-i\frac{\sqrt{3}}{2}\eta\right) 2\cos\left(\frac{1}{2}\xi\right) \right],$$

$$H_{36} = \exp\left(i\left(2\frac{\sqrt{3}}{3}\eta - 2z_0\zeta\right)\right)$$

$$\times \left[I^{44}(\tau_3;\tau_6) + I^{44}(\tau_3;-\mathbf{a}_2 + \tau_6)\exp\left(-i\frac{\sqrt{3}}{2}\eta\right) 2\cos\left(\frac{1}{2}\xi\right) \right],$$

$$H_{44} = I^{44}(\tau_4;\tau_4) + I^{44}(\tau_4;\mathbf{a}_1 + \tau_4)\exp\left(i\frac{\sqrt{3}}{2}\eta\right) 2\cos\left(\frac{1}{2}\xi\right),$$

$$H_{45} = \exp\left(i\left(2\frac{\sqrt{3}}{3}\eta + 2z_0\zeta\right)\right)$$

$$\times \left[I^{44}(\tau_4;\tau_5) + I^{44}(\tau_4;-\mathbf{a}_2 + \tau_5)\exp\left(-i\frac{\sqrt{3}}{2}\eta\right) 2\cos\left(\frac{1}{2}\xi\right) \right],$$

where we use the following brief notation: $\xi = ak_x$, $\eta = ak_y$ and $\zeta = ck_z$.

The interplanar distance in $2H$-$TaSe_2$ is much larger than the distance between Ta atoms in the same plane (6.350 Å vs. 3.436 Å). For this reason, and also for comparison with the ARPES data, we consider the dispersion law only in the $k_z = 0$ plane of the Brillouin zone. The matrix elements of the Hamiltonian in the $k_z = 0$ plane are given by

$$H_{11} = I^{11}(0; 0) + 2I^{11}(0; \mathbf{a}_1)\left(\cos\xi + 2\cos\left(\frac{1}{2}\xi\right)\cos\left(\frac{\sqrt{3}}{2}\eta\right)\right),$$

$$H_{12} = 2I^{11}(0; \tau_2),$$

$$H_{13} = H_{14} = H_{15} = H_{16} = 0,$$

$$H_{33} = I^{44}(\tau_3; \tau_3) + I^{44}(\tau_3; \mathbf{a}_1 + \tau_3)\exp\left(i\frac{\sqrt{3}}{2}\eta\right)2\cos\left(\frac{1}{2}\xi\right),$$

$$H_{34} = I^{44}(\tau_3; \tau_4) + I^{44}(\tau_3; \mathbf{a}_1 + \tau_4)\exp\left(i\frac{\sqrt{3}}{2}\eta\right)2\cos\left(\frac{1}{2}\xi\right),$$

$$H_{35} = \exp\left(i2\frac{\sqrt{3}}{3}\eta\right) \times \left[I^{44}(\tau_3; \tau_5) + I^{44}(\tau_3; -\mathbf{a}_2 + \tau_5)\exp\left(-i\frac{\sqrt{3}}{2}\eta\right)2\cos\left(\frac{1}{2}\xi\right)\right],$$

$$H_{36} = \exp\left(i2\frac{\sqrt{3}}{3}\eta\right) \times \left[I^{44}(\tau_3; \tau_6) + I^{44}(\tau_3; -\mathbf{a}_2 + \tau_6)\exp\left(-i\frac{\sqrt{3}}{2}\eta\right)2\cos\left(\frac{1}{2}\xi\right)\right],$$

$$H_{44} = I^{44}(\tau_4; \tau_4) + I^{44}(\tau_4; \mathbf{a}_1 + \tau_4)\exp\left(i\frac{\sqrt{3}}{2}\eta\right)2\cos\left(\frac{1}{2}\xi\right),$$

$$H_{45} = \exp\left(i2\frac{\sqrt{3}}{3}\eta\right) \times \left[I^{44}(\tau_4; \tau_5) + I^{44}(\tau_4; -\mathbf{a}_2 + \tau_5)\exp\left(-i\frac{\sqrt{3}}{2}\eta\right)2\cos\left(\frac{1}{2}\xi\right)\right].$$

For brevity, we introduce shorthand notation for the energy integrals (Table 15.6) and rewrite the matrix elements of the Hamiltonian in the $k_z = 0$ plane using this notation

$$H_{11} = x_1 + 2x_2\left(\cos\xi + 2\cos\left(\frac{1}{2}\xi\right)\cos\left(\frac{\sqrt{3}}{2}\eta\right)\right),$$

$$H_{12} = x_3,$$

Table 15.6 Shorthand notation for the energy integrals in the $2H$-TaSe$_2$ structure

x_i	Integral[a]	x_i	Integral[a]	x_i	Integral[a]
x_1	$I^{11}(0;0)$	x_6	$I^{44}(\boldsymbol{\tau}_3;\boldsymbol{\tau}_4)$	x_{11}	$I^{44}(\boldsymbol{\tau}_3;-\mathbf{a}_2+\boldsymbol{\tau}_6)$
x_2	$I^{11}(0;\mathbf{a}_1)$	x_7	$I^{44}(\boldsymbol{\tau}_3;\mathbf{a}_1+\boldsymbol{\tau}_4)$	x_{12}	$I^{44}(\boldsymbol{\tau}_4;\boldsymbol{\tau}_4)$
x_3	$I^{11}(0;\boldsymbol{\tau}_2)$	x_8	$I^{44}(\boldsymbol{\tau}_3;\boldsymbol{\tau}_5)$	x_{13}	$I^{44}(\boldsymbol{\tau}_4;\mathbf{a}_1+\boldsymbol{\tau}_4)$
x_4	$I^{44}(\boldsymbol{\tau}_3;\boldsymbol{\tau}_3)$	x_9	$I^{44}(\boldsymbol{\tau}_3;-\mathbf{a}_2+\boldsymbol{\tau}_5)$	x_{14}	$I^{44}(\boldsymbol{\tau}_4;\boldsymbol{\tau}_5)$
x_5	$I^{44}(\boldsymbol{\tau}_3;\mathbf{a}_1+\boldsymbol{\tau}_3)$	x_{10}	$I^{44}(\boldsymbol{\tau}_3;\boldsymbol{\tau}_6)$	x_{15}	$I^{44}(\boldsymbol{\tau}_4;-\mathbf{a}_2+\boldsymbol{\tau}_5)$

[a] Ref. [19]

$$
\begin{aligned}
&H_{13}=H_{14}=H_{15}=H_{16}=0,\\
&H_{33}=x_4+x_5\exp\left(i\frac{\sqrt{3}}{2}\eta\right)2\cos\left(\frac{1}{2}\xi\right),\\
&H_{34}=x_6+x_7\exp\left(i\frac{\sqrt{3}}{2}\eta\right)2\cos\left(\frac{1}{2}\xi\right),\\
&H_{35}=\exp\left(i2\frac{\sqrt{3}}{3}\eta\right)\left[x_8+x_9\exp\left(-i\frac{\sqrt{3}}{2}\eta\right)2\cos\left(\frac{1}{2}\xi\right)\right],\\
&H_{36}=\exp\left(i2\frac{\sqrt{3}}{3}\eta\right)\left[x_{10}+x_{11}\exp\left(-i\frac{\sqrt{3}}{2}\eta\right)2\cos\left(\frac{1}{2}\xi\right)\right],\\
&H_{44}=x_{12}+x_{13}\exp\left(i\frac{\sqrt{3}}{2}\eta\right)2\cos\left(\frac{1}{2}\xi\right),\\
&H_{45}=\exp\left(i2\frac{\sqrt{3}}{3}\eta\right)\left[x_{14}+x_{15}\exp\left(-i\frac{\sqrt{3}}{2}\eta\right)2\cos\left(\frac{1}{2}\xi\right)\right].
\end{aligned}
$$

Thus, the matrix of the Hamiltonian (15.18) in the $k_z=0$ becomes

$$
\begin{pmatrix}
H_{11} & H_{12} & 0 & 0 & 0 & 0\\
H_{12} & H_{11} & 0 & 0 & 0 & 0\\
0 & 0 & H_{33} & H_{34} & H_{35} & H_{36}\\
0 & 0 & H_{34}^* & H_{44} & H_{45} & H_{35}\\
0 & 0 & H_{35}^* & H_{45}^* & H_{44} & H_{34}\\
0 & 0 & H_{36}^* & H_{35}^* & H_{34}^* & H_{33}
\end{pmatrix}. \tag{15.19}
$$

Finally, we consider the form of the matrix of the Hamiltonian in $2H$-TaSe$_2$ and expressions for the matrix elements at high symmetry points in the Brillouin zone.

15.5 Hamiltonian Matrix at High Symmetry Points

Let us consider the explicit form of (nonzero) matrix elements of matrix (15.19) at the high symmetry points in the Brillouin zone indicated in Table 15.2. We confine ourselves to the points Γ, K, M, T, Σ[6], where the fundamental differences between the LDA calculation (see Fig. 15.3) and dispersion law obtained from the ARPES experiment (see Fig. 15.2) are most apparent:

$$
\begin{aligned}
\Gamma\colon & H_{11} = x_1 + 6x_2, \quad H_{12} = 2x_3, \quad H_{33} = x_4 + 2x_5,\\
& H_{34} = x_6 + 2x_7, \quad H_{35} = x_8 + 2x_9, \quad H_{36} = x_{10} + 2x_{11},\\
& H_{44} = x_{12} + 2x_{13}, \quad H_{45} = x_{14} + 2x_{15}.\\
K\colon & H_{11} = x_1 - 3x_2, \quad H_{12} = 2x_3, \quad H_{33} = x_4 - x_5,\\
& H_{34} = x_6 - x_7, \quad H_{35} = x_8 - x_9, \quad H_{36} = x_{10} - x_{11},\\
& H_{44} = x_{12} - x_{13}, \quad H_{45} = x_{14} - x_{15}.\\
M\colon & H_{11} = x_1 - 2x_2, \quad H_{12} = 2x_3, \quad H_{33} = x_4,\\
& H_{34} = x_6, \quad H_{35} = \gamma_1 x_8, \quad H_{36} = \gamma_1 x_{10}, \quad H_{44} = x_{12}, \quad H_{45} = \gamma_1 x_{14}.\\
T\colon & H_{11} = x_1 + x_2, \quad H_{12} = 2x_3, \quad H_{33} = x_4 + x_5, \quad H_{34} = x_6 + x_7,\\
& H_{35} = x_8 + x_9, \quad H_{36} = x_{10} + x_{11}, \quad H_{44} = x_{12} + x_{13}, \quad H_{45} = x_{14} + x_{15}.\\
\Sigma\colon & H_{11} = x_1 + 2x_2, \quad H_{12} = 2x_3, \quad H_{33} = x_4 + (1+i)x_5,\\
& H_{34} = x_6 + (1+i)x_7, \quad H_{35} = \gamma_2\left[x_8 + (1-i)x_9\right],\\
& H_{36} = \gamma_2\left[x_{10} + (1-i)x_{11}\right], \quad H_{44} = x_{12} + (1+i)x_{13},\\
& H_{45} = \gamma_2\left[x_{14} + (1-i)x_{15}\right].
\end{aligned}
$$

Here $\gamma_1 = \exp\left(i\frac{2\pi}{3}\right) = -\frac{1}{2} + i\frac{\sqrt{3}}{2}$ and $\gamma_2 = \exp\left(i\frac{\pi}{3}\right) = \frac{1}{2} + i\frac{\sqrt{3}}{2}$. The Hamiltonian matrix has the form (15.19) at all the above points.

15.6 Interim Conclusion

In this chapter, we have applied consistent treatment of symmetry in the tight-binding method, described in Chap. 13, to the $2H$-TaSe$_2$ structure. We studied the crystal structure, reciprocal lattice and Brillouin zone of $2H$-TaSe$_2$. We obtained an analytical form of the matrix elements of the $2H$-TaSe$_2$ Hamiltonian, which can also be used for any other compound having the same structure.

[6] Points T and Σ are chosen at $\mu = 1/2$.

We have shown that it is the hybridization of Ta d-states in the conduction band that is responsible for the direct interaction of Ta layers, rather than the interaction through Se p_z-orbitals, as originally assumed in [14].

Once the independent energy integrals x_i are calculated, the expressions for the matrix elements of the Hamiltonian can be used, for example, to calculate the Fermi velocity $v_F = \partial\varepsilon(\mathbf{k})/\partial\mathbf{k}|_{\varepsilon=\varepsilon_F}$ and, most importantly, to obtain the spectral function $A(\mathbf{k}, \omega)$, which is directly measured in the ARPES experiment. The independent energy integrals x_i can be obtained from the LDA calculation. The simplest version of the 2D interpolation of the experimental $\varepsilon(k)$ by the tight-binding method is presented in [10] (for the Slater-Koster interpolation see, e.g. [20]).

For simplicity, we have demonstrated the calculation without spin. Since Ta is a heavy element (atomic number $Z = 73$), relativistic effects must be taken into account to obtain more accurate quantitative results. It is possible to calculate all matrix elements and interpolate the experimental $\varepsilon(k)$ in the tight-binding method taking the spin–orbit interaction into account following [12, 21]. For spin-polarized calculations in LDA see, e.g. [10, 17]; in particular, they remove the accidental degeneracy at the point K.

References

1. J.A. Wilson, F.J. Di Salvo, S. Mahajan, Adv. Phys. **24:2**, 117 (1975)
2. L.F. Mattheiss, Phys. Rev. B **8**, 3719 (1973)
3. D.E. Moncton, J.D. Axe, F.J. Di Salvo, Phys. Rev. B **16**, 801 (1977)
4. K. Rossnagel, J. Phys. Condens. Matter **23**, 213001 (2011)
5. S.V. Borisenko, A.A. Kordyuk, A.N. Yaresko, V.B. Zabolotnyy, D.S. Inosov, R. Schuster, B. Buchner, R. Weber, R. Follath, L. Patthey, H. Berger, Phys. Rev. Lett. **100**, 196402 (2008)
6. E.Z. Kuchinskii, I.A. Nekrasov, M.V. Sadovskii, JETP **114**, 671 (2012)
7. D.S. Inosov, D.V. Evtushinsky, V.B. Zabolotnyy, A.A. Kordyuk, B. Buchner, R. Follath, H. Berger, S.V. Borisenko, Phys. Rev. B **79**, 125112 (2009)
8. G. Wexler, A.M. Woolley, J. Phys. C: Solid State Phys. **9**, 1185 (1976)
9. H.E. Brauer, H.I. Starnberg, L.J. Holleboom, H.P. Hughes, V.N. Strocov, J. Phys. Condens. Matter **13**, 9879 (2001)
10. J. Laverock, D. Newby, E. Abreu, R. Averitt, K.E. Smith, R.P. Singh, G. Balakrishnan, J. Adell, T. Balasubramanian, Phys. Rev. B **88**, 035108 (2013)
11. J.C. Slater, *Quantum Theory of Molecules and Solids, Vol.2: Symmetry and Energy Bands in Crystals* (McGraw-Hill, New York, 1965)
12. R.F. Egorov, B.I. Reser, V.P. Shirokovskii, Phys. Stat. Sol. **26**, 391 (1968)
13. G.F. Koster, Solid State Phys. **5**, 173 (1957)
14. E.O. Zavarygin, *Consistent Treatment of Symmetry in the Tight Binding Method for Conduction Band of* $2H$-*TaSe*$_2$, Bachelor's Thesis, Ural Federal University [in Russian] (2012)
15. O.K. Andersen, Phys. Rev. B **12**, 3060 (1975)
16. U. Barth, L.J. Hedin, J. Phys. C: Solid State Phys. **5**, 1629 (1972)
17. M.D. Johannes, I.I. Mazin, C.A. Howells, Phys. Rev. B **73**, 205102 (2006)
18. G.F. Koster, J.O. Dimmock, R.G. Wheeler, H. Statz, *Properties of the Thirty-Two Point Groups* (M.I.T. Press, Cambridge, Mass, 1963)
19. N.B. Melnikov, B.I. Reser, Phys. Scr. (submitted) 2022

20. Y.W. Li, J. Jiang, H.F. Yang, D. Prabhakaran, Z.K. Liu, L.X. Yang, Y.L. Chen, Phys. Rev. B **97**, 115118 (2018)
21. R.F. Egorov, B.I. Reser, V.P. Shirokovskii, Trudy Instituta Fiziki Metallov [in Russian] **27**, 98 (1968)

Chapter 16
Symmetrized Plane Waves

... symmetry and group theory took years to develop. But eventually this work had quite fundamental applications. It spread the basic truth that laws of nature have simple invariance properties. (The Recollections of Eugene P. Wigner as told to Andrew Szanton, Plenum, New York, 1992)

In many problems of solid-state physics, including calculation of electron energy bands, it is necessary to construct symmetrized plane waves (SPWs).

A SPW is a linear combination of plane waves that belongs to a row of an irreducible representation of the group of the wavevector $G_{\mathbf{k}}$. Use of SPWs greatly simplifies the calculation of the electron energy bands. Indeed, by Wigner's theorem (Sect. 7.2.2), in the presence of symmetry the Hamiltonian matrix elements for any two functions belonging to different rows of the same representation are equal to zero, and the matrix elements for any two functions belonging to the same row of the same representation are equal to each other. Then the Hamiltonian matrix is divided into blocks corresponding to individual lines of each representation, which makes it possible to effectively increase the number of the coordinate functions without increasing the amount of calculations. Moreover, the blocks corresponding to different rows of the same representation are identical (Appendix C.1), and hence it is sufficient to consider symmetrized functions that belong to one, for example, the first, row of each irreducible representation. Note that the use of symmetrized functions not only simplifies the calculation of the energy bands, but also makes it possible to classify electronic states according to symmetry.

Simple illustrative examples of constructing SPWs by hand can be found in textbooks (see, e.g. [1–3]). A set of tables of coefficients for constructing SPWs for all 73 *symmorphic* space groups has been given in [4]. SPWs for some nonsymmorphic space groups are present in the literature (see, e.g. [5–7]). Here we describe an efficient method of calculating SPWs [8], which is applicable to both symmorphic nonsymmorphic space groups. In energy band calculations, SPWs are useful in the orthogonalized plane waves (OPW) and augmented plane waves (APW) methods (for the use in the OPW method, see Chap. 9). Another application of SPWs is calculation of integrals over the Brillouin zone (see, e.g. [9]).

N. B. Melnikov and B. I. Reser, *Space Group Representations*,
https://doi.org/10.1007/978-3-031-13991-8_16

16.1 Projection Operator

To construct SPWs for a symmetry vector **k** in the Brillouin zone, we use the projection operator (8.16), which generates a function belonging to a ρ-th row of a j-th small representation:

$$O^{\mathbf{k}j}_{\rho\sigma} = \frac{l_j}{n(G_{\mathbf{k}})} \sum_{\{\alpha|\mathbf{t}\}\in G_{\mathbf{k}}} d^{\mathbf{k}j*}_{\rho\sigma}(\{\alpha|\mathbf{t}\})\, P_{\{\alpha|\mathbf{t}\}} \quad (\sigma \text{ is fixed}). \tag{16.1}$$

Here $d^{\mathbf{k}j}$ is the j-th irreducible (small) representation of the group of the wavevector $G_{\mathbf{k}}$; l_j is the dimension of $d^{\mathbf{k}j}$; $n(G_{\mathbf{k}})$ is the number of elements of the group $G_{\mathbf{k}}$. The small representation can be written as

$$d^{\mathbf{k}j}(\{\alpha|\mathbf{t}\}) = d^{\mathbf{k}j}(\{\alpha|\mathbf{R}_n + \boldsymbol{\tau}_\alpha\}) = \mathrm{e}^{-\mathrm{i}\mathbf{k}\mathbf{R}_n}\, \Delta^{\mathbf{k}j}(\{\alpha|\boldsymbol{\tau}_\alpha\})\,, \tag{16.2}$$

where $\mathbf{R}_n$ is the lattice vector, $\boldsymbol{\tau}_\alpha$ is the nonprimitive translation and $\Delta^{\mathbf{k}j}$ is the j-th irreducible projective representation of the *point* group of the wavevector $G^0_{\mathbf{k}}$. Then the action of the operator (16.1) on the Bloch functions (including plane waves) is given by formula (8.20),

$$\tilde{O}^{\mathbf{k}j}_{\rho\sigma} = \frac{l_j}{n(G^0_{\mathbf{k}})} \sum_{\alpha\in G^0_{\mathbf{k}}} \Delta^{\mathbf{k}j*}_{\rho\sigma}(\{\alpha|\boldsymbol{\tau}_\alpha\})\, P_{\{\alpha|\boldsymbol{\tau}_\alpha\}}, \tag{16.3}$$

where $n(G^0_{\mathbf{k}})$ is the number of elements of the group $G^0_{\mathbf{k}}$ (for details, see Sect. 8.2). The operator (16.3) is determined solely by the irreducible projective representations of the *point* group of the wavevector $G^0_{\mathbf{k}}$, and the number of terms $n(G^0_{\mathbf{k}})$ in the sum (16.3) is much smaller than the number of terms $n(G_{\mathbf{k}})$ in the sum (16.1).

Matrix elements between the SPWs belonging to the same row of the same representation are equal to each other (see Sect. 8.2.2). Hence it is sufficient to calculate SPWs that belong to only one row of each irreducible representation $\Delta^{\mathbf{k}j}$. All SPWs belonging to the *first* row of the j-th irreducible representation of the group $G_{\mathbf{k}}$ can be constructed using the projection operator

$$\tilde{O}^{\mathbf{k}j}_{11} = \frac{l_j}{n(G^0_{\mathbf{k}})} \sum_{\alpha\in G^0_{\mathbf{k}}} \Delta^{\mathbf{k}j*}_{11}(\{\alpha|\boldsymbol{\tau}_\alpha\})\, P_{\{\alpha|\boldsymbol{\tau}_\alpha\}}\,. \tag{16.4}$$

In what follows, we will proceed from the assumption that the unitary matrices $\Delta^{\mathbf{k}j}$ have only one nonzero element in each row. For brevity, this type of matrices will be called canonical here. If the theory of induced representations is used to obtain irreducible representations, the matrices immediately have the canonical form. In other cases, it is easy to reduce the matrix to the canonical form using the corresponding unitary transformation.[1]

Let $\{e^{i\mathbf{k}_\mu \mathbf{r}}\}$ be the set of initial (nonsymmetrized) plane waves labeled by $\mathbf{k}_\mu = \mathbf{k} + \mathbf{K}_\mu$ ($\mathbf{K}_\mu$ is the reciprocal lattice vector). Then the action of the operator $P_{\{\alpha|\boldsymbol{\tau}_\alpha\}}$ on a plane wave is written as[2]

$$P_{\{\alpha|\boldsymbol{\tau}_\alpha\}}\, e^{i\mathbf{k}_\mu \mathbf{r}} = e^{-i\alpha\mathbf{k}_\mu \boldsymbol{\tau}_\alpha}\, e^{i\alpha\mathbf{k}_\mu \mathbf{r}}. \tag{16.5}$$

Recalling the coset decomposition (5.19) of the group of the wavevector $G_{\mathbf{k}}$ with respect to the translation group T,

$$G_{\mathbf{k}} = \bigcup_m T\{\alpha_m|\boldsymbol{\tau}_{\alpha_m}\}, \qquad m = 1, 2, \ldots, n(G^0_{\mathbf{k}}) = n(G_{\mathbf{k}})/n(T),$$

we see from formulas (16.4) and (16.5) that the SPW $\tilde{O}^{\mathbf{k}j}_{11} e^{i\mathbf{k}_\mu \mathbf{r}}$ is a linear combination of no more than $n(G^0_{\mathbf{k}})$ plane waves $e^{i\alpha_m \mathbf{k}_\mu \mathbf{r}}$.

The SPWs constructed by applying the projection operator (16.4) to *all* plane waves $e^{i\mathbf{k}_\mu \mathbf{r}}$ may be linearly dependent. In particular, the action of the projection operator $\tilde{O}^{\mathbf{k}j}_{11}$ on a plane wave $e^{i\alpha_m \mathbf{k}_\mu \mathbf{r}}$ that is present in the SPW $\tilde{O}^{\mathbf{k}j}_{11} e^{i\mathbf{k}_\mu \mathbf{r}}$ gives the same SPW (up to a phase factor). As we show, a similar result holds for symmetrized functions corresponding to any irreducible representation $\mathcal{D}$ of an arbitrary group G.

Let φ be an arbitrary function and $\{\varphi_m\}$ is the set of functions such that $\varphi_m = c_m \hat{R}_m \varphi$, where c_m is a nonzero constant and $\hat{R}_m$ is the shorthand notation for the symmetry operator corresponding to the element R_m of the group G. If the function φ_m is present in the symmetrized function

$$O_{11}\varphi = \frac{l}{n(G)} \sum_{R_i \in G} D^*_{11}(R_i)\hat{R}_i \varphi = \frac{l}{n(G)} \sum_{R_i \in G} D^*_{11}(R_i) c_i^{-1} \varphi_i,$$

then the coefficient $D^*_{11}(R_m)c_m^{-1}$ of the m-th term must be nonzero, i.e. $D_{11}(R_m) \neq 0$. From the assumption that the matrix $D(R_m)$ has the canonical form, it follows that $D_{11}(R_m)$ is the only nonzero matrix element in the first row: $D_{1\sigma}(R_m) = D_{11}(R_m)\,\delta_{1\sigma}$. Now we consider the symmetrized function obtained from φ_m:

---

[1] In some nonsymorphic groups there may be irreducible representations such that all matrices $\Delta^{\mathbf{k}j}$ cannot be simultaneously reduced to the canonical form. This case requires special consideration.

[2] Recall that the action of the operator $P_{\{\alpha|\mathbf{t}\}}$ on an arbitrary function $f(\mathbf{r})$ is determined by the relation

$$P_{\{\alpha|\mathbf{t}\}} f(\mathbf{r}) = f(\{\alpha|\mathbf{t}\}^{-1}\mathbf{r}) = f(\alpha^{-1}\mathbf{r} - \alpha^{-1}\mathbf{t}).$$

$$O_{11}\varphi_m = \frac{l}{n(G)} c_m \sum_i D^*_{11}(R_i)\hat{R}_i\hat{R}_m\varphi = \frac{l}{n(G)} c_m \sum_l D^*_{11}(R_l R_m^{-1})\hat{R}_l\varphi$$

$$= \frac{l}{n(G)} c_m \sum_l \sum_\sigma D^*_{1\sigma}(R_l)\, D^*_{\sigma 1}(R_m^{-1})\hat{R}_l\varphi = c_m \sum_\sigma D_{1\sigma}(R_m)\, O_{1\sigma}\varphi\,.$$

Using $D_{1\sigma}(R_m) = D_{11}(R_m)\,\delta_{1\sigma}$, we reduce the latter to

$$O_{11}\varphi_m = c_m D_{11}(R_m)\, O_{11}\varphi\,, \tag{16.6}$$

i.e. the two symmetrized functions coincide (up to a phase factor).

In our case, G is the group of the wavevector $G_{\mathbf{k}}$, the group element R_m is $\{\alpha_m|\mathbf{R}_n + \boldsymbol{\tau}_{\alpha_m}\}$, the matrix $D(R_m)$ is $d^{\mathbf{k}j}(\{\alpha_m|\mathbf{R}_n + \boldsymbol{\tau}_{\alpha_m}\})$, $\{\varphi_m\}$ is the set of the plane waves $\{\mathrm{e}^{\mathrm{i}\alpha_m\mathbf{k}_\mu\mathbf{r}}\}$ and φ is the plane wave $\mathrm{e}^{\mathrm{i}\mathbf{k}_\mu\mathbf{r}}$. From the condition $\varphi_m = c_m\hat{R}_m\varphi$ we easily obtain that $c_m = \mathrm{e}^{\mathrm{i}\alpha_m\mathbf{k}_\mu(\mathbf{R}_n+\boldsymbol{\tau}_{\alpha_m})}$. Then relation (16.6) is written in the form

$$O^{\mathbf{k}j}_{11}\,\mathrm{e}^{\mathrm{i}\alpha_m\mathbf{k}_\mu\mathbf{r}} = \mathrm{e}^{\mathrm{i}\alpha_m\mathbf{k}_\mu(\mathbf{R}_n+\boldsymbol{\tau}_{\alpha_m})}\, d^{\mathbf{k}j}_{11}(\{\alpha_m|(\mathbf{R}_n + \boldsymbol{\tau}_{\alpha_m})\})\; O^{\mathbf{k}j}_{11}\,\mathrm{e}^{\mathrm{i}\mathbf{k}_\mu\mathbf{r}}\,. \tag{16.7}$$

Taking (16.2) and (16.3) into account, we finally obtain

$$\tilde{O}^{\mathbf{k}j}_{11}\,\mathrm{e}^{\mathrm{i}\alpha_m\mathbf{k}_\mu\mathbf{r}} = \mathrm{e}^{\mathrm{i}\alpha_m\mathbf{k}_\mu\boldsymbol{\tau}_{\alpha m}}\,\Delta^{\mathbf{k}j}_{11}(\{\alpha_m|\boldsymbol{\tau}_{\alpha_m}\})\,\tilde{O}^{\mathbf{k}j}_{11}\,\mathrm{e}^{\mathrm{i}\mathbf{k}_\mu\mathbf{r}}\,. \tag{16.8}$$

Clearly, if we apply the projection operator (16.4) to a plane wave $\mathrm{e}^{\mathrm{i}\mathbf{k}_\nu\mathbf{r}}$ that is not one of the plane waves $\mathrm{e}^{\mathrm{i}\alpha_m\mathbf{k}_\mu\mathbf{r}}$ appearing in the already constructed SPWs, we obtain a linearly independent SPW.

16.2 Construction of Symmetrized Plane Waves

As we have shown, the SPW obtained from $\mathrm{e}^{\mathrm{i}\mathbf{k}_\mu\mathbf{r}}$ is a linear combination of the plane waves $\mathrm{e}^{\mathrm{i}\alpha_m\mathbf{k}_\mu\mathbf{r}}$, where α_m are the representatives of the coset decomposition of the group of the wavevector $G_{\mathbf{k}}$ over the translation subgroup T. To describe the algorithm for constructing the SPWs, we first need an expression for the symmetrization coefficients.

From relations (16.4) and (16.5) it follows that an SPW belonging to the first row of the j-th irreducible representations of the group of the wavevector $G_{\mathbf{k}}$ can be written as

$$\varphi^j_{\mathbf{k}_\mu} = \tilde{O}^{\mathbf{k}j}_{11}\,\mathrm{e}^{\mathrm{i}\mathbf{k}_\mu\mathbf{r}} = \frac{l_j}{n(G^0_{\mathbf{k}})} \sum_{\{\alpha_m|\boldsymbol{\tau}_{\alpha m}\}} c^j_{\alpha_m\mathbf{k}_\mu}\,\varphi_{\alpha_m\mathbf{k}_\mu}\,, \tag{16.9}$$

where

$$c^{j}_{\alpha_m \mathbf{k}_\mu} = \Delta^{\mathbf{k}j*}_{11}(\{\alpha_m|\boldsymbol{\tau}_{\alpha_m}\})\, \mathrm{e}^{-\mathrm{i}\alpha_m \mathbf{k}_\mu \boldsymbol{\tau}_{\alpha m}}\,, \tag{16.10}$$

$$\varphi_{\alpha_m \mathbf{k}_\mu} = \mathrm{e}^{\mathrm{i}\alpha_m \mathbf{k}_\mu \mathbf{r}}\,. \tag{16.11}$$

Some transformations α_p can map the vector $\mathbf{k}_\mu$ to the same resulting vector $\alpha_p \mathbf{k}_\mu$. Therefore, it is convenient to rewrite expression (16.9) in the form

$$\varphi^{j}_{\mathbf{k}_\mu} = \sum_{\{\alpha_n|\boldsymbol{\tau}_{\alpha_n}\}} C^{j}_{\alpha_n \mathbf{k}_\mu}\, \varphi_{\alpha_n \mathbf{k}_\mu}\,, \tag{16.12}$$

where the summation is carried out over the elements $\{\alpha_n|\boldsymbol{\tau}_{\alpha_n}\}$ for which all vectors $\alpha_n \mathbf{k}_\mu$ are different, and the corresponding symmetrization coefficients

$$C^{j}_{\alpha_n \mathbf{k}_\mu} = \frac{l_j}{n(G^0_{\mathbf{k}})} \sum_{\{\alpha_{p(n)}|\boldsymbol{\tau}_{\alpha_{p(n)}}\}} c^{j}_{\alpha_{p(n)} \mathbf{k}_\mu} \tag{16.13}$$

are nonzero.

If it is necessary to obtain only several symmetrized functions, the construction of SPWs can be performed directly. If it is required to find tens or hundreds of symmetrized functions, then the direct computation becomes excessively complicated and error-prone. There is a number of algorithms for constructing SPWs (see, e.g. [4, 10–13]) that are applicable only to symmorphic groups. In the case of nonsymmorphic groups, the problem becomes much more complicated. The point is that, for symmorphic groups, it is possible to classify SPWs according to the types of vectors $\mathbf{k}_\mu$, and the number of these types is small [4]. For nonsymmorphic groups, because of the factor $\mathrm{e}^{-\mathrm{i}\alpha_m \mathbf{k}_\mu \boldsymbol{\tau}_{\alpha m}}$, the classification of SPWs by type becomes very difficult [14]. The algorithm we present here is suitable for both symmorphic and nonsymorphic groups [8][3]. It consists of three stages.

(a) *Arrange the initial set of plane waves in ascending order of* $|\mathbf{k}_\mu|$

First, from all reciprocal lattice vectors, we select the vectors $\mathbf{K}_p$ such that $(\mathbf{K}_p)_x \geq (\mathbf{K}_p)_y \geq (\mathbf{K}_p)_z \geq 0$ and $|\mathbf{K}_{p_1}| \leq |\mathbf{K}_{p_2}|$ for $p_1 < p_2$ (any other reciprocal lattice vector $\mathbf{K}_\mu$ can be obtained from $\mathbf{K}_p$ by a rotation α_i from the crystal class G^0). For example, for the fcc lattice, using the fact that the components $(\mathbf{K}_p)_x$, $(\mathbf{K}_p)_y$, $(\mathbf{K}_p)_z$ are all even or all odd, we can immediately write any number of vectors $\mathbf{K}_p$: (000), (111), (200), etc.[4] Next, we construct all distinct vectors $\mathbf{k}_{pi} = \mathbf{k} + \alpha_i \mathbf{K}_p$ $(\alpha_i \in G^0)$, arrange them in ascending order of $|\mathbf{k}_{pi}|$ and divide them into sets of vectors with the same modulus—sets of equal modulus for short. We arrange the vectors $\mathbf{k}_{pi}$ into the

[3] Another computer algorithm that can be used for constructing SPWs for symmorphic and nonsymorphic groups SYMPW was presented and implemented by [15] (see also [1]).

[4] The components of the vector $\mathbf{K}_p$ are written in units of $2\pi/a$, where a is the lattice constant of the cubic lattice.

sets of equal modulus because an SPW can contain only the plane waves with the vectors $\mathbf{k}_{pi}$ having the same modulus. Note that a set of equal modulus may contain vectors $\mathbf{k}_{pi}$ that are not related to each other by any transformation α_m from $G^0_{\mathbf{k}}$.

(b) *Decompose reducible representations corresponding to the sets of equal modulus into irreducible components*

The plane waves corresponding to a set of equal modulus form a basis of a reducible projective representation $\mathcal{D}^{\mathbf{k}}$. Each representation $\mathcal{D}^{\mathbf{k}}$ decomposes into irreducible projective representations. The number of times an irreducible projective representation $\Delta^{\mathbf{k}j}$ enters the representation $\mathcal{D}^{\mathbf{k}}$ (the multiplicity of occurrence a_j) is calculated by the formula

$$a_j = \frac{1}{n(G^0_{\mathbf{k}})} \sum_{\{\alpha_m|\boldsymbol{\tau}_{\alpha m}\}} \chi^{\mathbf{k}}(\{\alpha_m|\boldsymbol{\tau}_{\alpha_m}\})\, \chi^{\mathbf{k}j*}(\{\alpha_m|\boldsymbol{\tau}_{\alpha_m}\}). \tag{16.14}$$

Here $\chi^{\mathbf{k}}(\{\alpha_m|\boldsymbol{\tau}_{\alpha_m}\})$ and $\chi^{\mathbf{k}j}(\{\alpha_m|\boldsymbol{\tau}_{\alpha_m}\})$ are the characters of the matrices $D^{\mathbf{k}}(\{\alpha_m|\boldsymbol{\tau}_{\alpha_m}\})$ and $\Delta^{\mathbf{k}j}(\{\alpha_m|\boldsymbol{\tau}_{\alpha_m}\})$, respectively. In the case of plane waves, for $\chi^{\mathbf{k}}(\{\alpha_m|\boldsymbol{\tau}_{\alpha_m}\})$, using formula (16.5), we can obtain

$$\chi^{\mathbf{k}}(\{\alpha_m|\boldsymbol{\tau}_{\alpha_m}\}) = \sum_{\mathbf{k}_\mu} \mathrm{e}^{-\mathrm{i}\alpha_m \mathbf{k}_\mu \boldsymbol{\tau}_{\alpha m}}\, \delta_{\alpha_m \mathbf{k}_\mu, \mathbf{k}_\mu}\,, \tag{16.15}$$

where the summation is carried out over all vectors $\mathbf{k}_\mu$ from the set of equal modulus.

The multiplicity of occurrence a_j indicates the number of SPWs transforming according to the j-th representation, which can be constructed from the plane waves of a given set of equal modulus.

Since the dimension of the reducible representation $\mathcal{D}^{\mathbf{k}}$ must equal to the sum of products of the dimensions l_j of the irreducible representations $\Delta^{\mathbf{k}j}$ and their multiplicities of the occurrence a_j, calculating the value $\sum_j a_j l_j$ and comparing it with the number of vectors in the sets of equal modulus allows us to make sure that the obtained sets are complete, i.e. that each of them contains all vectors $\mathbf{k}_\mu$ with the given modulus.

(c) *Construct the SPWs that belong to the j-th representation of the group of the wavevector*

The construction of the SPWs is carried out separately for each set of plane waves corresponding to a set of $\mathbf{k}_\mu$ with the given modulus.

According to (16.12) and (16.13), an SPW is completely determined by indicating the vectors $\alpha_n \mathbf{k}_\mu$ and the coefficients $C^j_{\alpha_n \mathbf{k}_\mu}$ of the plane waves that are present in the linear combination. It is convenient to organize calculations as follows.

We take the first vector of a set of equal modulus $\mathbf{k}_1$. If the matrix element $\Delta_{11}(\{\alpha_1|\boldsymbol{\tau}_{\alpha_1}\})$ is nonzero, we calculate the vector $\alpha_1\mathbf{k}_1$ and, using formula (16.10), we calculate the coefficient $C_{\alpha_1\mathbf{k}_1}$[5]. We store these values. Then we calculate $\alpha_2\mathbf{k}_1$

[5] This step can be superfluous, as usually α_1 is the identity element.

and $C_{\alpha_2\mathbf{k}_1}$. Compare the resulting vectors $\alpha_1\mathbf{k}_1$ and $\alpha_2\mathbf{k}_1$; if they coincide, add up the coefficients $C_{\alpha_1\mathbf{k}_1}$ and $C_{\alpha_2\mathbf{k}_1}$; otherwise, store $\alpha_2\mathbf{k}_1$ and $C_{\alpha_2\mathbf{k}_1}$. Next, we act on the vector $\mathbf{k}_1$ with the remaining elements α_m, and each time the newly obtained vector $\alpha_m\mathbf{k}_1$ is compared with the previously obtained ones and, if it coincides with any vector, we add up the corresponding coefficients, and omit the vector $\alpha_m\mathbf{k}_1$ itself. After all the elements $\{\alpha_m|\tau_{\alpha_m}\}$ have been exhausted, we obtain the set of symmetrization coefficients $C^j_{\alpha_n\mathbf{k}_1} \neq 0$ and vectors $\alpha_n\mathbf{k}_1$, which determine the first SPW $\varphi^j_{\mathbf{k}_1}$.

To construct the next SPW, we take the second vector of the set of equal modulus $\mathbf{k}_2$ and compare it with the vectors included in $\varphi^j_{\mathbf{k}_1}$; if $\mathbf{k}_2$ is already contained in $\varphi^j_{\mathbf{k}_1}$, we omit the second vector and go to the third, otherwise, we do the same with it as with the first one, and so on, until we obtain all a_j of the SPWs that transform according to the j-th representation. Note that the described procedure does not ensure that the SPW will be normalized.

The above algorithm was implemented in [8]. The appendix to [8] contains the code, a list of identifiers, and a test problem. The program is written for cubic structures. This restriction is related to two circumstances: (1) all wavevectors are expressed in units of $2\pi/a$, and nonprimitive translations are expressed in units of a; (2) the rotational parts of the symmetry transformations are given in a form that takes into account only the permutation and change in signs of the vector components. If necessary, a minor modification of the program allows one to remove these restrictions. For reasons of computational convenience, the program has a scale coefficent (SC), multiplication by which makes the components of the reduced wavevector integer. As the output, the program returns the following information about an SPW obtained from the vectors of a particular set of equal modulus: an array of vectors arranged in the order of their appearance in the SPW; an array of the corresponding symmetrization coefficients; number of the SPWs; an array indicating the number of vectors in each SPW.

The test problem presented in [8] is the construction of the SPWs for the group of the wavevector $G_{\mathbf{k}}$ at the point W of the Brillouin zone for the diamond structure. The matrices of irreducible projective representations are given in Appendix F. They are obtained using the method summarized in Chap. 5 and tables presented in Chap. 6.

References

1. M. El-Batanouny, F. Wooten, *Symmetry and Condensed Matter Physics: A Computational Approach* (Cambridge Univ. Press, Cambridge, 2008)
2. M. Dresselhaus, G. Dresselhaus, A. Jorio, *Group Theory: Application to the Physics of Condensed Matter* (Springer, Berlin, 2008)
3. W. Hergert, R.M. Geilhufe, *Group Theory in Solid State Physics and Photonics: Problem Solving with Mathematica* (Wiley, New York, 2018)
4. A. Luehrmann, Adv. Phys. **17**, 1 (1968)
5. W. Gorzkowski, Phys. Stat. Sol. **3**, 599 (1963)
6. J.C. Slater, Rev. Mod. Phys. **37**, 68 (1965)
7. R. Dirl, K. Payer, B.L. Davies, Comput. Phys. Commun. **98**, 98 (1996)

8. B.I. Reser, R.F. Egorov, *Construction of Symmetrized Plane Waves* [in Russian] (1973). Dep. VINITI, No. 6658–73
9. N.O. Folland, J. Osman, J. Math. Phys. **23**, 1538 (1982)
10. H. Schlosser, J. Phys. Chem. Solids **23**, 963 (1962)
11. V. Hoffstein, O. Moller, Computer Phys. Commun. **2**, 11 (1971)
12. V. Hoffstein, O. Moller, Computer Phys. Commun. **2**, 17 (1971)
13. V. Hoffstein, O. Moller, Computer Phys. Commun. **2**, 26 (1971)
14. B.I. Reser, *The OPW method and interpolation based on the tight binding in calculating the energy band structure of crystals* [in Russian]. Diss. for the degree of cand. of phys. and math. sciences, Inst. of Metal Phys., Ural Science Center, USSR Acad. of Sciences, Sverdlovsk (1972). 153 pp
15. E. Blokker, S. Flodmark, *Description of the Symmetry Projection. Programs SYMPRJ and SYMPW* (1972). Univ. Stockholm, Inst. Phys. (USIP) Rep. 72–13

Chapter 17
Symmetrized Bloch Sums

One record more, the last of all—and then
This chronicle of mine is done, fulfilled.
(A. Pushkin, The uncensored Boris Godunov (1825), transl. by
A. Wood, Univ. Wisconsin Press, Madison, WI, 2006)

Use of symmetrized functions simplifies the calculation of electronic states and makes it possible to classify them by symmetry. Since only certain combinations of plane waves and Bloch sums are always used as initial functions, it is necessary to construct only symmetrized plane waves and symmetrized Bloch sums (SBS). Construction of the symmetrized plane waves was discussed in the previous chapter. Here we focus on the construction of the SBS (for applications of SBS to energy band interpolation, see Chap. 14).

Construction of the SBS has a long history (see, e.g. [1]). Originally construction of the SBS was considered as a purely "manual" task. Often tables of symmetrized functions for a particular crystal structure made the subject of an entire publication (see, e.g. [2, 3]). The first mention of a solution to this problem on a computer and a link to an unpublished program can be found in [4]. The program using this algorithm was implemented in [5] and necessary corrections were described in [6]. Another computer program for constructing the SBS was published by [7], but it was designed mainly without taking into account the general theory of the projection operators developed in [4], and, in our opinion, is not efficient enough. In this chapter, we describe a simple and economical program for constructing the SBS [8].

17.1 Theory

17.1.1 Projection Matrix

We consider a group G of symmetry transformations g and the isomorphic group of symmetry operators P_g defined by the equality

N. B. Melnikov and B. I. Reser, *Space Group Representations*,
https://doi.org/10.1007/978-3-031-13991-8_17

$$P_g f(\mathbf{r}) = f(g^{-1}\mathbf{r}), \tag{17.1}$$

where $f(\mathbf{r})$ is an arbitrary function. Let

$$\mathbf{\Phi} = (\Phi_1, \Phi_2, \ldots, \Phi_n) \tag{17.2}$$

be the initial set of basis functions of some, in the general case, reducible representation $\mathcal{D}(g)$ of the group G, i.e. for all $g \in G$ we have

$$P_g \Phi_\nu = \sum_{\mu=1}^{n} \Phi_\mu D_{\mu\nu}(g). \tag{17.3}$$

To construct functions that belong to a σ-th row of a j-th irreducible representation, we apply the projection operator (see Sect. 7.1.3) to all the functions (17.2). With a minor change of notation, we write the projection operator as

$$\hat{S}^j_{\sigma\sigma} = \frac{l_j}{n(G)} \sum_{g \in G} D^{j*}_{\sigma\sigma}(g)\, P_g, \tag{17.4}$$

where $D^j(g)$ is the matrix of the irreducible representation of the group G, l_j is the dimension of the representation and $n(G)$ is the order of the group. Applying the operator (17.4) to the function Φ_ν and taking (17.3) into account, we have

$$\begin{aligned} \hat{S}^j_{\sigma\sigma}\Phi_\nu &= \frac{l_j}{n(G)} \sum_{g \in G} D^{j*}_{\sigma\sigma}(g)\, P_g \Phi_\nu = \frac{l_j}{n(G)} \sum_{g \in G} D^{j*}_{\sigma\sigma}(g) \sum_{\mu} \Phi_\mu D_{\mu\nu}(g) \\ &= \sum_{\mu} \Phi_\mu \left[\frac{l_j}{n(G)} \sum_{g \in G} D^{j*}_{\sigma\sigma}(g) D_{\mu\nu}(g) \right] \equiv \sum_{\mu} \Phi_\mu (S^j_{\sigma\sigma})_{\mu\nu}, \end{aligned} \tag{17.5}$$

where

$$(S^j_{\sigma\sigma})_{\mu\nu} = \frac{l_j}{n(G)} \sum_{g \in G} D^{j*}_{\sigma\sigma}(g) D_{\mu\nu}(g). \tag{17.6}$$

Hence

$$\hat{S}^j_{\sigma\sigma}\, \mathbf{\Phi} = \mathbf{\Phi}\, \mathbf{S}^j_{\sigma\sigma}, \tag{17.7}$$

where $\mathbf{S}^j_{\sigma\sigma}$ is the matrix with the elements (17.6), which we call the *projection matrix*.

Formulas (17.7) and (17.6) in principle solve the problem of constructing symmetrized functions, but in practical terms they are clearly unsatisfactory. After rather laborious calculations by formula (17.6) of all $n \times n$ matrix elements $(S^j_{\sigma\sigma})_{\mu\nu}$, in the general case, we obtain the matrix $\mathbf{S}^{j}_{\sigma\sigma}$ such that some of its columns will turn out to be zero, since not all functions Φ_{ν} have a nonzero projection onto the space of the j-th representation, but the rest of the columns will turn out to be linearly dependent, mutually nonorthogonal and not normalized.

Following Flodmark [4], we show that the projection matrix $\mathbf{S}^{j}_{\sigma\sigma}$ of dimension $n \times n$ can be replaced by a rectangular matrix of dimension $n \times m$ $(m < n)$ with nonzero orthonormal columns, which immediately gives linearly independent symmetrized functions.[1]

Like the projection operator $\hat{S}^{j}_{\sigma\sigma}$, the projection matrix $\mathbf{S}^{j}_{\sigma\sigma}$ is hermitian and idempotent[2]:

$$\mathbf{S}^{\dagger} = \mathbf{S}, \tag{17.8}$$

$$\mathbf{S}^{2} = \mathbf{S}. \tag{17.9}$$

From relation (17.8) it follows that the matrix $\mathbf{S}$ can be reduced to a diagonal form, then from relation (17.9) it follows that its eigenvalues can be equal only to 1 or 0. Hence, there exists a unitary transformation $\mathbf{U}$ such that

$$\mathbf{U}^{-1}\mathbf{S}\mathbf{U} = \begin{pmatrix} \mathbf{1} & \mathbf{0} \\ \mathbf{0} & \mathbf{0} \end{pmatrix} = \mathbf{E}_0 \tag{17.10}$$

(here $\mathbf{1}$ and $\mathbf{0}$ denote the unit and zero blocks of the matrix, respectively), i.e.

$$\mathbf{S}\mathbf{U} = \mathbf{U}\mathbf{E}_0 = (\mathbf{U}_r, \mathbf{0}), \tag{17.11}$$

where $\mathbf{U}_r$ is the rectangular matrix that consists of columns corresponding to the unit eigenvalues

$$\mathbf{S}\mathbf{U}_r = \mathbf{U}_r. \tag{17.12}$$

The product $\mathbf{\Phi}\mathbf{U}_r$ gives a set of linearly independent symmetrized functions.

It remains to find the rectangular matrix $\mathbf{U}_r$. It can be done, using a standard method for diagonalizing a Hermitian matrix. However, the matrix $\mathbf{U}_r$ can also be found directly using the idempotency property (17.9) of the projection matrix. Let us introduce the notation $S_{\cdot\mu}$ for the μ-th column and $S_{\mu\cdot}$ for the μ-th row of the matrix $\mathbf{S}$. If we choose the columns of the matrix $\mathbf{U}_r$ in the form $S_{\cdot\mu}$, relation (17.12) will be fulfilled due to (17.9); however, the columns of $S_{\cdot\mu}$ in the general case will not satisfy the orthonormality condition, which should hold for the columns of the matrix $\mathbf{U}_r$:

$$(U_{\cdot\mu}, U_{\cdot\nu}) = \delta_{\mu\nu}. \tag{17.13}$$

[1] These functions will be orthonormal if such were the initial functions Φ_{ν}.

[2] Since the indices j and σ are fixed, we will not write them in this section for brevity.

To satisfy this condition as well, we orthonormalize the columns $S_{\cdot\mu}$ using the Gram–Schmidt process; i.e. we choose the columns of the matrix $\mathbf{U}_r$ as follows:

$$U_{\cdot\mu} = V_{\cdot\mu} / \sqrt{(V_{\cdot\mu}, V_{\cdot\mu})}, \tag{17.14}$$

$$V_{\cdot 1} = S_{\cdot 1}, \tag{17.15}$$

$$V_{\cdot\mu} = S_{\cdot\mu} - \sum_{\nu=1}^{\mu-1} (U_{\cdot\nu}, S_{\cdot\mu})\, U_{\cdot\nu}, \qquad \mu > 1. \tag{17.16}$$

Using properties (17.8) and (17.9) of the projection matrix, it is easy to prove that

$$(U_{\cdot\nu}, S_{\cdot\mu}) = U^*_{\mu\nu} \tag{17.17}$$

and

$$(V_{\cdot\mu}, V_{\cdot\mu}) = V_{\mu\mu}. \tag{17.18}$$

These equalities simplify the expressions (17.14)–(17.16) for $V_{\cdot\mu}$ and $U_{\cdot\mu}$:

$$U_{\cdot\mu} = V_{\cdot\mu} / \sqrt{V_{\mu\mu}}, \tag{17.19}$$

$$V_{\cdot 1} = S_{\cdot 1}, \tag{17.20}$$

$$V_{\cdot\mu} = S_{\cdot\mu} - \sum_{\nu=1}^{\mu-1} U^*_{\mu\nu}\, U_{\cdot\nu}, \qquad \mu > 1. \tag{17.21}$$

According to (17.10) and (17.11) the number of columns m of the matrix $\mathbf{U}_r$ is equal to

$$m = \mathrm{Tr}(\mathbf{E}_0) = \mathrm{Tr}(\mathbf{S}). \tag{17.22}$$

Let us investigate the properties of the matrix $\mathbf{S}$ in order to determine which of its columns are needed to construct the matrix $\mathbf{U}_r$. From relations (17.8) and (17.9) it follows that

$$S_{\nu\nu} - S^2_{\nu\nu} = \sum_{\mu\neq\nu} |S_{\mu\nu}|^2. \tag{17.23}$$

Hence

$$0 \leq S_{\nu\nu} \leq 1. \tag{17.24}$$

If the diagonal element $S_{\nu\nu}$ is 0 or 1, all the off-diagonal elements in the corresponding row and column are zero. Consequently, for $S_{\nu\nu} = 0$ the original basis function Φ_ν does not appear in the symmetrized functions (with fixed σ and j), and for $S_{\nu\nu} = 1$ it is itself a linearly independent symmetrized function.

Next we consider the columns of the matrix **S** for which $S_{\nu\nu}$ is not equal to 0 or 1. Having previously calculated the sum of m' diagonal elements of such (i-th) columns, we determine the corresponding columns of the matrix $\mathbf{U}_r$ as follows. As $U_{\cdot 1}$, we choose the i_1-th column of the matrix **S**, normalized by formula (17.19). For $m' > 1$, the i_2-th column of the matrix **S** is made orthogonal to the i_1-th column by formula (17.21), and, if the obtained vector is zero, we do the same with the i_3-th column, etc., until we get $V_{\cdot 2} \neq 0$, which is then normalized by formula (17.19). This process is continued until all m' columns of $\mathbf{U}_r$ have been constructed.

17.1.2 Projection Matrix on Bloch Functions

Let us apply the above general theory to the space groups. As the initial functions Φ_i, we take Bloch sums of the form[3]

$$\Phi^{\mathbf{k}st}_{lm}(\mathbf{r}) = \sum_p \mathrm{e}^{i\mathbf{k}\mathbf{R}_p} \varphi^s_{lm}(\mathbf{r} - \mathbf{R}_p - \boldsymbol{\tau}^s_t), \tag{17.25}$$

$$\varphi^s_{lm}(\mathbf{r}) = R^s_l(r) Y_{lm}(\mathbf{r}), \tag{17.26}$$

where $R^s_l(r)$ is a function depending on the modulus of the vector **r**, Y_{lm} is the spherical function, $\mathbf{R}_p$ is the crystal lattice vector, $\boldsymbol{\tau}^s_t$ is the basis vector indicating the position of the t-th atom of the s-th sort in the Wigner–Seitz cell and **k** is the reduced wavevector (in the Brillouin zone).

The projection operator (17.4), which generates functions that belong to the σ-th row of the j-th irreducible representation of the group of the wavevector $G_\mathbf{k}$, can be written in the form

$$\hat{S}^{\mathbf{k}j}_{\sigma\sigma} = \frac{l_j}{n(G_\mathbf{k})} \sum_{\{\alpha|\mathbf{R}_p+\boldsymbol{\tau}_\alpha\}\in G_\mathbf{k}} d^{\mathbf{k}j*}_{\sigma\sigma}(\{\alpha|\mathbf{R}_p+\boldsymbol{\tau}_\alpha\})\, P_{\{\alpha|\mathbf{R}_p+\boldsymbol{\tau}_\alpha\}}, \tag{17.27}$$

[3] Here, following Flodmark, we omit $\boldsymbol{\tau}^s_t$ in the exponential of the Bloch sum. The main quantum number n is also omitted, since it does not influence the symmetrization procedure.

where $d^{\mathbf{k}j}$ is the j-th irreducible representation of the group $G_{\mathbf{k}}$, α is the proper ($\alpha = \alpha'$) or improper ($\alpha = i\alpha'$) rotation, $\boldsymbol{\tau}_\alpha$ is the nonprimitive translation vector. We take into account

$$d^{\mathbf{k}j}(\{\alpha|\mathbf{R}_p + \boldsymbol{\tau}_\alpha\}) = \mathrm{e}^{-i\mathbf{k}\mathbf{R}_p}\,\Delta^{\mathbf{k}j}(\{\alpha|\boldsymbol{\tau}_\alpha\}), \tag{17.28}$$

where $\Delta^{\mathbf{k}j}$ is the irreducible projective representation of the factor group $G_{\mathbf{k}}/T$ (T is the translation group), which is isomorphic to the point group of the wavevector $G^0_{\mathbf{k}}$, consisting of the elements α. Then the action of the projection operator (17.27) on the Bloch sum (17.25) is reduced to

$$\hat{S}^{\mathbf{k}j}_{\sigma\sigma}\,\Phi^{\mathbf{k}st}_{lm}(\mathbf{r}) = \frac{l_j}{n(G^0_{\mathbf{k}})}\sum_{\{\alpha|\boldsymbol{\tau}_\alpha\}}\Delta^{\mathbf{k}j*}_{\sigma\sigma}(\{\alpha|\boldsymbol{\tau}_\alpha\})\,P_{\{\alpha|\boldsymbol{\tau}_\alpha\}}\,\Phi^{\mathbf{k}st}_{lm}(\mathbf{r}). \tag{17.29}$$

We write function (17.25) as

$$\Phi^{\mathbf{k}st}_{lm}(\mathbf{r}) = \sum_p \mathrm{e}^{i\mathbf{k}\mathbf{R}_p}\,P_{\{\varepsilon|\mathbf{R}_p+\boldsymbol{\tau}^s_t\}}\,\varphi^s_{lm}(\mathbf{r}), \tag{17.30}$$

where $P_{\{\varepsilon|\mathbf{R}_p+\boldsymbol{\tau}^s_t\}}$ is the operator in the space of functions defined by (17.1). Recall that[4]

$$P_{\{\alpha|\mathbf{t}\}}\,P_{\{\beta|\mathbf{t}'\}} = P_{\{\alpha\beta|\alpha\mathbf{t}'+\mathbf{t}\}} \tag{17.31}$$

where $\mathbf{t}$ and $\mathbf{t}'$ are arbitrary translations. Now, applying the operator $P_{\{\alpha|\boldsymbol{\tau}_\alpha\}}$ to the function (17.30) and taking into account (17.31), we obtain

$$\begin{aligned} P_{\{\alpha|\boldsymbol{\tau}_\alpha\}}\,\Phi^{\mathbf{k}st}_{lm}(\mathbf{r}) &= \sum_p \mathrm{e}^{i\mathbf{k}\mathbf{R}_p}\,P_{\{\alpha|\boldsymbol{\tau}_\alpha\}}\,P_{\{\varepsilon|\mathbf{R}_p+\boldsymbol{\tau}^s_t\}}\,\varphi^s_{lm}(\mathbf{r}) \\ &= \sum_p \mathrm{e}^{i\mathbf{k}\mathbf{R}_p}\,P_{\{\varepsilon|\alpha\mathbf{R}_p+\alpha\boldsymbol{\tau}^s_t+\boldsymbol{\tau}_\alpha\}}\,P_{\{\alpha|0\}}\,\varphi^s_{lm}(\mathbf{r}). \end{aligned} \tag{17.32}$$

Using the fact that the spherical functions $Y_{lm}(\mathbf{r})$ are the basis functions of the irreducible representation $\mathcal{D}^{l\pm}$ of the full rotation group (plus sign for even l, minus sign for odd ones), i.e. they satisfy the relation

$$P_{\{\alpha|0\}}\,Y_{lm}(\mathbf{r}) = \sum_{m'} D^{l\pm}_{m'm}(\alpha)\,Y_{lm'}(\mathbf{r}), \tag{17.33}$$

---

[4] The operator $P_{\{\alpha|\mathbf{R}_p+\boldsymbol{\tau}_\alpha\}}$, corresponding to a crystal symmetry transformation, is a special case of the operators $P_{\{\alpha|\mathbf{t}\}}$, corresponding to a linear space transformation.

we obtain

$$P_{\{\alpha|\boldsymbol{\tau}_\alpha\}}\, \Phi^{\mathbf{k}st}_{lm}(\mathbf{r}) = \sum_{m'} D^{l\pm}_{m'm}(\alpha) \sum_{p} \mathrm{e}^{i\mathbf{k}\mathbf{R}_p}\, P_{\{\varepsilon|\alpha\mathbf{R}_p+\alpha\boldsymbol{\tau}^s_t+\boldsymbol{\tau}_\alpha\}}\, \varphi^s_{lm'}(\mathbf{r}). \tag{17.34}$$

The vector $\alpha\mathbf{R}_p + \alpha\boldsymbol{\tau}^s_t + \boldsymbol{\tau}_\alpha$ can be represented as $\{\alpha|\boldsymbol{\tau}_\alpha\}(\mathbf{R}_p + \boldsymbol{\tau}^s_t)$. Since $\{\alpha|\boldsymbol{\tau}_\alpha\}$ is an element of the crystal symmetry group, this vector characterizes the position of some v-th atom of some u-th cell:

$$\alpha\mathbf{R}_p + \alpha\boldsymbol{\tau}^s_t + \boldsymbol{\tau}_\alpha = \mathbf{R}_{u(s,t,\alpha)} + \boldsymbol{\tau}^s_{v(s,t,\alpha)}. \tag{17.35}$$

Hence, the values $u(s,t,\alpha)$ and $v(s,t,\alpha)$ are determined directly:

$$\mathbf{R}_{u(s,t,\alpha)} = \alpha\mathbf{R}_p + \mathbf{R}(s,t,\alpha), \tag{17.36}$$

$$\boldsymbol{\tau}^s_{v(s,t,\alpha)} = \alpha\boldsymbol{\tau}^s_t + \boldsymbol{\tau}_\alpha - \mathbf{R}(s,t,\alpha), \tag{17.37}$$

where $\mathbf{R}(s,t,\alpha)$ is the translation vector chosen so that the vector on the right-hand side of equality (17.37) lies within the Wigner–Seitz cell (basis vector). Since α transforms the crystal lattice into itself, the summation over p in expression (17.34) can be replaced by the summation over u. Expressing $\mathbf{R}_u$ from (17.35), we write (17.34) as (we omit the arguments (s,t,α) of u,v and $\mathbf{R}$ where it does not lead to confusion)

$$\begin{aligned} P_{\{\alpha|\boldsymbol{\tau}_\alpha\}}\, \Phi^{\mathbf{k}st}_{lm}(\mathbf{r}) &= \sum_{m'} D^{l\pm}_{m'm}(\alpha) \sum_{u} \mathrm{e}^{i\mathbf{k}\alpha^{-1}(\mathbf{R}_u-\mathbf{R})}\, P_{\{\varepsilon|\mathbf{R}_u+\boldsymbol{\tau}^s_v\}}\, \varphi^s_{lm'}(\mathbf{r}) \\ &= \sum_{m'} D^{l\pm}_{m'm}(\alpha)\, \mathrm{e}^{-i\mathbf{k}\mathbf{R}}\, \Phi^{\mathbf{k}sv}_{lm'}(\mathbf{r}) = \sum_{t'm'} D^{l\pm}_{m'm}(\alpha)\, \mathrm{e}^{-i\mathbf{k}\mathbf{R}}\, \delta_{t',v}\, \Phi^{\mathbf{k}st'}_{lm'}(\mathbf{r}). \end{aligned} \tag{17.38}$$

Hence the Bloch sums $\Phi^{\mathbf{k}st}_{lm}(\mathbf{r})$ for fixed $\mathbf{k}$, s and l form a basis of some (generally reducible) representation of the group $G_{\mathbf{k}}$. The matrix elements of this representation are

$$D^{\mathbf{k}sl}_{t'm',tm}(\{\alpha|\boldsymbol{\tau}_\alpha\}) = D^{l\pm}_{m'm}(\alpha)\, \mathrm{e}^{-i\mathbf{k}\mathbf{R}(s,t,\alpha)}\, \delta_{t',v(s,t,\alpha)}. \tag{17.39}$$

Taking (17.29) and (17.38) into account, the elements of the projection matrix (17.6) on Bloch sums take the form[5]

$$S^{\mathbf{k}jsl}_{t'm',tm} = \frac{l_j}{n(G^0_{\mathbf{k}})} \sum_{\{\alpha|\boldsymbol{\tau}_\alpha\}} \Delta^{\mathbf{k}j*}_{\sigma\sigma}(\{\alpha|\boldsymbol{\tau}_\alpha\})\, D^{l\pm}_{m'm}(\alpha)\, \mathrm{e}^{-i\mathbf{k}\mathbf{R}(s,t,\alpha)}\, \delta_{t',v(s,t,\alpha)}. \tag{17.40}$$

[5] For brevity, we omit the fixed row number σ everywhere except for $\Delta^{\mathbf{k}j}_{\sigma\sigma}$.

17.1.3 Irreducible Representations of the Rotation Group

Here we use spherical functions $Y_{lm}(\vartheta, \varphi)$ of the form

$$Y_{lm}(\vartheta, \varphi) = \begin{cases} (-1)^m N_{l|m|} P_l^{|m|}(\cos\vartheta) e^{im\varphi}, & m = 0, \ldots, l\,, \\ N_{l|m|} P_l^{|m|}(\cos\vartheta) e^{im\varphi}, & m = -l, \ldots, -1\,, \end{cases} \tag{17.41}$$

where

$$N_{l|m|} = \sqrt{\frac{2l+1}{4\pi} \frac{(l-|m|)!}{(l+|m|)!}} \tag{17.42}$$

is the normalization factor, and

$$P_l^{|m|}(\cos\vartheta) = \frac{1}{2^l l!} \sin^{|m|}\vartheta \frac{d^{l+|m|}}{(d\cos\vartheta)^{l+|m|}} (\cos^2\vartheta - 1) \tag{17.43}$$

is the associated Legendre polynomial. The choice of the phases of the spherical functions Y_{lm} here corresponds to the widely used definition adopted by Condon and Shortley [9].

We define the proper rotation by the Euler angles α, β, γ and denote it by $\{\alpha, \beta, \gamma\}$[6]. The rotation $\{\alpha, \beta, \gamma\}$ here is a composition of three positive rotations [7]: first the space rotation around the z-axis through the angle α, then rotation around the y-axis through the angle β and then another rotation around the z-axis through the angle γ. Angles α, β, γ vary within $0 \leq \alpha, \gamma < 2\pi$ and $0 \leq \beta \leq \pi$. We emphasize that the rotations, like all transformations in this book, are active rotations, i.e. it is the points in space that rotate, and not the coordinate system.

A formula for the representation matrix elements of the group of proper rotations that transform the spherical functions (17.41)–(17.43) are given in Wigner's monograph [10, formula (15.27)]. However, Wigner considered rotations as passive, i.e. it is the coordinate system that rotates, not the points in space, so we cannot use his formula directly.

We denote the transformation matrix associated with some rotation by $\mathbf{R}$ and equip it with the index a if it corresponds to the active transformation, and the index p if it corresponds to the passive one. Then, according to our definition of the Euler angles, we have

$$\mathbf{R}^a_{\{\alpha\beta\gamma\}} = \mathbf{R}^a_\gamma \mathbf{R}^a_\beta \mathbf{R}^a_\alpha\,, \tag{17.44}$$

---

[6] In what follows, when it is necessary to distinguish between the rotation α and the Euler angle α that characterizes it, we will supply the latter with the index e.

[7] Recall that a positive rotation around a certain axis is the rotation of the right-handed screw moving in the positive direction of this axis.

while Wigner's convention leads to

$$\mathbf{R}^p_{\{\alpha\beta\gamma\}} = \mathbf{R}^p_\alpha \mathbf{R}^p_\beta \mathbf{R}^p_\gamma. \tag{17.45}$$

An analogue of Wigner's formula [10, (15.27)] for active rotations can be obtained by repeating all the reasoning of [10, Chap. 15], but taking into account the following two circumstances.

1. Since $\mathbf{R}^a = (\mathbf{R}^p)^{-1} = \tilde{\mathbf{R}}^p$, the unitary matrices $\mathbf{u}^a_1(\alpha)$ and $\mathbf{u}^a_2(\beta)$ must have the form (compare with [10, formulas (15.14a) and (15.14b)])

$$\mathbf{u}^a_1(\alpha) = \begin{pmatrix} e^{i\frac{\alpha}{2}} & 0 \\ 0 & e^{-i\frac{\alpha}{2}} \end{pmatrix}, \qquad \mathbf{u}^a_2(\beta) = \begin{pmatrix} \cos\frac{\beta}{2} & \sin\frac{\beta}{2} \\ -\sin\frac{\beta}{2} & \cos\frac{\beta}{2} \end{pmatrix}.$$

2. According to (17.44), the active rotation with the Euler angles α, β, γ is the product $\mathbf{u}^a_1(\gamma)\,\mathbf{u}^a_2(\beta)\,\mathbf{u}^a_1(\alpha)$. Hence for the coefficients a and b of the transformation $\mathbf{u}$, we obtain (compare with [10, formulas (15.9) and (15.15)])

$$a = e^{i\frac{\alpha}{2}} \cos\frac{\beta}{2}\, e^{i\frac{\gamma}{2}}, \qquad b = e^{-i\frac{\alpha}{2}} \sin\frac{\beta}{2}\, e^{i\frac{\gamma}{2}}.$$

Instead, we obtain the formula for the representation matrix elements of the group of active rotations directly from Wigner's formula [10, (15.27)], based on the following easily proved (see, e.g. [1]) assertion.

There is a one-to-one correspondence [8] $g^p = (g^a)^{-1}$ *between the elements* g^a *of the group of active transformations* G^a *and elements* g^p *of the group of passive transformations* G^p*. If* $D^a(g^a)$ *is a representation of the group* G^a *with the representation basis functions* Φ_μ*, then the set of matrices* $D^p(g^p) = D^a((g^a)^{-1})$ *forms a representation of the group* G^p *with the same basis functions* Φ_μ*.*

Applying this assertion and taking (17.45) into account, we have

$$D^a(\{\alpha,\beta,\gamma\}^a) = D^p((\{\alpha,\beta,\gamma\}^p)^{-1}) = D^p(\{-\gamma,-\beta,-\alpha\}^p). \tag{17.46}$$

Then from Wigner's formula [10, (15.27)], replacing α by $-\gamma$, β by $-\beta$ and γ by $-\alpha$, we immediately obtain

$$D^l_{m'm}(\{\alpha,\beta,\gamma\}) = e^{-im'\gamma}\, d^l_{m'm}(\beta)\, e^{-im\alpha}. \tag{17.47}$$

Here

[8] This correspondence is not an isomorphism.

$$d^l_{m'm}(\beta) = \sum_k (-1)^{k+m'-m} \frac{\sqrt{(l+m)!(l-m)!(l+m')!(l-m')!}}{(l-m'-k)!(l+m-k)!k!(k+m'-m)!} \times \cos^{2l+m-m'-2k}\frac{\beta}{2}\,\sin^{2k+m'-m}\frac{\beta}{2}, \quad (17.48)$$

where the summation over k goes from the largest of the numbers 0 and $m-m'$ to the smallest of the numbers $l-m'$ and $l+m$.

The matrix elements $d^l_{m'm}(\beta)$ satisfy the relations

$$d^l_{m'm}(\beta) = d^l_{-m,-m'}(\beta) = (-1)^{m-m'}\, d^l_{mm'}(\beta). \quad (17.49)$$

The first equality is verified by direct substitution of $-m$ instead of m' and $-m'$ instead of m in formula (17.48). The second is proved as follows. Since the matrix $D^l(\{\alpha,\beta,\gamma\})$ is unitary, taking (17.44) into account, we have

$$D^l_{m'm}(\{\alpha,\beta,\gamma\}) = D^{l*}_{mm'}(\{\alpha,\beta,\gamma\}^{-1}) = D^{l*}_{mm'}(\{-\alpha,-\beta,-\gamma\}).$$

Hence

$$d^l_{m'm}(\beta) = d^l_{mm'}(-\beta) = (-1)^{m-m'}\, d^l_{mm'}(\beta).$$

Due to relations (17.49), it is sufficient to calculate the matrix elements $d^l_{m'm}(\beta)$ within the limits $-l \le m' \le 0,\ m' \le m \le -m'$.

17.2 Algorithm

We wish to construct the symmetrized Bloch sum, which belongs to a certain row of the j-th irreducible representation of the group of wavevector $\mathbf{k}$ and is obtained from the Bloch sum corresponding to the localized function of the t-th atom of the s-th sort with the quantum numbers l and m. It is convenient to represent this kind of function in the form

$$F^{\mathbf{k}jsl}_{tm}(\mathbf{r}) = \sum_{t'm'} \Phi^{\mathbf{k}sl}_{t'm'}(\mathbf{r})\, U^{\mathbf{k}jsl}_{t'm',tm} = \sum_{t'm'}\sum_{\mathbf{R}_p} \mathrm{e}^{i\mathbf{k}\mathbf{R}_p} \varphi^s_{lm'}(\mathbf{r}-\mathbf{R}_p-\boldsymbol{\tau}^s_{t'})\, U^{\mathbf{k}jsl}_{t'm',tm} \equiv \sum_{\mathbf{R}_p} \mathrm{e}^{i\mathbf{k}\mathbf{R}_p} f^{\mathbf{k}jsl}_{tm}(\mathbf{r}-\mathbf{R}_p), \quad (17.50)$$

where

$$f^{\mathbf{k}jsl}_{tm}(\mathbf{r}) = \sum_{t'm'} \varphi^s_{lm'}(\mathbf{r}-\boldsymbol{\tau}^s_{t'})\, U^{\mathbf{k}jsl}_{t'm',tm} \quad (17.51)$$

is the part of the function $F_{tm}^{\mathbf{k}jsl}(\mathbf{r})$ consisting of the localized atomic functions $\varphi_{lm'}^{s}(\mathbf{r}-\boldsymbol{\tau}_{t'}^{s})$ centered in the Wigner–Seitz cell. This representation of symmetrized Bloch sums allows us to restrict ourselves to considering the functions $f_{tm}^{\mathbf{k}jsl}(\mathbf{r})$, which we will call the symmetrized localized functions.

Construction of functions (17.51) is reduced to calculation of the matrix elements $U_{t'm',tm}^{\mathbf{k}jsl}$. According to the general theory presented in Sect. 17.1, the latter are expressed in terms of the matrix elements $S_{t'm',tm}^{\mathbf{k}jsl}$ given by formula (17.40).

Let us describe in more detail the calculation of the matrix elements $S_{t'm',tm}^{\mathbf{k}jsl}$. We write the matrices $D^{l\pm}(\alpha)$ of the irreducible representations of the full rotation group in the form

$$D^{l\pm}(\alpha)=c_\alpha\, D^{l}(\alpha'), \tag{17.52}$$

where

$$c_\alpha=\begin{cases}1, & \text{if }\ \alpha=\alpha',\\ (-1)^l, & \text{if }\ \alpha=i\alpha'.\end{cases} \tag{17.53}$$

Substituting (17.52) to (17.40) and taking (17.47) into account, for the elements of the projection matrix we obtain the following expression

$$\begin{aligned} S_{t'm',tm}^{\mathbf{k}jsl}=\frac{l_j}{n(G_{\mathbf{k}}^0)}&\sum_{\{\alpha|\boldsymbol{\tau}_\alpha\}}\Delta_{\sigma\sigma}^{\mathbf{k}j*}(\{\alpha|\boldsymbol{\tau}_\alpha\})c_\alpha \mathrm{e}^{-im'\gamma}d_{m'm}^{l}(\beta)\mathrm{e}^{-im\alpha_e}\\ &\times \mathrm{e}^{-i\mathbf{k}\mathbf{R}(s,t,\alpha)}\delta_{t',v(s,t,\alpha)}. \end{aligned} \tag{17.54}$$

Since we sum up the *products* of complex numbers, it is convenient to represent expression (17.54) in the form

$$S_{t'm',tm}^{\mathbf{k}jsl}=\frac{l_j}{n(G_{\mathbf{k}}^0)}\sum_{\{\alpha|\boldsymbol{\tau}_\alpha\}}Z_{m'm}^{\mathbf{k}jl}(\{\alpha|\boldsymbol{\tau}_\alpha\})\,\mathrm{e}^{i\Psi_{m'tm}^{\mathbf{k}js}(\{\alpha|\boldsymbol{\tau}_\alpha\})}\,\delta_{t',v(s,t,\alpha)}, \tag{17.55}$$

where

$$Z_{m'm}^{\mathbf{k}jl}(\{\alpha|\boldsymbol{\tau}_\alpha\})=|\Delta_{\sigma\sigma}^{\mathbf{k}j}(\{\alpha|\boldsymbol{\tau}_\alpha\})|\,c_\alpha\, d_{m'm}^{l}(\beta), \tag{17.56}$$

$$\Psi_{m'tm}^{\mathbf{k}js}(\{\alpha|\boldsymbol{\tau}_\alpha\})=-\arg\Delta_{\sigma\sigma}^{\mathbf{k}j}(\{\alpha|\boldsymbol{\tau}_\alpha\})-m'\gamma-m\alpha_{\mathrm{e}}-\mathbf{k}\mathbf{R}(s,t,\alpha). \tag{17.57}$$

We calculate the vectors $\mathbf{R}(s,t,\alpha)$ and values $v(s,t,\alpha)$ defined by relation (17.37) as follows. We fix the vector $\boldsymbol{\tau}_t^s$. Acting on it by the transformation $\{\alpha|\boldsymbol{\tau}_\alpha\}$, we obtain $\alpha\boldsymbol{\tau}_t^s+\boldsymbol{\tau}_\alpha$. From this vector we subtract the basis vectors $\boldsymbol{\tau}_{t'}^s$ one by one and each time check whether $\mathbf{a} = \alpha\boldsymbol{\tau}_t^s + \boldsymbol{\tau}_\alpha - \boldsymbol{\tau}_{t'}^s$ is a lattice vector. To do this, we calculate the scalar products of the vector $\mathbf{a}$ with the basic vectors of the reciprocal lattice $\mathbf{b}_i$: $\mathbf{ab}_i = 2\pi\mu_i$ $(i = 1, 2, 3)$. If all μ_i are integers, then $\mathbf{R}(s, t, \alpha) = \mathbf{a}$, $v(s, t, \alpha) = t'$.

Since for all crystallographic point groups, the Euler angle β takes only three different values [1]: $\beta = 0, \frac{\pi}{2}, \pi$, it is convenient to calculate the matrix elements $d^l_{m'm}(\beta)$ for these three values in advance. If $\beta = 0$, we have a rotation at an angle $\alpha + \gamma$ about the z-axis. In this case (see, e.g. [10, p. 154])

$$D^l_{m'm}(\{\alpha, 0, \gamma\}) = \mathrm{e}^{-im(\alpha+\gamma)}\,\delta_{m'm},$$

and hence

$$d^l_{m'm}(0) = \delta_{m'm}. \tag{17.58}$$

If $\beta = \pi$, it can be seen from formula (17.48) that the terms such that $2l + m - m' - 2k = 0$ can be nonzero, i.e. the terms with $k = \frac{2l+m-m'}{2} = \frac{l+m+l-m'}{2}$. However, only for $m = -m'$ does the value $k = l - m'$ satisfy the condition $k \leq \min(l - m', l + m)$. The corresponding term in formula (17.48) is equal to $(-1)^{l+m'}$. From this we obtain

$$d^l_{m'm}(\pi) = (-1)^{l+m'}\,\delta_{m',-m}. \tag{17.59}$$

Thus, it is necessary to calculate only $d^l_{m'm}(\frac{\pi}{2})$. Formula (17.48) with $\beta = \frac{\pi}{2}$ has the form

$$d^l_{m'm}\left(\frac{\pi}{2}\right) = 2^{-l}\sum_k (-1)^{k+m'-m}\frac{\sqrt{(l+m)!(l-m)!(l+m')!(l-m')!}}{(l-m'-k)!(l+m-k)!k!(k+m'-m)!}. \tag{17.60}$$

All other details of calculating the symmetrized localized functions should be clear from the program, given in [8], itself. As a test example, the program constructs symmetrized Bloch sums for the Cr_3Si structure with the space group O_h^3, discussed in Chap. 12.

17.3 Summary

Let us summarize the main advantages of the above algorithm for constructing symmetrized Bloch sums.

1. The algorithm takes into account that the projection matrix is Hermitian (17.8) and idempotent (17.9). This has allowed us
   (a) to calculate not the entire projection matrix, but only a part of it that gives linearly independent symmetrized functions;

(b) not to calculate the scalar products of columns of the projection matrix during the Gram–Schmidt orthogonalization process;
(c) obtain the multiplicity of occurrence as a trace of the projection matrix.

2. Calculation of matrix elements $D^{\mathbf{k}sl}_{t'm',tm}(\{\alpha|\boldsymbol{\tau}_\alpha\})$ of the reducible representation is mainly reduced to the calculation of the matrix elements $D^l_{m'm}(\alpha')$ of the irreducible representation of the rotation group. (The vectors $\mathbf{R}(s, t, \alpha)$ and the values $v(s, t, \alpha)$, which depend only on the crystal structure, are determined in advance).
3. In turn, by specifying rotations by the Euler angles α_e, β, γ the calculation of the matrix elements $D^l_{m'm}(\alpha')$ is reduced to the calculation of the matrix elements $d^l_{m'm}(\beta)$, the number of which is made four times smaller using relation (17.49).
4. Next, taking into account the fact that in crystallographic point groups β takes only the values 0, $\frac{\pi}{2}$, π, we have substantially simplified the calculation of the remaining matrix elements $d^l_{m'm}(\beta)$ (see formulas (17.58)–(17.60)).
5. Finally, we have simplified computations of the elements $S^{\mathbf{k}jsl}_{t'm',tm}$ of the projection matrix by using exponential form of complex numbers [see formulas (17.55)–(17.57)].

This allowed us to create a simple and efficient algorithm and software for constructing symmetrized Bloch sums.

References

1. C.J. Bradley, A.P. Cracknell, *The Mathematical Theory of Symmetry in Solids. Representation Theory for Point Groups and Space Groups* (Clarendon, Oxford, 1972). Reprinted: Clarendon, Oxford, 2010
2. S.L. Altmann, A.P. Cracknell, Rev. Mod. Phys. **37**, 19 (1965)
3. S.L. Altmann, C.J. Bradley, Rev. Mod. Phys. **37**, 33 (1965)
4. S. Flodmark, Phys. Rev. **132**, 1343 (1963)
5. E. Blokker, J. Comput. Phys. **12**, 471 (1973)
6. S. Flodmark, J. Comput. Phys. **25**, 314 (1977)
7. L.M. Sandratskii, R.F. Egorov, *Construction of Symmetrized Spherical Harmonics for Lattices with a Basis* [in Russian] (1978). Dep. VINITI, No. 274–78
8. B.I. Reser, *Construction of Symmetrized Bloch Sums* [in Russian] (1979). Dep. VINITI, No. 3258–79
9. E.U. Condon, G.H. Shortley, *The theory of atomic spectra* (Cambridge Univ. Press, Cambridge, 1959)
10. E.P. Wigner, *Group Theory and its Application to the Quantum Mechanics of Atomic Spectra*, 2nd edn. (Academic, New York, 1959)

Chapter 18
Discussion and Outlook

But all things come to an end, even troublesome and tedious things ... (A.D. Sakharov, Memoir, Knopf, New York, 1990)

18.1 Representation Theory of the Space Groups

The irreducible representations of symmorphic groups were studied in the celebrated paper by Bouckaert, Smoluchowski and Wigner [1] in the example of cubic crystals. In this case, it was shown that the problem of finding small representations of the group of the wavevector $G_{\mathbf{k}}$ reduces to finding vector representations of the point group $G^0_{\mathbf{k}}$. Furthermore, relations were deduced between small representations of the groups $G_{\mathbf{k}}$ when the vector $\mathbf{k}$ changes continuously (compatibility relations). For the symmorphic space groups, all high symmetry points in the Brillouin zone were found and character tables of small representations were constructed.

Nonsymmorphic groups were first analyzed by Herring [2] in the example of the diamond and hexagonal close-packed structures. He used the fact that small representations of the group $G_{\mathbf{k}}$ can be determined from vector representations of the factor group $G_{\mathbf{k}}/T_{\mathbf{k}}$. Here, $T_{\mathbf{k}}$ is the translation subgroup with $\mathbf{kR}_n = 2\pi m$ (m is an integer) and, hence, the matrices of the small representations are unit matrices. Once the irreducible representations of this auxiliary group are obtained, the true small representations can be deduced merely from the multiplication table. Herring [2] obtained only characters, but the same procedure can be used to deduce matrices of the small representation as well. The above method, though rather laborious, was widely used later on and made its way to textbooks [3, 4]. An analogous method was independently presented in [5], also in the example of the diamond structure. Döring and Zehler [5] were probably the first ones who pointed out a simple procedure for deducing the small representations for vectors inside the Brillouin zone and noted completeness of the projective representation system of the factor group $G_{\mathbf{k}}/T$ (in our terminology).

N. B. Melnikov and B. I. Reser, *Space Group Representations*,
https://doi.org/10.1007/978-3-031-13991-8_18

A more clear description of the irreducible representations of the space groups, which followed the paper of Seitz [6], was given by Koster [7] (see also [8]). In the coming years, several alternative approaches appeared in the literature [9–13].

This book presents a methodologically transparent and practically efficient scheme based on [12]. It allows us to construct the irreducible representations of the space groups of any crystal structure both single- and double-valued.

18.2 Tables of the Space Group Representations

In practical applications of the space group representations people mostly use tables. Let us mention some of them.

Character tables of the space group representations for simple cubic structures were given in the paper [1]. Characters of representations for two nonsymmorphic space groups, with the hexagonal close-packed and diamond structures were presented in the paper [2].

Complete *character* tables of the projective representations for the 32 crystallographic *point* groups were deduced by Döring [14]. He also explained how to obtain characters of the space group representations using these tables. A logical continuation of the paper [14] was the work by Hurley [15], who presented *matrices* of projective representations of the 32 crystallographic point groups. Unfortunately, use of these tables is hindered by a number of reasons. Firstly, Hurley does not describe all possible realizations of the point groups, and hence, in each particular case, one needs to find the correspondence between the abstract group description and its realization by an enumerative technique. Secondly, there are no formulas for the character factors; i.e. one needs to convert the initial defining relations to the form given in the tables. Thirdly, Hurley does not write how to choose numerical coefficients in front of the matrices of the generators. Finally, the tables in [15] are not complete because the vector representations are missing.

The *characters* of representations of all *space* groups are given in tables by Zak et al. [11]. Unlike the tables in [14], these tables do not require any additional calculations and are very convenient in use. Unfortunately, our discouraging experience of using the tables [11] (see the footnote at p. 49 in [16]) casts doubts about their exactness.

In many physical problems the group characters are insufficient, and one needs representation matrices as well. The monograph of Bir and Pikus [17] presents matrices of projective *representations* of the 32 crystallographic *point* groups. Unlike the work of Hurley [15], the monograph [17] gives the matrices for the groups generators only and uses the minimal set of the generators. The tables of the projective representations are extremely compact but constructing the space group representations with their help is very laborious. Of the three critical remarks related to the work of Hurley [15], only the first one is relevant to the monograph [17]. But there is one more point of criticism that is relevant to both of them: The description of the construction method for the representation matrices is missing, which makes the results of [15] and [17] irreproducible.

Representation matrices of all 230 *space* groups were first obtained by Kovalev [18]. In fact, he tabulated projective representations of the point groups for a fixed choice of the coordinate system in each space group and fixed vector **k** in the Brillouin zone. In some cases these restrictions are undesirable. Furthermore, the method of obtaining the matrices is not described, which makes them irreproducible. This is a serious problem because there are many inaccuracies and misprints [13, 19] and also part of the space group representations, whose point groups do not coincide with a holohedry, are missing altogether.

Faddeyev [9] constructed the tables of the space group representations for wavevectors whose point groups coincide with crystal classes.[1] In principle these tables allow to obtain all representations, but in practice one faces a lot of technical difficulties. Moreover, the book tabulates the matrices of only space group generators and uses terminology and notation that is not usual for the solid-state theory.

Slater [20] presented convenient tables of representations of the 20 space groups corresponding to the crystal structures that are the most widespread in nature. But using these tables, one should keep in mind that the author adopted an unconventional definition of the symmetry operator. To obtain a representation matrix in the usual sense one needs to replace each matrix by its inverse. Due to the unitary property, that means one needs to transpose and take the complex conjugate of each matrix.

Representations of all space groups can be found in the tables by Miller and Love [21] and in the monograph by Bradley and Cracknell [13]. But the representation matrices are given only for the group generators. The authors do not give explicit expressions for the representatives $\{\alpha_m|\tau_{\alpha_m}\}$ in the decomposition of $G_\mathbf{k}$ with respect to the translation group T, and one needs to obtain them by means of fairly laborious calculations. Therefore, constructing the representation from the tables [13, 21] is by no means an easy task. Moreover, the tables [21], obtained by a computer and presented in the form of a computer listing, are heavily codified.

As for the tables presented in this book, they are preceded by a method description, which makes them reproducible. A clear advantage is our tables are not related to either a specific choice of coordinate system in a particular space group, nor a specific choice of the **k** vector in the Brillouin zone.[2] Finally, the presence of all possible realizations of the point group and necessary formulas for the character and gauge factors makes our tables easy to use.

Before finishing our brief description of the existing tables of the space group representations, let us note that the choice of one set of tables or the another is to a large extent a matter of taste. One should keep in mind that the representation

[1] In [9], these representations are called basic.

[2] The choice of the coordinate system in the crystal determines the symmetry elements with nonelementary translations $\{\alpha_m|\tau_{\alpha_m}\}$. They, in turn, determine the character factors ω_{ij} and gauge factors δ_i, which define the final form of small representations of the group of the wavevector $G_\mathbf{k}$ (and the space group G).

matrices obtained by different authors can differ.[3] There is a number of reasons for these discrepancies, for example different choices of the arm of the star, different definitions of the symmetry operator, different choices of the sign in the exponent that determines the translation subgroup representation, etc. Noncoincidence of group representations can also come from two different ways of looking at a symmetry operator: either as a transformation of the points of space or as a transformation of the coordinate axes.

18.3 Applications of the Space Group Representations

A brief note on the applications. The crystal (ligand) field theory uses *point* group symmetry to study *ionic* crystals and is completely beyond the scope of this book (see, e.g. [22–25]).[4]

The range of applications of the representations theory of the *space* group is also much wider than we could cover. It includes crystal lattice vibrations [28–32], strain effects in semiconductors [17, 33], structural and magnetic phase transitions [34, 35], X-ray and neutron scattering [36, 37], superconductivity [38, 39], and much more. We hope that other applications of symmetry to physical problems will be made by our readers. Our purpose was to prepare them.

And one last point. In writing this book, we proceeded from the conviction that a simple theory will take roots, while a complex one will die out. To what extent we have succeeded in simplifying the theory is for the reader to judge. What is important is that our tables are complete and compact, and applications are still relevant.

References

1. L.P. Bouckaert, R. Smoluchowski, E. Wigner, Phys. Rev. **50**, 58 (1936)
2. C. Herring, J. Franklin Inst. **233**, 525 (1942)
3. V. Heine, *Group Theory in Quantum Mechanics* (Pergamon, Oxford, 1960). Reprinted: Dover, New York, 2009
4. L.D. Landau, E.M. Lifshitz, *Statistical Physics (Course on Theoretical Physics)*, vol. 5, 3rd edn. (Pergamon, Oxford, 1985)
5. W. Döring, V. Zehler, Ann. Physik **448**, 214 (1953)
6. F. Seitz, Ann. Math. **37**, 17 (1936)
7. G.F. Koster, Solid State Phys. **5**, 173 (1957)
8. A.V. Sokolov, V.P. Shirokovskii, Sov. Phys. Usp. **3**, 551 (1961)
9. D.K. Faddeev, *Tables of the Principal Unitary Representations of Fedorov Groups* [in Russian] (Acad. Sci. USSR, Moscow, 1961). Engl. transl.: Pergamon, Oxford, 1964
10. I.V.V. Raghavacharyulu, Can. J. Phys. **39**, 830 (1961)
11. J. Zak, A. Casher, M. Glück, Y. Gur, *The Irreducible Representations of Space Groups* (Benjamin, New York, 1969)

[3] We are not speaking about differences that come from unitary transformations.

[4] For the use of group theoretical analysis in crystal field theory see, e.g. [26, 27].

12. R.F. Egorov, V.K. Zvezdin, B.I. Reser, *Representations of Space Groups* [in Russian] (1971). Dep. VINITI, No. 2865–71
13. C.J. Bradley, A.P. Cracknell, *The Mathematical Theory of Symmetry in Solids. Representation Theory for Point Groups and Space Groups* (Clarendon, Oxford, 1972). Reprinted: Clarendon, Oxford, 2010
14. W. Döring, Z. Naturforschg. **14 a**, 343 (1959)
15. A.C. Hurley, Phil. Trans. Roy. Soc. London **A260**, 1 (1966)
16. B.I. Reser, R.F. Egorov, V.K. Zvezdin, *Representations of the Space Groups. II. Tables* [in Russian] (1979). Dep. VINITI, No. 568–79
17. G.L. Bir, G.E. Pikus, *Symmetry and Strain-induced Effects in Semiconductors* [in Russian] (Nauka, Moscow, 1972). Engl. transl.: Wiley, New York, 1974
18. O.V. Kovalev, *Irreducible Representations of Space Groups* [in Russian] (Acad. Sci. Ukrain. SSR, Kiev, 1961). Engl. transl.: Gordon and Breach, New York, 1965
19. J.L. Warren, Rev. Mod. Phys. **40**, 38 (1968)
20. J.C. Slater, *Quantum Theory of Molecules and Solids, Vol.2: Symmetry and Energy Bands in Crystals* (McGraw-Hill, New York, 1965)
21. S.C. Miller, W.F. Love, *Tables of Irreducible Representations of Space Groups and Co-representations of Magnetic Space Groups* (Pruett Press, Boulder, CO, 1967)
22. H. Bethe, Ann. Phys. **395**, 133 (1929)
23. J.H. Van Vleck, *The Theory of Electric and Magnetic Susceptibilities* (Oxford Univ. Press, Oxford, 1932)
24. C.J. Ballhausen, *Introduction to Ligand Field Theory* (McGraw-Hill, New York, 1962)
25. M.A. Hitchman, B.N. Figgis, *Ligand Field Theory and Its Applications* (Wiley-VCH, New York, 2000)
26. B.I. Reser, V.I. Cherepanov, Solid State Phys. (USSR) **7**, 2024 (1965)
27. V.I. Bobrovskii, Sov. Phys.—Phys. Metal. Metalloved. **65**, 669 (1988)
28. H. Poulet, J.P. Mathieu, *Vibration Spectra and Symmetry of Crystals* (Gordon & Breach, New York, 1970)
29. J.L. Birman, *Theory of Crystal Space Groups and Lattice Dynamics: Infra-Red and Raman Optical Processes of Insulating Crystals* (Springer, Berlin, 1974)
30. F. Bassani, G. Pastori-Parravicini, *Electronic States and Optical Transitions in Solids* (Pergamon, Oxford, 1975)
31. V.I. Bobrovskii, *Group-Theoretical Analysis of the Vibrational Spectrum of Crystals* [in Russian] (1976). Dep. VINITI, No. 2928–76
32. B. Di Bartolo, R.C. Powell, *Crystal Symmetry, Lattice Vibrations and Optical Spectroscopy of Solids: A Group Theoretical Approach* (World Scientific, New Jersey, 2014)
33. Y. Sun, S.E. Thompson, T. Nishida, *Strain Effect in Semiconductors: Theory and Device Applications* (Springer, New York, 2010)
34. H.F. Franzen, *Second-Order Phase Transitions and the Irreducible Representation of Space Groups* (Springer, Berlin, 1982)
35. Y.A. Izyumov, V. Syromyatnikov, *Phase Transitions and Crystal Symmetry* (Kluwer Academic, Dordrecht, 1990)
36. Y.A. Izyumov, V.E. Naish, R.P. Ozerov, *Neutron Diffraction of Magnetic Materials* (Springer, Berlin, 1991)
37. M. Ladd, R. Palmer, *Structure Determination by X-ray Crystallography: Analysis by X-rays and Neutrons*, 5th edn. (Springer, New York, 2013)
38. N. Plakida, *High-Temperature Cuprate Superconductors* (Springer, Berlin, 2010)
39. Y. Izyumov, E. Kurmaev, *High-T_c Superconductors Based on FeAs Compounds* (Springer, Berlin, 2010)

Appendix A
Frobenius-Schur Theorem and Its Applications

A.1 Frobenius-Schur Theorem for Projective Representations

Theorem A.1 *Let $\mathcal{D}$ be an irreducible projective representation of a group G. Then*

$$\sum_{a\in G} \omega_{a,a}\chi(a^2) = \begin{cases} n(G), & \mathcal{D}\sim\mathcal{D}^*, \quad SS^* = E, \\ 0, & \mathcal{D}\nsim\mathcal{D}^*, \\ -n(G), & \mathcal{D}\sim\mathcal{D}^*, \quad SS^* = -E. \end{cases} \tag{A.1}$$

Here the matrix S is such that $\mathcal{D}^* = S^{-1}\mathcal{D}S$.

Proof By the definition of the projective representation (4.1), we have $\omega_{a,a}D(a^2) = D(a)D(a)$. Hence

$$\sum_{a\in G} \omega_{a,a}\chi(a^2) = \sum_{a\in G} \omega_{a,a} \operatorname{Tr} D(a^2) = \sum_{a\in G} \operatorname{Tr}(\omega_{a,a}D(a^2)) = \sum_{a\in G} \operatorname{Tr}(D(a)D(a)). \tag{A.2}$$

From this it follows that

$$\begin{aligned} \sum_{a\in G} \omega_{a,a}\chi(a^2) &= \sum_{a\in G}\sum_{i} (D(a)D(a))_{ii} \\ &= \sum_{a\in G}\sum_{i}\sum_{j} D_{ij}(a)D_{ji}(a) \\ &= \sum_{i}\sum_{j}\sum_{a\in G} D_{ij}(a)[D^*_{ji}(a)]^*. \end{aligned} \tag{A.3}$$

If $\mathcal{D}\nsim\mathcal{D}^*$, then the sum vanishes according to the orthogonality relations (4.15). If $\mathcal{D}\sim\mathcal{D}^*$, i.e. $\mathcal{D}^* = S^{-1}\mathcal{D}S$, then

N. B. Melnikov and B. I. Reser, *Space Group Representations*,
https://doi.org/10.1007/978-3-031-13991-8

$$\sum_{a\in G}\omega_{a,a}\chi(a^2)=\sum_i\sum_j\sum_{a\in G}D_{ij}(a)(S^{-1}D(a)S)^*_{ji}$$
$$=\sum_{ij}\sum_{a\in G}\sum_{mn}D_{ij}(a)(S^{-1})^*_{jm}D^*_{mn}(a)S^*_{ni}.$$

Taking the orthogonality relations (4.15) into account again, for $SS^*=\pm E$, we obtain

$$\sum_{a\in G}\omega_{a,a}\chi(a^2)=\sum_{ij}\sum_{mn}(S^{-1})^*_{jm}S^*_{ni}\sum_{a\in G}D_{ij}(a)D^*_{mn}(a)$$
$$=\frac{n(G)}{l}\sum_{ij}(S^{-1})^*_{ji}S^*_{ji}=\frac{n(G)}{l}\sum_{ij}S_{ij}S^*_{ji} \tag{A.4}$$
$$=\frac{n(G)}{l}\,\mathrm{Sp}(SS^*)=\pm n(G).$$

Thus, relation (A.1) is proved.

Note that the above theorem includes the Frobenius-Schur theorem for vector representations as a special case when $\omega_{a,a}=1$.

A.2 Reality Criterion for Space Groups

Theorem A.2 *Let $d^{\mathbf{k}j}$ be the $(\mathbf{k}j)$-th small representation of the group of the wavevector $G_\mathbf{k}$ of a space group G. Then*

$$\sum_{\beta\mathbf{k}\approx-\mathbf{k}}\chi_d^{\mathbf{k}j}(\{\beta|\tau_\beta\}^2)=\begin{cases}n(G^0_\mathbf{k}), & \textit{case a: } \mathcal{D}^{\mathbf{k}j} \textit{ is real,}\\ 0, & \textit{case b: } \mathcal{D}^{\mathbf{k}j} \textit{ is complex,}\\ -n(G^0_\mathbf{k}), & \textit{case c: } \mathcal{D}^{\mathbf{k}j} \textit{ is pseudoreal,}\end{cases} \tag{A.5}$$

where $\chi_d^{\mathbf{k}j}$ is the character of the small representation $d^{\mathbf{k}j}$ of the group of the wavevector $G_\mathbf{k}$, $\mathcal{D}^{\mathbf{k}j}$ is the regular continuation of $d^{\mathbf{k}j}$ to the space group G, and $G^0_\mathbf{k}$ is the point group of the wavevector.

Proof We consider an irreducible representation $\mathcal{D}^{\mathbf{k}j}$ of the space group G. The representation $\mathcal{D}^{\mathbf{k}j}$ is the regular continuation of the $(\mathbf{k}j)$-th small representation of the group of the wavevector $G_\mathbf{k}$. Recalling formulas (3.58) and (5.31) for the elements of the block-matrix $D^{\mathbf{k}j}$, we write

$$\sum_{a\in G}\chi_D^{\mathbf{k}j}(a^2)=\sum_{a\in G}\mathrm{Tr}(D^{\mathbf{k}j}(a^2))$$
$$=\sum_{a\in G}\sum_{Ll\lambda}D^{\mathbf{k}j}_{Ll\lambda,Ll\lambda}(a^2)=\sum_{a\in G}\sum_{l\lambda}\sum_{g_La^2g_L^{-1}\in G_{\mathbf{k}}}d^{\mathbf{k}j}_{l\lambda,l\lambda}(g_La^2g_L^{-1}) \tag{A.6}$$
$$=\sum_{a\in G}\sum_{g_La^2g_L^{-1}\in G_{\mathbf{k}}}\mathrm{Tr}(d^{\mathbf{k}j}(g_La^2g_L^{-1})).$$

Introducing $b=g_Lag_L^{-1}$, we have $b^2=g_La^2g_L^{-1}$. Then

$$b^2=\{\beta|\mathbf{R}_n+\boldsymbol{\tau}_\beta\}^2=\{\beta^2|\beta\boldsymbol{\tau}_\beta+\boldsymbol{\tau}_\beta+\beta\mathbf{R}_n+\mathbf{R}_n\}.$$

The condition $b^2\in G_{\mathbf{k}}$ implies $\beta^2\mathbf{k}=\mathbf{k}+\mathbf{K}_\mu$. From (5.21) it follows that

$$d^{\mathbf{k}j}(b^2)=\mathrm{e}^{-i\mathbf{k}(\beta\mathbf{R}_n+\mathbf{R}_n)}\,\Delta^{\mathbf{k}j}(\{\beta^2|\beta\boldsymbol{\tau}_\beta+\boldsymbol{\tau}_\beta\}).$$

Then (A.6) is written as

$$\sum_{a\in G}\chi_D^{\mathbf{k}j}(a^2)=\frac{n(G)}{n(G_{\mathbf{k}})}\sum_{\substack{\beta\\ \{\beta^2\mathbf{k}=\mathbf{k}+\mathbf{K}_\mu\}}}\sum_{\mathbf{R}_n}\mathrm{e}^{-i\mathbf{k}(\beta\mathbf{R}_n+\mathbf{R}_n)}\,\mathrm{Tr}(\Delta^{\mathbf{k}j}(\{\beta^2|\beta\boldsymbol{\tau}_\beta+\boldsymbol{\tau}_\beta\})).$$

Taking into account

$$\sum_{\mathbf{R}_n}\mathrm{e}^{-i\mathbf{k}(\beta\mathbf{R}_n+\mathbf{R}_n)}=\sum_{\mathbf{R}_n}\mathrm{e}^{-i(\beta^{-1}\mathbf{k}+\mathbf{k})\mathbf{R}_n}=n(T)\sum_{\mathbf{K}_\mu}\delta_{\beta\mathbf{k}+\mathbf{k},\mathbf{K}_\mu},$$

where $n(T)$ is the number of lattice sites $\mathbf{R}_n$ in the crystal, we obtain

$$\sum_{a\in G}\chi_D^{\mathbf{k}j}(a^2)=\frac{n(T)n(G)}{n(G_{\mathbf{k}})}\sum_{\substack{\beta\\ \{\beta\mathbf{k}=-\mathbf{k}+\mathbf{K}_\mu\}}}\mathrm{Tr}(\Delta^{\mathbf{k}j}(\{\beta^2|\beta\boldsymbol{\tau}_\beta+\boldsymbol{\tau}_\beta\})).$$

Clearly, if $\beta\mathbf{k}=-\mathbf{k}+\mathbf{K}_\mu$ then $\beta^2\in G^0_{\mathbf{k}}$ (but not necessarily $\beta\in G^0_{\mathbf{k}}$). Since $G^0_{\mathbf{k}}$ is isomorphic to $G_{\mathbf{k}}/T$, we finally obtain

$$\sum_{a\in G}\chi_D^{\mathbf{k}j}(a^2)=\frac{n(G)}{n(G^0_{\mathbf{k}})}\sum_{\substack{\beta\\ \{\beta\mathbf{k}=-\mathbf{k}+\mathbf{K}_\mu\}}}\mathrm{Tr}(d^{\mathbf{k}j}(b^2))=\frac{n(G)}{n(G^0_{\mathbf{k}})}\sum_{\substack{\beta\\ \{\beta\mathbf{k}=-\mathbf{k}+\mathbf{K}_\mu\}}}\chi_d^{\mathbf{k}j}(b^2), \tag{A.7}$$

where $b=\{\beta|\mathbf{R}_n+\boldsymbol{\tau}_\beta\}$. Here $\mathbf{R}_n$ can be chosen arbitrarily, because of the condition $\beta\mathbf{k}=-\mathbf{k}+\mathbf{K}_\mu$. Substituting (A.7) in the Frobenius-Schur theorem,

$$\sum_{a \in G} \chi(a^2) = \begin{cases} n(G), & \text{case a}, \\ 0, & \text{case b}, \\ -n(G), & \text{case c}, \end{cases}$$

we come to relation (A.5).

Appendix B
Construction of Projective Representations of the Point Group T

According to Eqs. (6.30) and (6.28), the relations defining projective representation of the group T have the form[1]

$$\begin{aligned} &A^2 = E, && && \\ &B^2 = E, && BA = \omega^2 AB, && \\ &C^3 = E, && CA = BC, && CB = \omega ABC, \end{aligned} \tag{B.1}$$

$$\omega^4 = 1. \tag{B.2}$$

Let us consider the matrices A, B, C and $\Omega \equiv \omega E$ as the generators of the covering group $\tilde{T}$. An element of this group is

$$\Omega^\mu A^\alpha B^\beta C^\gamma, \qquad 1 \leq \mu \leq 4, \qquad 1 \leq \alpha, \beta \leq 2, \qquad 1 \leq \gamma \leq 3. \tag{B.3}$$

As an invariant subgroup H of the covering group $\tilde{T}$ we choose the group consisting of

$$\Omega^\mu A^\alpha B^\beta, \qquad 1 \leq \mu \leq 4, \qquad 1 \leq \alpha, \beta \leq 2. \tag{B.4}$$

Then the expansion the group $\tilde{T}$ with respect to H is

$$\tilde{T} = HE \oplus HC \oplus HC^2. \tag{B.5}$$

The fact that H is an invariant subgroup is easy to verify. Indeed, taking (B.1) and (B.2) into account, we have

[1] For brevity we omit the primes on the matrices and indices of the character factor.

N. B. Melnikov and B. I. Reser, *Space Group Representations*,
https://doi.org/10.1007/978-3-031-13991-8

$$
\begin{aligned}
C\Omega^{\mu} A^{\alpha} B^{\beta} C^{-1} &= \Omega^{\mu} B^{\alpha} C B^{\beta} C^{-1} = \Omega^{\mu+\beta+2(\beta-1)} B^{\alpha} A^{\beta} B^{\beta} C^{-1} \\
&= \Omega^{\mu+\beta+2(\beta-1)+2\alpha\beta} A^{\beta} B^{\alpha+\beta}, \qquad \text{(B.6)}\\
C^2\Omega^{\mu} A^{\alpha} B^{\beta} (C^2)^{-1} &= C\Omega^{\mu+\beta+2(\beta-1)+2\alpha\beta} A^{\beta} B^{\alpha+\beta} C^{-1} \\
&= \Omega^{\mu+\alpha+2(\alpha-1)+2\alpha\beta} A^{\alpha+\beta} B^{\alpha}. \qquad \text{(B.7)}
\end{aligned}
$$

In particular, from formula (B.6) it follows that

$$
CAC^{-1} = B, \qquad CBC^{-1} = \Omega AB, \tag{B.8}
$$

and from formula (B.7) it follows that

$$
C^2 A (C^2)^{-1} = \Omega AB, \qquad C^2 B (C^2)^{-1} = A. \tag{B.9}
$$

Representations of the invariant subgroup H are easily obtained from the representations of the covering group $\tilde{D}_2$. (We assume that the representations of $\tilde{D}_2$ are already known by the time we consider the group T (see Sect. 6.6).) In Table B.1 the representations of the invariant subgroup H are given in columns 2–4. Columns 5 and 6 indicate the star and normalizer of each representation, which are determined as follows. Taking (B.6) and (B.7) into account, we write down the character of the conjugate representation

$$
\chi^{i_c}(\Omega^{\mu} A^{\alpha} B^{\beta}) = \chi^{i}(\Omega^{\mu+\beta+2(\beta-1)+2\alpha\beta} A^{\beta} B^{\alpha+\beta}), \tag{B.10}
$$

$$
\chi^{i_{c^2}}(\Omega^{\mu} A^{\alpha} B^{\beta}) = \chi^{i}(\Omega^{\mu+\alpha+2(\alpha-1)+2\alpha\beta} A^{\alpha+\beta} B^{\alpha}), \tag{B.11}
$$

and determine the representation of the invariant subgroup H with the same character. The set of indices of the conjugate representations, including i, gives the star of the representation, and the set of classes HC^{γ} $(1 \le \gamma \le 3)$ with C^{γ} such that $\chi^{i_{c^{\gamma}}} = \chi^{i}$ gives the normalizer of the representation.

Let us proceed to construction of representations of the covering group $\tilde{T}$.

First, we consider the representations d^i whose normalizer coincides with the invariant subgroup. Choosing the representation d^3 from the orbit[2] d^2, d^3, d^4, and the representation d^5 from the orbit d^5, d^6, d^7, using the regular representation formula (3.58) and taking (B.8) and (B.9) into account, we immediately obtain

[2] This is our historically established choice.

Table B.1 Representations of the invariant subgroup H of the covering group $\tilde{T}$, their stars and normalizers

i	Ω	A	B	Star	Normalizer
1		1	1	1	$\tilde{T}$
2	1	1	-1	2, 3, 4	H
3		-1	1	2, 3, 4	H
4		-1	-1	2, 3, 4	H
5		1	1	5, 6, 7	H
6	-1	1	-1	5, 6, 7	H
7		-1	1	5, 6, 7	H
8		-1	-1	8	$\tilde{T}$
9	$i\begin{pmatrix}1&0\\0&1\end{pmatrix}$	$\begin{pmatrix}1&0\\0&-1\end{pmatrix}$	$\begin{pmatrix}0&1\\1&0\end{pmatrix}$	9	$\tilde{T}$
10	$-i\begin{pmatrix}1&0\\0&1\end{pmatrix}$	$\begin{pmatrix}1&0\\0&-1\end{pmatrix}$	$\begin{pmatrix}0&1\\1&0\end{pmatrix}$	10	$\tilde{T}$

$$
\begin{aligned}
D^i(\Omega) &= d^i(\Omega)\begin{pmatrix}1&0&0\\0&1&0\\0&0&1\end{pmatrix},\\
D^i(A) &= \begin{pmatrix}d^i(A)&0&0\\0&d^i(B)&0\\0&0&d^i(\Omega AB)\end{pmatrix},\\
D^i(B) &= \begin{pmatrix}d^i(B)&0&0\\0&d^i(\Omega AB)&0\\0&0&d^i(A)\end{pmatrix},\\
D^i(C) &= \begin{pmatrix}0&1&0\\0&0&1\\1&0&0\end{pmatrix}, \qquad i = 3, 5
\end{aligned}
\tag{B.12}
$$

Now we consider the representations d^i whose normalizer coincides with the whole group $\tilde{T}$. The factor group $\tilde{T}/H$ of the normalizer of these representations $\tilde{T}$ over the invariant subgroup H is isomorphic to the point group C_3.

To construct one-dimensional representations d^1 and d^8, we need *vector* representations of the group C_3, because the matrices of the unitary transformation $U(C^\gamma)$ can be chosen to be unity and elements C^γ form a group (see Sect. 3.7). The vector representations of the group C_3 are presented in Table B.2. By formula (3.41) we obtain

$$
\begin{gathered}
D^{ij}(\Omega) = d^i(\Omega),\ D^{ij}(A) = d^i(A),\ D^{ij}(B) = d^i(B),\ D^{ij}(C) = \Delta^j(C),\\
i = 1, 8; \qquad j = 1, 2, 3.
\end{gathered}
\tag{B.13}
$$

Table B.2 Vector representations of the group C_3

	C
Δ^1	1
Δ^2	ε
Δ^3	ε^2

To construct two-dimensional representations d^9 and d^{10}, we need to obtain the matrices of the unitary transformation $U(C^\gamma)$ as solutions of the system of matrix equations (3.24). Let us consider the system of equations for the matrix $U(C)$:

$$U(C)\, d^i(A) = d^i(CAC^{-1})\, U(C), \tag{B.14}$$

$$U(C)\, d^i(B) = d^i(CBC^{-1})\, U(C). \tag{B.15}$$

Taking relations (B.8) into account, we write this system of equations as

$$U(C)\, d^i(A) = d^i(B)\, U(C), \tag{B.16}$$

$$U(C)\, d^i(B) = d^i(\Omega AB)\, U(C). \tag{B.17}$$

Equation (B.16),

$$\begin{pmatrix} u_{11} & u_{12} \\ u_{21} & u_{22} \end{pmatrix} \begin{pmatrix} 1 & 0 \\ 0 & -1 \end{pmatrix} = \begin{pmatrix} 0 & 1 \\ 1 & 0 \end{pmatrix} \begin{pmatrix} u_{11} & u_{12} \\ u_{21} & u_{22} \end{pmatrix},$$

yields

$$U(C) = \begin{pmatrix} u_{11} & u_{12} \\ u_{11} & -u_{12} \end{pmatrix}.$$

Substituting this matrix in equation (B.17)

$$\begin{pmatrix} u_{11} & u_{12} \\ u_{11} & -u_{12} \end{pmatrix} \begin{pmatrix} 0 & 1 \\ 1 & 0 \end{pmatrix} = \omega \begin{pmatrix} 0 & 1 \\ -1 & 0 \end{pmatrix} \begin{pmatrix} u_{11} & u_{12} \\ u_{11} & -u_{12} \end{pmatrix},$$

and taking $\omega^2 = -1$ into account, we obtain

$$U(C) = u_{11} \begin{pmatrix} 1 & \omega \\ 1 & -\omega \end{pmatrix}.$$

Finally, since U is unitary, i.e. from the equation

$$u_{11} \begin{pmatrix} 1 & \omega \\ 1 & -\omega \end{pmatrix} u_{11}^* \begin{pmatrix} 1 & 1 \\ \omega^* & -\omega^* \end{pmatrix} = \begin{pmatrix} 1 & 0 \\ 0 & 1 \end{pmatrix},$$

we have

$$U(C) = \frac{1}{\sqrt{2}} \begin{pmatrix} 1 & \omega \\ 1 & -\omega \end{pmatrix} \tag{B.18}$$

up to an arbitrary phase factor. Perfectly analogously, we consider the system of equations

$$U(C^2)\, d^i(A) = d^i(C^2 A (C^2)^{-1})\, U(C^2), \tag{B.19}$$
$$U(C^2)\, d^i(B) = d^i(C^2 B (C^2)^{-1})\, U(C^2). \tag{B.20}$$

Taking (B.9) into account, we rewrite the above as follows:

$$U(C^2)\, d^i(A) = d^i(\Omega A B)\, U(C^2), \tag{B.21}$$
$$U(C^2)\, d^i(B) = d^i(A)\, U(C^2). \tag{B.22}$$

From this it follows that

$$U(C^2) = \frac{1}{\sqrt{2}} \begin{pmatrix} 1 & 1 \\ -\omega & \omega \end{pmatrix}. \tag{B.23}$$

Using the explicit form (B.18) and (B.23) of matrices $U(C)$ and $U(C^2)$, we calculate the factor system of the projective representation of the group C_3. The relation [see (3.27)]

$$U(C^2) = U(C)U(C)\, \omega_{C,C} \tag{B.24}$$

gives

$$\omega_{C,C} = \frac{1-\omega}{\sqrt{2}}, \tag{B.25}$$

and the relation

$$U(E) = U(C^2)U(C)\, \omega_{C^2,C} \tag{B.26}$$

gives

$$\omega_{C^2,C} = 1. \tag{B.27}$$

Taking (B.25) and (B.27) into account, for the matrix $\Delta^{ij}(C)$ of the projective representation of the group C_3 we have

$$\Delta^{ij}(C) = (\omega_{C^3})^{1/3} \Delta^j(C) = (\omega_{C,C}\omega_{C^2,C})^{1/3} \Delta^j(C) = \left(\frac{1-\omega}{\sqrt{2}}\right)^{1/3} \Delta^j(C),$$
$$i = 9, 10; \qquad j = 1, 2, 3. \tag{B.28}$$

Now by formula (3.41), taking (B.18) and (B.28) into account, we obtain

$$D^{ij}(\Omega) = d^i(\Omega), \qquad D^{ij}(A) = d^i(A), \qquad D^{ij}(B) = d^i(B),$$
$$D^{ij}(C) = U^i(C) \otimes \Delta^{ij}(C) = \frac{1}{\sqrt{2}}\begin{pmatrix} 1 & \omega \\ 1 & -\omega \end{pmatrix}\left(\frac{1-\omega}{\sqrt{2}}\right)^{1/3} \Delta^j(C),$$
$$i = 9, 10; \qquad j = 1, 2, 3. \tag{B.29}$$

Finally,

$$D^{9,j}(C) = \frac{1}{\sqrt{2}}\begin{pmatrix} 1 & i \\ 1 & -i \end{pmatrix}\left(\frac{1-i}{\sqrt{2}}\right)^{1/3} \Delta^j(C) = \frac{1}{\sqrt{2}}\begin{pmatrix} 1 & i \\ 1 & -i \end{pmatrix} e^{-i\pi/12}\Delta^j(C),$$
$$D^{10,j}(C) = \frac{1}{\sqrt{2}}\begin{pmatrix} 1 & -i \\ 1 & i \end{pmatrix}\left(\frac{1+i}{\sqrt{2}}\right)^{1/3} \Delta^j(C) = \frac{1}{\sqrt{2}}\begin{pmatrix} 1 & -i \\ 1 & i \end{pmatrix} e^{i\pi/12}\Delta^j(C),$$
$$j = 1, 2, 3. \tag{B.30}$$

Appendix C
More on Symmetry in Quantum Mechanics

C.1 Discussion of Wigner's Theorem

Wigner's theorem was formulated and proved in Chap. 7. This theorem helps to classify the states of a physical system according to the symmetry group of the Hamiltonian. Here we discuss some consequences of Wigner's theorem for the matrix of the Hamiltonian (for details, see [1]).

We consider an orthonormal basis consisting of the basis functions of irreducible representations. In this basis, the matrix S of a symmetry operator P_a is block-diagonal,

$$S = \begin{pmatrix} S_1 & 0 & 0 & \dots \\ 0 & S_2 & 0 & \dots \\ 0 & 0 & S_3 & \dots \\ \dots & \dots & \dots & \dots \end{pmatrix},$$

where $S_1, S_2, \dots$ are the *representation matrices*. The matrix of the Hamiltonian has the form

$$H = \begin{pmatrix} H_{11} & H_{12} & H_{13} & \dots \\ H_{21} & H_{22} & H_{23} & \dots \\ H_{31} & H_{32} & H_{33} & \dots \\ \dots & \dots & \dots & \dots \end{pmatrix}.$$

Here H_{ij} are *submatrices*. Most of them are rectangular, because different representations do not necessarily have the same dimensions. If we form the products HS and SH, multiplying the corresponding submatrices as matrix elements, we obtain

$$HS = \begin{pmatrix} H_{11}S_1 & H_{12}S_2 & \dots \\ H_{21}S_1 & H_{22}S_2 & \dots \\ H_{31}S_1 & H_{32}S_2 & \dots \\ \dots & \dots & \dots \end{pmatrix}. \quad SH = \begin{pmatrix} S_1H_{11} & S_1H_{12} & S_1H_{13} & \dots \\ S_2H_{21} & S_2H_{22} & S_2H_{23} & \dots \\ S_3H_{31} & S_3H_{32} & S_3H_{33} & \dots \\ \dots & \dots & \dots & \dots \end{pmatrix}.$$

Since S is a symmetry operator, it commutes with H,

N. B. Melnikov and B. I. Reser, *Space Group Representations*,
https://doi.org/10.1007/978-3-031-13991-8

$$SH - HS = 0,$$

or equivalently,

$$S_i H_{ij} - H_{ij} S_j = 0.$$

By Schur's lemma we come to the following conclusion.

(i) When S_i and S_j are non-equivalent, we have $H_{ij} = 0$.
(ii) When S_i and S_j are equivalent, H_{ij} is a multiple of the unit matrix E, i.e. $H_{ij} = H'_{ij} E$, where H'_{ij} is a number.

Let us arrange the eigenspaces V_p of irreducible representations $\mathcal{D}_p$ in such a way that the symmetry operator is written as

$$S = \left(\begin{array}{c|c|c} \begin{matrix} S_1^{(1)} & 0 & \cdots \\ 0 & S_2^{(1)} & \cdots \\ \cdots & \cdots & \cdots \end{matrix} & & 0 \\ \hline & \begin{matrix} S_1^{(2)} & 0 & \cdots \\ 0 & S_2^{(2)} & \cdots \\ \cdots & \cdots & \cdots \end{matrix} & \\ \hline 0 & & \ddots \end{array} \right),$$

where the matrix $S_m^{(p)}$ corresponds to the m-th irreducible representation $\mathcal{D}_p$. Then, H is also block-diagonal but consists of larger blocks $H^{(p)}$ (possibly infinite) corresponding to *non-equivalent* irreducible representations $\mathcal{D}_p$,

$$H = \left(\begin{array}{c|c|c} \begin{matrix} H_{11}^{(1)} & H_{12}^{(1)} & \cdots \\ H_{21}^{(1)} & H_{22}^{(1)} & \cdots \\ \cdots & \cdots & \cdots \end{matrix} & & 0 \\ \hline & \begin{matrix} H_{11}^{(2)} & H_{12}^{(2)} & \cdots \\ H_{21}^{(2)} & H_{22}^{(2)} & \cdots \\ \cdots & \cdots & \cdots \end{matrix} & \\ \hline 0 & & \ddots \end{array} \right),$$

where $H_{ij}^{(p)} = H_{ij}^{(p)\prime} E$ is a scalar submatrix. Thus, the function space is decomposed in the invariant subspaces V_p with respect to the group G and the operator H at the same time. Each of them corresponds to one of the non-equivalent irreducible representations $\mathcal{D}_p$ of G. The orthogonal functions $f_{m\kappa}^{(p)}$ that lead to this decomposition are represented by symbols with three indices

p labels the big square matrices;
m labels the small square matrices;
κ labels the rows of the small square matrices.

These functions are not yet the Schrödinger wave functions for which H is diagonal. But the initial eigenvalue problem is split into smaller ones, based solely on the symmetry of H with respect to the group G.

C.2 Projection Operator Corresponding to G and $G_{\mathbf{k}}$

Let $\mathcal{D}^{\mathbf{k}j}$ be the irreducible representation of the space group G obtained by the continuation of the j-th small representation $d^{\mathbf{k}j}$ of the group of the wavevector $G_{\mathbf{k}}$. According to (7.3), we define the projection operator corresponding to a particular row of the representation $\mathcal{D}^{\mathbf{k}j}$ as

$$\mathcal{O}^{\mathbf{k}j}_{R\rho,S\sigma} = \frac{l_{\mathbf{k}j}}{n(G)} \sum_{a \in G} D^{\mathbf{k}j*}_{R\rho,S\sigma}(a) P_a$$

(index $(S\sigma)$ is fixed)[3]. This operator can be used to construct symmetrized functions with respect to the space group G (Sect. 8.2).

Next, we construct the symmetrized functions with respect to the group of the wavevector $G_{\mathbf{k}}$. This is obtained by using the operator of the form

$$O^{\mathbf{k}j}_{\rho\sigma} \equiv \mathcal{O}^{\mathbf{k}j}_{1\rho,1\sigma} = \frac{l_{\mathbf{k}j}}{n(G)} \sum_{a \in G} D^{\mathbf{k}j*}_{1\rho,1\sigma}(a) P_a.$$

According to (5.5), any element $a = \{\alpha|\mathbf{t}\}$ of the space group G can be written as

$$\{\alpha|\mathbf{t}\} = \{\alpha|\mathbf{R}_n + \boldsymbol{\tau}_\alpha\} = \{\varepsilon|\mathbf{R}_n\}\{\alpha|\boldsymbol{\tau}_\alpha\}.$$

Then

$$O^{\mathbf{k}j}_{\rho\sigma} = \frac{l_{\mathbf{k}j}}{n(G)} \sum_{\{\varepsilon|\mathbf{R}_n\}\{\alpha|\boldsymbol{\tau}_\alpha\}} D^{\mathbf{k}j*}_{1\rho,1\sigma}(\{\varepsilon|\mathbf{R}_n\}\{\alpha|\boldsymbol{\tau}_\alpha\}) P_{\{\varepsilon|\mathbf{R}_n\}} P_{\{\alpha|\boldsymbol{\tau}_\alpha\}}\,. \tag{C.1}$$

According to (5.33), we have

$$D^{\mathbf{k}j}_{1\rho,1\sigma}(\{\varepsilon|\mathbf{R}_n\}\{\alpha|\boldsymbol{\tau}_\alpha\}) = d^{\mathbf{k}j}_{\rho\sigma}(\{\varepsilon|\mathbf{R}_n\}\{\alpha|\boldsymbol{\tau}_\alpha\})\, \delta_{G_{\mathbf{k}}\{\varepsilon|\mathbf{R}_n\}\{\alpha|\boldsymbol{\tau}_\alpha\},G_{\mathbf{k}}},$$

---

[3] It is important to distinguish the operator $\mathcal{O}^{\mathbf{k}j}$ with respect to the space group G, operator $O^{\mathbf{k}j}$ with respect to the group of the wavevector $G_{\mathbf{k}}$ and operator $\tilde{O}^{\mathbf{k}j}$ with respect to the point group of the wavevector $G^0_{\mathbf{k}}$. The operator $\mathcal{O}^{\mathbf{k}j}$ is used in the studies of the phonon spectra. In the electron band calculations, we deal only with the operators $O^{\mathbf{k}j}$ and $\tilde{O}^{\mathbf{k}j}$.

where $d^{\mathbf{k}j}$ is the j-th small representation of the group of the wavevector $G_{\mathbf{k}}$. By (5.32) we have $l_{\mathbf{k}j} = n(G)\, l_j / n(G_{\mathbf{k}})$, where l_j is the dimension of $d^{\mathbf{k}j}$. Hence the operator (C.1) takes the form

$$O^{\mathbf{k}j}_{\rho\sigma} = \frac{l_j}{n(G_{\mathbf{k}})} \sum_{\{\varepsilon|\mathbf{R}_n\}\{\alpha_m|\boldsymbol{\tau}_{\alpha_m}\}} d^{\mathbf{k}j*}_{\rho\sigma}(\{\varepsilon|\mathbf{R}_n\}\{\alpha_m|\boldsymbol{\tau}_{\alpha_m}\})\, P_{\{\varepsilon|\mathbf{R}_n\}} P_{\{\alpha_m|\boldsymbol{\tau}_{\alpha_m}\}}. \tag{C.2}$$

Taking (5.21) into account,

$$d^{\mathbf{k}j}(\{\varepsilon|\mathbf{R}_n\}\{\alpha_m|\boldsymbol{\tau}_{\alpha_m}\}) = \mathrm{e}^{-i\mathbf{k}\mathbf{R}_n}\, \Delta^{\mathbf{k}j}(\{\alpha_m|\boldsymbol{\tau}_{\alpha_m}\}),$$

we can write this operator as the product of two operators

$$O^{\mathbf{k}j}_{\rho\sigma} = \left(\frac{l_j}{n(G^0_{\mathbf{k}})} \sum_{\{\alpha_m|\boldsymbol{\tau}_{\alpha_m}\}} \Delta^{\mathbf{k}j*}_{\rho\sigma}(\{\alpha_m|\boldsymbol{\tau}_{\alpha_m}\})\, P_{\{\alpha_m|\boldsymbol{\tau}_{\alpha_m}\}} \right) \left(\frac{1}{n(T)} \sum_{\{\varepsilon|\mathbf{R}_n\}} \mathrm{e}^{i\mathbf{k}\mathbf{R}_n} P_{\{\varepsilon|\mathbf{R}_n\}} \right). \tag{C.3}$$

Here $\Delta^{\mathbf{k}j}$ is the j-th irreducible projective representation of the factor group $G_{\mathbf{k}}/T$, which is isomorphic to the point group $G^0_{\mathbf{k}}$ consisting of rotational parts α_m of the representatives $\{\alpha_m|\boldsymbol{\tau}_{\alpha_m}\}$ in the decomposition of the group $G_{\mathbf{k}}$ into cosets with respect to T (see Sect. 5.6).

Usually, one chooses Bloch functions as the basis functions

$$\psi_{\mathbf{k}}(\mathbf{r}) = \mathrm{e}^{i\mathbf{k}\mathbf{r}} u_{\mathbf{k}}(\mathbf{r}). \tag{C.4}$$

It is easy to check that the action of the second operator on the right-hand side of (C.3) on a Bloch function does not change the function. Hence the projection operator $\tilde{O}^{\mathbf{k}j}_{\rho\sigma}$ takes the form

$$\tilde{O}^{\mathbf{k}j}_{\rho\sigma} = \frac{l_j}{n(G^0_{\mathbf{k}})} \sum_{\{\alpha_m|\boldsymbol{\tau}_{\alpha_m}\}} \Delta^{\mathbf{k}j*}_{\rho\sigma}(\{\alpha_m|\boldsymbol{\tau}_{\alpha_m}\})\, P_{\{\alpha_m|\boldsymbol{\tau}_{\alpha_m}\}}. \tag{C.5}$$

In this case, one can restrict oneself to obtaining only irreducible projective representations of the group $G^0_{\mathbf{k}}$.

According to Wigner's theorem (Sect. 7.2.2), to calculate the matrix elements of the Hamiltonian it is sufficient to obtain only *one* symmetrized basis function corresponding to each irreducible representation. For instance, we can construct symmetrized basis functions belonging to the first row of each irreducible representation of the space group. This is obtained by using the projection operator of the form

$$O^{\mathbf{k}j} \equiv O^{\mathbf{k}j}_{11,11} = \frac{l_{\mathbf{k}j}}{n(G)} \sum_{a \in G} D^{\mathbf{k}j*}_{11,11}(a)\, P_a.$$

Using (C.3) we obtain

$$O^{\mathbf{k}j} = \left(\frac{l_j}{n(G^0_{\mathbf{k}})} \sum_{\{\alpha_m|\tau_{\alpha_m}\}} \Delta^{\mathbf{k}j*}_{1,1}(\{\alpha_m|\tau_{\alpha_m}\}) P_{\{\alpha_m|\tau_{\alpha_m}\}} \right) \left(\frac{1}{n(T)} \sum_{\{\varepsilon|\mathbf{R}_n\}} \mathrm{e}^{i\mathbf{k}\mathbf{R}_n} P_{\{\varepsilon|\mathbf{R}_n\}} \right).$$

If we choose Bloch functions (C.4) as the initial basis, then $O^{\mathbf{k}j}$ takes the form

$$\tilde{O}^{\mathbf{k}j} = \frac{l_j}{n(G^0_{\mathbf{k}})} \sum_{\{\alpha_m|\tau_{\alpha_m}\}} \Delta^{\mathbf{k}j*}_{1,1}(\{\alpha_m|\tau_{\alpha_m}\}) P_{\{\alpha_m|\tau_{\alpha_m}\}}.$$

In this case, one can restrict oneself to obtaining only irreducible projective representations of the group $G^0_{\mathbf{k}}$.

Reference

1. P.H.E. Meijer, E. Bauer, *Group Theory: The Application to Quantum Mechanics*, 2nd edn. (North-Holland, Amsterdam, 1962)

Appendix D
Factor System of Double-Valued Representations

D.1 Composition of Two Rotations

The rotation α by an angle θ around the axis with a unit vector $\mathbf{n}$ can be represented by the two-dimensional matrix (8.26),

$$S(\alpha) = \mathrm{e}^{-i(\theta/2)\boldsymbol{\sigma}\cdot\mathbf{n}} = \cos\frac{\theta}{2}\,E - i\sin\frac{\theta}{2}\,(\boldsymbol{\sigma}\cdot\mathbf{n}) \tag{D.1}$$

(see textbooks on classical mechanics [1, 2] and quantum mechanics [3, 4]). As we see below, it is convenient to choose θ such that $-\pi < \theta \le \pi$ (rather than $0 < \theta \le 2\pi$), where a positive angle is given by a counter clockwise rotation as viewed down from the endpoint of the vector $\mathbf{n}$. Furthermore, the pairs of the parameters $(\theta, \mathbf{n})$ and $(-\theta, -\mathbf{n})$ describe the same rotation. For the sake of definiteness, for all rotations such that $0 \le \theta \le \pi$, we take $\mathbf{n}$ in the *positive hemisphere*,

$$\text{(i)}\ \ z > 0; \qquad \text{(ii)}\ \ z = 0,\quad x > 0; \qquad \text{(iii)}\ \ z = 0,\quad x = 0,\quad y > 0,$$

and for all rotations such that $-\pi < \theta < 0$, we take $\mathbf{n}$ in the *negative hemisphere*. With this convention it is sufficient to consider only nonnegative angles

$$0 \le \theta \le \pi. \tag{D.2}$$

Each rotation α now has a unique parametric point, and the parametric points defined this way are called *standard parametric points*.

Theorem D.1 *The composition of two matrices $S(\alpha)$ and $S(\beta)$ is given by*

$$S(\alpha)S(\beta) = \omega_{\alpha,\beta}S(\alpha\beta),$$

where $\omega_{\alpha,\beta}$ is given by

N. B. Melnikov and B. I. Reser, *Space Group Representations*,
https://doi.org/10.1007/978-3-031-13991-8

$$
\omega_{\alpha,\beta} = \begin{cases} \dfrac{\cos\frac{\theta_\alpha}{2}\cos\frac{\theta_\beta}{2} - \sin\frac{\theta_\alpha}{2}\sin\frac{\theta_\beta}{2}(\mathbf{n}_\alpha \cdot \mathbf{n}_\beta)}{\cos\frac{\theta_{\alpha\beta}}{2}}, & \theta_{\alpha\beta} \neq \pi, \\ \sin\frac{\theta_\alpha}{2}\cos\frac{\theta_\beta}{2}(\mathbf{n}_\alpha \cdot \mathbf{n}_{\alpha\beta}) + \cos\frac{\theta_\alpha}{2}\sin\frac{\theta_\beta}{2}(\mathbf{n}_\beta \cdot \mathbf{n}_{\alpha\beta}) & \\ + \sin\frac{\theta_\alpha}{2}\sin\frac{\theta_\beta}{2}([\mathbf{n}_\alpha \times \mathbf{n}_\beta] \cdot \mathbf{n}_{\alpha\beta}), & \theta_{\alpha\beta} = \pi. \end{cases} \tag{D.3}
$$

Proof From the well-known property of the Pauli matrices

$$
(\boldsymbol{\sigma} \cdot \mathbf{a})(\boldsymbol{\sigma} \cdot \mathbf{b}) = (\mathbf{a} \cdot \mathbf{b})\, E + i(\boldsymbol{\sigma} \cdot (\mathbf{a} \times \mathbf{b})),
$$

it follows that the result is also a rotation

$$
S(\alpha)S(\beta) = \cos\frac{\theta'_{\alpha\beta}}{2}\, E - i \sin\frac{\theta'_{\alpha\beta}}{2}\, (\boldsymbol{\sigma} \cdot \mathbf{n}'_{\alpha\beta}),
$$

where the rotation angle $\theta'_{\alpha\beta}$ and axis $\mathbf{n}'_{\alpha\beta}$ are determined by[4]

$$
\cos\frac{\theta'_{\alpha\beta}}{2} = \cos\frac{\theta_\alpha}{2}\cos\frac{\theta_\beta}{2} - \sin\frac{\theta_\alpha}{2}\sin\frac{\theta_\beta}{2}\,(\mathbf{n}_\alpha \cdot \mathbf{n}_\beta), \tag{D.4}
$$

$$
\begin{aligned} \sin\frac{\theta'_{\alpha\beta}}{2}\,\mathbf{n}'_{\alpha\beta} &= \sin\frac{\theta_\alpha}{2}\cos\frac{\theta_\beta}{2}\,\mathbf{n}_\alpha + \sin\frac{\theta_\beta}{2}\cos\frac{\theta_\alpha}{2}\,\mathbf{n}_\beta \\ &+ \sin\frac{\theta_\alpha}{2}\sin\frac{\theta_\beta}{2}\,(\mathbf{n}_\alpha \times \mathbf{n}_\beta). \end{aligned} \tag{D.5}
$$

The prime indicates that, while θ_α and θ_β must be standard parameters, the rotation angle $\theta'_{\alpha\beta}$ may not be so. By $\theta_{\alpha\beta}$ and $\mathbf{n}_{\alpha\beta}$ we denote the angle and axis such that condition (D.2) is satisfied. Therefore, if $\theta'_{\alpha\beta} \neq \pi$ and hence $\theta_{\alpha\beta} \neq \pi$, then by relation (D.4) we obtain that $\omega_{\alpha,\beta}$ is equal to the ratio $\cos(\theta'_{\alpha\beta}/2)/\cos(\theta_{\alpha\beta}/2)$. This is the first case in formula (D.3). If $\theta'_{\alpha\beta}$ and hence $\theta_{\alpha\beta}$ are equal to π, then by relation (D.5) we obtain that $\omega_{\alpha,\beta}$ is equal to $\mathbf{n}'_{\alpha\beta} \cdot \mathbf{n}_{\alpha\beta}$. This is the second case in formula (D.3). The proof is complete.

To sum up the above discussion of formula (D.3), $\omega_{\alpha,\beta} = 1$ if $(\theta'_{\alpha\beta}, \mathbf{n}'_{\alpha\beta})$ are standard parameters, and $\omega_{\alpha,\beta} = -1$ if they are not. When $\theta'_{\alpha\beta} \neq \pi$, this simply means that

$$
\omega_{\alpha,\beta} = \begin{cases} 1, & 0 < \theta'_{\alpha\beta} < \pi, \\ -1, & \pi < \theta'_{\alpha\beta} < 2\pi, \end{cases} \tag{D.6}
$$

where the angle $\theta'_{\alpha\beta}$ is given by (D.4). In particular, if α and β are rotations about the same axis then the composition is also the rotation about this axis and the rotation angles add up. Indeed, if $\mathbf{n}_\alpha = \mathbf{n}_\beta = \mathbf{n}$, then by formulas (D.4) and (D.5) we imme-

[4] This is the *composition rule of two spatial rotations*, obtained by Rodrigues (1840); his geometrical construction is presented by [1, 5]. A different way of deriving the composition rule was presented by Gibbs [6, Sect. 128, p. 347]. Here we give a more compact derivation based on representing the rotations by the Pauli matrices.

diately obtain $\theta'_{\alpha\beta} = \theta_\alpha + \theta_\beta$ and $\mathbf{n}'_{\alpha\beta} = \mathbf{n}$. Thus, if α and β are rotations about the same axis (or rotations plus inversion), formula (D.6) immediately yields $\omega_{\alpha,\beta}$. For instance, the projective factor for the composition of any rotation by the angle $2\pi/3$ with itself is -1.

If $\theta'_{\alpha\beta} = \pi$, one needs to look at the vector $\mathbf{n}'_{\alpha\beta}$. Let us illustrate this by the example of the group D_2, which consists of four elements: the identity element and three rotations by the angle π around each coordinate axis $\mathbf{x}$, $\mathbf{y}$ and $\mathbf{z}$. The two-by-two matrices corresponding to these rotations are $S(\pi, \mathbf{x})$, $S(\pi, \mathbf{y})$ and $S(\pi, \mathbf{z})$. Then by formulas (D.4) and (D.5) we have

$$
\begin{aligned}
S(\pi, \mathbf{z})S(\pi, \mathbf{x}) &= S(\pi, \mathbf{z} \times \mathbf{x}) = S(\pi, \mathbf{y}), \\
S(\pi, \mathbf{x})S(\pi, \mathbf{z}) &= S(\pi, \mathbf{x} \times \mathbf{z}) = S(\pi, -\mathbf{y}).
\end{aligned}
$$

Using formula (D.3), we obtain the projective factors,

$$
\begin{aligned}
S(\pi, \mathbf{z})S(\pi, \mathbf{x}) &= S(\pi, \mathbf{y}), \\
S(\pi, \mathbf{x})S(\pi, \mathbf{z}) &= -S(\pi, \mathbf{y}).
\end{aligned}
$$

Another, even simpler, example is the case of two rotations about the same axis such that $\theta_\alpha + \theta_\beta = \pi$ and $\mathbf{n}_\alpha = \mathbf{n}_\beta = \mathbf{n}$. For instance, both elements α and β are equal to the rotation by the angle $\pi/2$ around the z-th axis, the element number 9 in Table 6.1. Then the right-hand side of formula (D.5) becomes

$$
\sin\frac{\theta_\alpha}{2}\cos\frac{\theta_\beta}{2}\,\mathbf{n}_\alpha + \sin\frac{\theta_\beta}{2}\cos\frac{\theta_\alpha}{2}\,\mathbf{n}_\beta + \sin\frac{\theta_\alpha}{2}\sin\frac{\theta_\beta}{2}\,(\mathbf{n}_\alpha \times \mathbf{n}_\beta) = \sin\frac{\theta_\alpha + \theta_\beta}{2}\,\mathbf{n} = \mathbf{n}.
$$

Hence formula (D.5) gives

$$
\sin\frac{\theta'_{\alpha\beta}}{2}\,\mathbf{n}'_{\alpha\beta} = \mathbf{n},
$$

i.e. either $\theta'_{\alpha\beta} = \pi$ and $\mathbf{n}'_{\alpha\beta} = \mathbf{n}$ or $\theta'_{\alpha\beta} = -\pi$ and $\mathbf{n}'_{\alpha\beta} = -\mathbf{n}$. According to our convention (D.2), we choose the former. Therefore, by formula (D.3) we have $\omega_{\alpha,\beta} = 1$ and $S(\alpha)S(\alpha) = S(\alpha^2)$, in full agreement with the multiplication Table 6.2.

D.2 Euler-Rodrigues Parameters

The method of calculating the projective factors can be made more transparent if we use different parameters for the rotation operators.

We replace θ and $\mathbf{n}$ by new parameters

$$
\lambda = \cos\frac{\theta}{2}, \qquad \boldsymbol{\Lambda} = \sin\frac{\theta}{2}\,\mathbf{n}, \tag{D.7}
$$

called the *Euler-Rodrigues parameters*. Then (D.1) becomes

$$S(\lambda, \mathbf{\Lambda}) = \lambda E - i\mathbf{\Lambda} \cdot \boldsymbol{\sigma}. \tag{D.8}$$

Standard Euler-Rodrigues parameters $\lambda \geq 0$ and $\Lambda > 0$ are those corresponding to standard parametric points.

From the well-known property of the Pauli matrices,

$$(\mathbf{a} \cdot \boldsymbol{\sigma})(\mathbf{b} \cdot \boldsymbol{\sigma}) = (\mathbf{a} \cdot \mathbf{b})\, E + i(\mathbf{a} \times \mathbf{b}) \cdot \boldsymbol{\sigma},$$

we obtain

$$S(\lambda_1, \mathbf{\Lambda}_1) S(\lambda_2, \mathbf{\Lambda}_2) = S(\lambda_3', \mathbf{\Lambda}_3'),$$

where

$$\lambda_3' = \lambda_1 \lambda_2 - \mathbf{\Lambda}_1 \cdot \mathbf{\Lambda}_2, \tag{D.9}$$

$$\mathbf{\Lambda}_3' = \lambda_1 \mathbf{\Lambda}_1 + \lambda_2 \mathbf{\Lambda}_2 + \mathbf{\Lambda}_1 \times \mathbf{\Lambda}_2. \tag{D.10}$$

By the prime we indicate that, while $(\lambda_1, \mathbf{\Lambda}_1)$ and $(\lambda_2, \mathbf{\Lambda}_2)$ must be standard Euler-Rodrigues parameters, $(\lambda_3', \mathbf{\Lambda}_3')$ may not be so. The following theorem say that $\omega_{\alpha,\beta} = 1$ if $(\lambda_3', \mathbf{\Lambda}_3')$ are standard parameters, and $\omega_{\alpha,\beta} = -1$ if they are not.

Theorem D.2 *The composition of two matrices* $S(\lambda_1, \mathbf{\Lambda}_1)$ *and* $S(\lambda_2, \mathbf{\Lambda}_2)$ *is given by*

$$S(\lambda_1, \mathbf{\Lambda}_1) S(\lambda_2, \mathbf{\Lambda}_2) = \omega_{\alpha,\beta}\; S(\lambda_3, \mathbf{\Lambda}_3),$$

where $(\lambda_3, \mathbf{\Lambda}_3)$ *are standard parameters and projective factors are given by*

$$\omega_{\alpha,\beta} = \begin{cases} 1, & \text{either } \lambda_3 > 0 \quad \text{or} \quad \lambda_3 = 0 \text{ and } \mathbf{\Lambda}_3 > 0, \\ -1, & \text{either } \lambda_3 < 0 \quad \text{or} \quad \lambda_3 = 0 \text{ and } \mathbf{\Lambda}_3 < 0. \end{cases} \tag{D.11}$$

Corollary D.1 *The composition of two symmetry operators* $P_{[\alpha|\mathbf{t}]}$ *and* $P_{[\beta|\mathbf{t}']}$ *is given by*

$$P_{[\alpha|\mathbf{t}]}\, P_{[\beta|\mathbf{t}']} = \omega_{\alpha,\beta}\, P_{[\alpha\beta|\alpha\mathbf{t}'+\mathbf{t}]},$$

where the projective factors are given by (D.11).

The proof can be found in [5]. We give two examples.

First, we obtain the table of projective factors for the group D_2. The group D_2 consists of four elements, the identity element and three rotations by the angle π around each coordinate axis, C_{2x}, C_{2y} and C_{2z}. The two-by-two matrices corresponding to these rotations are $S(0, \mathbf{x})$, $S(0, \mathbf{y})$ and $S(0, \mathbf{z})$. Then by formulas (D.9) and (D.10) we have

$$S(0, \mathbf{z})S(0, \mathbf{x}) = S(0, \mathbf{z} \times \mathbf{x}) = S(0, \mathbf{y}),$$
$$S(0, \mathbf{x})S(0, \mathbf{z}) = S(0, \mathbf{x} \times \mathbf{z}) = S(0, -\mathbf{y}).$$

Using formula (D.11), we obtain the projective factors,

$$S(0, \mathbf{z})S(0, \mathbf{x}) = S(0, \mathbf{y}),$$
$$S(0, \mathbf{x})S(0, \mathbf{z}) = -S(0, \mathbf{y}).$$

Similarly, we obtain +1 and −1 for the corresponding projective factors of other products in D_2.

As the second example, we consider the rotations by the angle π around the vectors $\frac{\sqrt{2}}{2}(1, 1, 0)$, $\frac{\sqrt{2}}{2}(1, 0, 1)$ and $\frac{\sqrt{2}}{2}(0, 1, 1)$. The two-by-two matrices corresponding to these rotations are $S(0, \frac{\sqrt{2}}{2}(\mathbf{x} + \mathbf{y}))$, $S(0, \frac{\sqrt{2}}{2}(\mathbf{x} + \mathbf{z})$ and $S(0, \frac{\sqrt{2}}{2}(\mathbf{y} + \mathbf{z}))$. Then by formulas (D.9) and (D.10) we have

$$S\left(0, \frac{\sqrt{2}}{2}(\mathbf{x} + \mathbf{y})\right) S\left(0, \frac{\sqrt{2}}{2}(\mathbf{x} + \mathbf{z})\right)$$
$$= S\left(-\frac{1}{2}(\mathbf{x} + \mathbf{y}) \cdot (\mathbf{x} + \mathbf{z}), \frac{1}{2}(\mathbf{x} + \mathbf{y}) \times (\mathbf{x} + \mathbf{z})\right) = S\left(-\frac{1}{2}, \frac{1}{2}(\mathbf{x} - \mathbf{y} - \mathbf{z})\right).$$

Using formula (D.11), we obtain the projective factor

$$S\left(0, \frac{\sqrt{2}}{2}(\mathbf{x} + \mathbf{y})\right) S\left(0, \frac{\sqrt{2}}{2}(\mathbf{x} + \mathbf{z})\right) = -S\left(\frac{1}{2}, \frac{1}{2}(-\mathbf{x} + \mathbf{y} + \mathbf{z})\right),$$

in full agreement with Table 6.2, i.e. the 19-th element times the 21-st element is equal to the minus 13-th element. Recall that the bar over the element $\alpha_i\alpha_j$ in the multiplication table (Table 6.2) means that $\omega_{\alpha_i,\alpha_j} = -1$.

References

1. A.I. Lurie, *Analytical Mechanics* (Nauka, Moscow, 1961). Engl. transl.: Springer, Berlin, 2002
2. H. Goldstein, *Classical Mechanics*, 2nd edn. (Addison-Wesley, Reading, MA, 1980)
3. L. Schiff, *Quantum Mechanics*, 3rd edn. (McGraw-Hill, New York, 1968)
4. L.D. Landau, E.M. Lifshitz, *Quantum Mechanics—Non-Relativistic Theory (Course of Theoretical Physics)*, vol. 3, 3rd edn. (Pergamon, Oxford, 1991)
5. S.L. Altmann, *Rotations, Quaternions, and Double Groups* (Oxford Univ. Press, New York, 1986)
6. J.W. Gibbs, E.B. Wilson, *Vector Analysis: A Textbook for the Use of Students of Mathematics and Physics, Founded upon the Lectures of JWG* (Yale Univ. Press, New Haven, CT, 1901)

Appendix E
Multiplication Table of the Point Group O_h

Here we present the multiplication table of the group O_h [1], which contains a number of corrections to the corresponding table of Kovalev [2] (Table E.1).

N. B. Melnikov and B. I. Reser, *Space Group Representations*,
https://doi.org/10.1007/978-3-031-13991-8

Table E.1 The multiplication table of the point group O_h

	j																							
i	1	2	3	4	5	6	7	8	9	10	11	12	13	14	15	16	17	18	19	20	21	22	23	24
1	1	2	3	4	5	6	7	8	9	10	11	12	13	14	15	16	17	18	19	20	21	22	23	24
2	2	1	4	3	6	5	8	7	10	9	12	11	14	13	16	15	18	17	20	19	22	21	24	23
3	3	4	1	2	7	8	5	6	11	12	9	10	15	16	13	14	19	20	17	18	23	24	21	22
4	4	3	2	1	8	7	6	5	12	11	10	9	16	15	14	13	20	19	18	17	24	23	22	21
5	5	8	6	7	9	12	10	11	1	4	2	3	17	20	18	19	21	24	22	23	13	16	14	15
6	6	7	5	8	10	11	9	12	2	3	1	4	18	19	17	20	22	23	21	24	14	15	13	16
7	7	6	8	5	11	10	12	9	3	2	4	1	19	18	20	17	23	22	24	21	15	14	16	13
8	8	5	7	6	12	9	11	10	4	1	3	2	20	17	19	18	24	21	23	22	16	13	15	14
9	9	11	12	10	1	3	4	2	5	7	8	6	21	23	24	22	13	15	16	14	17	19	20	18
10	10	12	11	9	2	4	3	1	6	8	7	5	22	24	23	21	14	16	15	13	18	20	19	17
11	11	9	10	12	3	1	2	4	7	5	6	8	23	21	22	24	15	13	14	16	19	17	18	20
12	12	10	9	11	4	2	1	3	8	6	5	7	24	22	21	23	16	14	13	15	20	18	17	19
13	13	15	14	16	21	23	22	24	17	19	18	20	1	3	2	4	9	11	10	12	5	7	6	8
14	14	16	13	15	22	24	21	23	18	20	17	19	2	4	1	3	10	12	9	11	6	8	5	7
15	15	13	16	14	23	21	24	22	19	17	20	18	3	1	4	2	11	9	12	10	7	5	8	6
16	16	14	15	13	24	22	23	21	20	18	19	17	4	2	3	1	12	10	11	9	8	6	7	5
17	17	18	20	19	13	14	16	15	21	22	24	23	5	6	8	7	1	2	4	3	9	10	12	11
18	18	17	19	20	14	13	15	16	22	21	13	24	6	5	7	8	2	1	3	4	10	9	11	12
19	19	20	18	17	15	16	14	13	23	24	22	21	7	8	6	5	3	4	2	1	11	12	10	9
20	20	19	17	18	16	15	13	14	24	23	21	22	8	7	5	6	4	3	1	2	12	11	9	10
21	21	24	23	22	17	20	19	18	13	16	15	14	9	12	11	10	5	8	7	6	1	4	3	2
22	22	23	24	21	18	19	20	17	14	15	16	13	10	11	12	9	6	7	8	5	2	3	4	1
23	23	22	21	24	19	18	17	20	15	14	13	16	11	10	9	12	7	6	5	8	3	2	1	4
24	24	21	22	23	20	17	18	19	16	13	14	15	12	19	10	11	8	5	6	7	4	1	2	3

The table presents the products $\alpha_i \alpha_j = \alpha_k$, where $\alpha_i, \alpha_j, \alpha_k \in O$. For the elements $\alpha_l \in O_h, l = 25, 26, \ldots, 48$, the multiplication rules are $\alpha_{i+24}\alpha_j = \alpha_i \alpha_{j+24} = \alpha_{k+24}$ and $\alpha_{i+24}\alpha_{j+24} = \alpha_k$. The numbering of the elements is as in monograph [2]

References

1. B.I. Reser, *The OPW method and interpolation based on the tight binding in calculating the energy band structure of crystals* [in Russian]. Diss. for the degree of cand. of phys. and math. sciences, Inst. of Metal Phys., Ural Science Center, USSR Acad. of Sciences, Sverdlovsk (1972), 153 pp
2. O.V. Kovalev, *Irreducible representations of space groups* [in Russian] (Acad. Sci. Ukrain. SSR, Kiev, 1961). Engl. transl.: Gordon and Breach, New York, 1965

Appendix F
Projective Representations and Compatibility Relations for O_h^7

Here we present matrices and characters of projective representations $\Delta^{\mathbf{k}j}$ of the point groups $G_{\mathbf{k}}^0$ and compatibility relations at the symmetry points for the group O_h^7 [1]. The numbering of the group elements is as in the monograph of Kovalev [2] (Tables F.1, F.2, F.3, F.4, F.5, F.6, F.7, F.8, F.9, F.10, F.11, F.12, F.13, F.14, F.15, F.16, F.17, F.18, F.19, F.20, F.21 and F.22).

Table F.1 Characters of representations of the group O_h of the point Γ

	1	2 3 4	5–12	13,16 17,18 21,23	14,15 19,20 22,24	25	26 27 28	29–36	37,40 41,42 45,47	38,39 43,44 46,48
Γ_1	1	1	1	1	1	1	1	1	1	1
Γ_2	1	1	1	−1	−1	1	1	1	−1	−1
Γ_{12}	2	2	−1	0	0	2	2	−1	0	0
Γ'_{15}	3	−1	0	−1	1	3	−1	0	−1	1
Γ'_{25}	3	−1	0	1	−1	3	−1	0	1	−1
Γ'_1	1	1	1	1	1	−1	−1	−1	−1	−1
Γ'_2	1	1	1	−1	−1	−1	−1	−1	1	1
Γ'_{12}	2	2	−1	0	0	−2	−2	1	0	0
Γ_{15}	3	−1	0	−1	1	−3	1	0	1	−1
Γ_{25}	3	−1	0	1	−1	−3	1	0	−1	1

N. B. Melnikov and B. I. Reser, *Space Group Representations*,
https://doi.org/10.1007/978-3-031-13991-8

Table F.2 Matrices of representations of the group O_h of the point Γ[a]

G	BSW	1	2	3	4	5	6	7	8	9	10	11	12
Γ_1^+	Γ_1	1	1	1	1	1	1	1	1	1	1	1	1
Γ_2^+	Γ_2	1	1	1	1	1	1	1	1	1	1	1	1
Γ_3^+	Γ_{12}	$\begin{matrix}1 & 0\\ 0 & 1\end{matrix}$	$\begin{matrix}1 & 0\\ 0 & 1\end{matrix}$	$\begin{matrix}1 & 0\\ 0 & 1\end{matrix}$	$\begin{matrix}1 & 0\\ 0 & 1\end{matrix}$	$\begin{matrix}\varepsilon & 0\\ 0 & \varepsilon^2\end{matrix}$	$\begin{matrix}\varepsilon & 0\\ 0 & \varepsilon^2\end{matrix}$	$\begin{matrix}\varepsilon & 0\\ 0 & \varepsilon^2\end{matrix}$	$\begin{matrix}\varepsilon & 0\\ 0 & \varepsilon^2\end{matrix}$	$\begin{matrix}\varepsilon^2 & 0\\ 0 & \varepsilon\end{matrix}$	$\begin{matrix}\varepsilon^2 & 0\\ 0 & \varepsilon\end{matrix}$	$\begin{matrix}\varepsilon^2 & 0\\ 0 & \varepsilon\end{matrix}$	$\begin{matrix}\varepsilon^2 & 0\\ 0 & \varepsilon\end{matrix}$
		2	2	2	2	−1	−1	−1	−1	−1	−1	−1	−1
Γ_4^+	Γ_{15}	$\begin{matrix}1 & 0 & 0\\ 0 & 1 & 0\\ 0 & 0 & 1\end{matrix}$	$\begin{matrix}1 & 0 & 0\\ 0 & -1 & 0\\ 0 & 0 & -1\end{matrix}$	$\begin{matrix}-1 & 0 & 0\\ 0 & 1 & 0\\ 0 & 0 & -1\end{matrix}$	$\begin{matrix}-1 & 0 & 0\\ 0 & -1 & 0\\ 0 & 0 & 1\end{matrix}$	$\begin{matrix}0 & 1 & 0\\ 0 & 0 & 1\\ 1 & 0 & 0\end{matrix}$	$\begin{matrix}0 & 1 & 0\\ 0 & 0 & -1\\ -1 & 0 & 0\end{matrix}$	$\begin{matrix}0 & -1 & 0\\ 0 & 0 & 1\\ -1 & 0 & 0\end{matrix}$	$\begin{matrix}0 & -1 & 0\\ 0 & 0 & -1\\ 1 & 0 & 0\end{matrix}$	$\begin{matrix}0 & 0 & 1\\ 1 & 0 & 0\\ 0 & 1 & 0\end{matrix}$	$\begin{matrix}0 & 0 & 1\\ -1 & 0 & 0\\ 0 & -1 & 0\end{matrix}$	$\begin{matrix}0 & 0 & -1\\ 1 & 0 & 0\\ 0 & -1 & 0\end{matrix}$	$\begin{matrix}0 & 0 & -1\\ -1 & 0 & 0\\ 0 & 1 & 0\end{matrix}$
		3	−1	−1	−1	0	0	0	0	0	0	0	0
Γ_5^+	Γ_{25}	$\begin{matrix}1 & 0 & 0\\ 0 & 1 & 0\\ 0 & 0 & 1\end{matrix}$	$\begin{matrix}1 & 0 & 0\\ 0 & -1 & 0\\ 0 & 0 & -1\end{matrix}$	$\begin{matrix}-1 & 0 & 0\\ 0 & 1 & 0\\ 0 & 0 & -1\end{matrix}$	$\begin{matrix}-1 & 0 & 0\\ 0 & -1 & 0\\ 0 & 0 & 1\end{matrix}$	$\begin{matrix}0 & 1 & 0\\ 0 & 0 & 1\\ 1 & 0 & 0\end{matrix}$	$\begin{matrix}0 & 1 & 0\\ 0 & 0 & -1\\ -1 & 0 & 0\end{matrix}$	$\begin{matrix}0 & -1 & 0\\ 0 & 0 & 1\\ -1 & 0 & 0\end{matrix}$	$\begin{matrix}0 & -1 & 0\\ 0 & 0 & -1\\ 1 & 0 & 0\end{matrix}$	$\begin{matrix}0 & 0 & 1\\ 1 & 0 & 0\\ 0 & 1 & 0\end{matrix}$	$\begin{matrix}0 & 0 & 1\\ -1 & 0 & 0\\ 0 & -1 & 0\end{matrix}$	$\begin{matrix}0 & 0 & -1\\ 1 & 0 & 0\\ 0 & -1 & 0\end{matrix}$	$\begin{matrix}0 & 0 & -1\\ -1 & 0 & 0\\ 0 & 1 & 0\end{matrix}$
		3	−1	−1	−1	0	0	0	0	0	0	0	0
G	BSW	13	14	15	16	17	18	19	20	21	22	23	24
Γ_1^+	Γ_1	1	1	1	1	1	1	1	1	1	1	1	1
Γ_2^+	Γ_2	−1	−1	−1	−1	−1	−1	−1	−1	−1	−1	−1	−1
Γ_3^+	Γ_{12}	$\begin{matrix}0 & 1\\ 1 & 0\end{matrix}$	$\begin{matrix}0 & 1\\ 1 & 0\end{matrix}$	$\begin{matrix}0 & 1\\ 1 & 0\end{matrix}$	$\begin{matrix}0 & 1\\ 1 & 0\end{matrix}$	$\begin{matrix}0 & \varepsilon\\ \varepsilon^2 & 0\end{matrix}$	$\begin{matrix}0 & \varepsilon\\ \varepsilon^2 & 0\end{matrix}$	$\begin{matrix}0 & \varepsilon\\ \varepsilon^2 & 0\end{matrix}$	$\begin{matrix}0 & \varepsilon\\ \varepsilon^2 & 0\end{matrix}$	$\begin{matrix}0 & \varepsilon^2\\ \varepsilon & 0\end{matrix}$	$\begin{matrix}0 & \varepsilon^2\\ \varepsilon & 0\end{matrix}$	$\begin{matrix}0 & \varepsilon^2\\ \varepsilon & 0\end{matrix}$	$\begin{matrix}0 & \varepsilon^2\\ \varepsilon & 0\end{matrix}$
		0	0	0	0	0	0	0	0	0	0	0	0
Γ_4^+	Γ_{15}	$\begin{matrix}0 & -1 & 0\\ -1 & 0 & 0\\ 0 & 0 & -1\end{matrix}$	$\begin{matrix}0 & -1 & 0\\ 1 & 0 & 0\\ 0 & 0 & 1\end{matrix}$	$\begin{matrix}0 & 1 & 0\\ -1 & 0 & 0\\ 0 & 0 & 1\end{matrix}$	$\begin{matrix}0 & 1 & 0\\ 1 & 0 & 0\\ 0 & 0 & -1\end{matrix}$	$\begin{matrix}-1 & 0 & 0\\ 0 & 0 & -1\\ 0 & -1 & 0\end{matrix}$	$\begin{matrix}-1 & 0 & 0\\ 0 & 0 & 1\\ 0 & 1 & 0\end{matrix}$	$\begin{matrix}1 & 0 & 0\\ 0 & 0 & -1\\ 0 & 1 & 0\end{matrix}$	$\begin{matrix}1 & 0 & 0\\ 0 & 0 & 1\\ 0 & -1 & 0\end{matrix}$	$\begin{matrix}0 & 0 & -1\\ 0 & -1 & 0\\ -1 & 0 & 0\end{matrix}$	$\begin{matrix}0 & 0 & -1\\ 0 & 1 & 0\\ 1 & 0 & 0\end{matrix}$	$\begin{matrix}0 & 0 & 1\\ 0 & -1 & 0\\ 1 & 0 & 0\end{matrix}$	$\begin{matrix}0 & 0 & 1\\ 0 & 1 & 0\\ -1 & 0 & 0\end{matrix}$
		−1	1	1	−1	−1	−1	1	1	−1	1	−1	1
Γ_5^+	Γ_{25}	$\begin{matrix}0 & 1 & 0\\ 1 & 0 & 0\\ 0 & 0 & 1\end{matrix}$	$\begin{matrix}0 & 1 & 0\\ -1 & 0 & 0\\ 0 & 0 & -1\end{matrix}$	$\begin{matrix}0 & -1 & 0\\ 1 & 0 & 0\\ 0 & 0 & -1\end{matrix}$	$\begin{matrix}0 & -1 & 0\\ -1 & 0 & 0\\ 0 & 0 & 1\end{matrix}$	$\begin{matrix}1 & 0 & 0\\ 0 & 0 & 1\\ 0 & 1 & 0\end{matrix}$	$\begin{matrix}1 & 0 & 0\\ 0 & 0 & -1\\ 0 & -1 & 0\end{matrix}$	$\begin{matrix}-1 & 0 & 0\\ 0 & 0 & 1\\ 0 & -1 & 0\end{matrix}$	$\begin{matrix}-1 & 0 & 0\\ 0 & 0 & -1\\ 0 & 1 & 0\end{matrix}$	$\begin{matrix}0 & 0 & 1\\ 0 & 1 & 0\\ 1 & 0 & 0\end{matrix}$	$\begin{matrix}0 & 0 & 1\\ 0 & -1 & 0\\ -1 & 0 & 0\end{matrix}$	$\begin{matrix}0 & 0 & -1\\ 0 & 1 & 0\\ -1 & 0 & 0\end{matrix}$	$\begin{matrix}0 & 0 & -1\\ 0 & -1 & 0\\ 1 & 0 & 0\end{matrix}$
		1	−1	−1	1	1	1	−1	−1	1	−1	1	−1

$\varepsilon = \exp(i2\pi/3)$

[a] To save space only matrices of representations of the group O are tabulated. Matrices of representations of the group $O_h = O \times C_{1h}$ are obtained by multiplying them by ± 1

Table F.3 Matrices and characters of representations of the group D_{4h} of the point X

	1	2	3	4	21	22	23	24	25	26	27	28	45	46	47	48
X_1	1 0	1 0	1 0	1 0	0 −1	0 1	0 −1	0 1	0 −1	0 1	0 −1	0 1	1 0	1 0	1 0	1 0
	0 1	0 −1	0 1	0 −1	−1 0	−1 0	−1 0	−1 0	−1 0	−1 0	−1 0	−1 0	0 1	0 −1	0 1	0 −1
	2	0	2	0	0	0	0	0	0	0	0	0	2	0	2	0
X_2	1 0	−1 0	1 0	−1 0	0 −1	0 −1	0 −1	0 −1	0 1	0 1	0 1	0 1	−1 0	1 0	−1 0	1 0
	0 1	0 1	0 1	0 1	−1 0	1 0	−1 0	1 0	1 0	−1 0	1 0	−1 0	0 −1	0 −1	0 −1	0 −1
	2	0	2	0	0	0	0	0	0	0	0	0	−2	0	−2	0
X_3	1 0	1 0	−1 0	−1 0	−1 0	1 0	1 0	−1 0	0 1	0 −1	0 −1	0 1	0 −1	0 −1	0 1	0 1
	0 1	0 −1	0 −1	0 1	0 −1	0 −1	0 1	0 1	1 0	1 0	−1 0	−1 0	−1 0	1 0	1 0	−1 0
	2	0	−2	0	−2	0	2	0	0	0	0	0	0	0	0	0
X_4	1 0	1 0	−1 0	−1 0	1 0	−1 0	−1 0	1 0	0 −1	0 1	0 1	0 −1	0 −1	0 −1	0 1	0 1
	0 1	0 −1	0 −1	0 1	0 1	0 1	0 −1	0 −1	−1 0	−1 0	1 0	1 0	−1 0	1 0	1 0	−1 0
	2	0	−2	0	2	0	−2	0	0	0	0	0	0	0	0	0

Table F.4 Matrices and characters of representations of the group D_{3h} of the point L

	1	5	9	13	17	21	25	29	33	37	41	45
L_1	1	1	1	1	1	1	1	1	1	1	1	1
L_2	1	1	1	−1	−1	−1	1	1	1	−1	−1	−1
L_3	$\begin{matrix}1 & 0\\ 0 & 1\end{matrix}$	$\begin{matrix}\varepsilon & 0\\ 0 & \varepsilon^2\end{matrix}$	$\begin{matrix}\varepsilon^2 & 0\\ 0 & \varepsilon\end{matrix}$	$\begin{matrix}0 & \varepsilon\\ \varepsilon^2 & 0\end{matrix}$	$\begin{matrix}0 & \varepsilon^2\\ \varepsilon & 0\end{matrix}$	$\begin{matrix}0 & 1\\ 1 & 0\end{matrix}$	$\begin{matrix}1 & 0\\ 0 & 1\end{matrix}$	$\begin{matrix}\varepsilon & 0\\ 0 & \varepsilon^2\end{matrix}$	$\begin{matrix}\varepsilon^2 & 0\\ 0 & \varepsilon\end{matrix}$	$\begin{matrix}0 & \varepsilon\\ \varepsilon^2 & 0\end{matrix}$	$\begin{matrix}0 & \varepsilon^2\\ \varepsilon & 0\end{matrix}$	$\begin{matrix}0 & 1\\ 1 & 0\end{matrix}$
	2	−1	−1	0	0	0	2	−1	−1	0	0	0
L_1'	1	1	1	1	1	1	−1	−1	−1	−1	−1	−1
L_2'	1	1	1	−1	−1	−1	−1	−1	−1	1	1	1
L_3'	$\begin{matrix}1 & 0\\ 0 & 1\end{matrix}$	$\begin{matrix}\varepsilon & 0\\ 0 & \varepsilon^2\end{matrix}$	$\begin{matrix}\varepsilon^2 & 0\\ 0 & \varepsilon\end{matrix}$	$\begin{matrix}0 & -\varepsilon\\ -\varepsilon^2 & 0\end{matrix}$	$\begin{matrix}0 & -\varepsilon^2\\ -\varepsilon & 0\end{matrix}$	$\begin{matrix}0 & -1\\ -1 & 0\end{matrix}$	$\begin{matrix}-1 & 0\\ 0 & -1\end{matrix}$	$\begin{matrix}-\varepsilon & 0\\ 0 & -\varepsilon^2\end{matrix}$	$\begin{matrix}-\varepsilon^2 & 0\\ 0 & -\varepsilon\end{matrix}$	$\begin{matrix}0 & \varepsilon\\ \varepsilon^2 & 0\end{matrix}$	$\begin{matrix}0 & \varepsilon^2\\ \varepsilon & 0\end{matrix}$	$\begin{matrix}0 & 1\\ 1 & 0\end{matrix}$
	2	−1	−1	0	0	0	−2	1	1	0	0	0

$\varepsilon = \exp(i2\pi/3)$

Table F.5 Matrices and characters of representations of the group D_{2d} of the point W

	1	2	17	18	27	28	43	44
W_1	$\begin{matrix}1 & 0\\ 0 & 1\end{matrix}$	$\begin{matrix}1 & 0\\ 0 & -1\end{matrix}$	$\begin{matrix}0 & -i\\ i & 0\end{matrix}$	$\begin{matrix}0 & i\\ i & 0\end{matrix}$	$\begin{matrix}0 & -1\\ i & 0\end{matrix}$	$\begin{matrix}0 & 1\\ i & 0\end{matrix}$	$\begin{matrix}1 & 0\\ 0 & i\end{matrix}$	$\begin{matrix}1 & 0\\ 0 & -i\end{matrix}$
	2	0	0	0	0	0	$1+i$	$1-i$
W_2	$\begin{matrix}1 & 0\\ 0 & 1\end{matrix}$	$\begin{matrix}1 & 0\\ 0 & -1\end{matrix}$	$\begin{matrix}0 & i\\ -i & 0\end{matrix}$	$\begin{matrix}0 & -i\\ -i & 0\end{matrix}$	$\begin{matrix}0 & -1\\ i & 0\end{matrix}$	$\begin{matrix}0 & 1\\ i & 0\end{matrix}$	$\begin{matrix}-1 & 0\\ 0 & -i\end{matrix}$	$\begin{matrix}-1 & 0\\ 0 & i\end{matrix}$
	2	0	0	0	0	0	$-1-i$	$-1+i$

Table F.6 Matrices and characters of representations of the group C_{2v} of the point K

	1	16	28	37
K_1	1	σ	σ	1
K_2	1	σ	$-\sigma$	−1
K_3	1	$-\sigma$	$-\sigma$	1
K_4	1	$-\sigma$	σ	−1

$\sigma = \exp(-i3\pi/4)$

Table F.7 Matrices and characters of representations of the group C_{2v} of the point U

	1	23	27	45
U_1	1	κ	κ	1
U_2	1	κ	$-\kappa$	−1
U_3	1	$-\kappa$	$-\kappa$	1
U_4	1	$-\kappa$	κ	−1

$\kappa = \exp(-i\pi/4)$

Table F.8 Matrices and characters of representations of the group C_{4v} of the point Δ

	1	3	22	24	26	28	45	47
Δ_1	1	1	ψ	ψ	ψ	ψ	1	1
Δ_2	1	1	$-\psi$	$-\psi$	ψ	ψ	-1	-1
Δ_2'	1	1	$-\psi$	$-\psi$	$-\psi$	$-\psi$	1	1
Δ_1'	1	1	ψ	ψ	$-\psi$	$-\psi$	-1	-1
Δ_5	1 0	-1 0	0 $-\psi$	0 ψ	0 $-\psi$	0 ψ	-1 0	1 0
	0 1	0 -1	ψ 0	$-\psi$ 0	$-\psi$ 0	ψ 0	0 1	0 -1
	2	-2	0	0	0	0	0	0

$\psi = \exp(-i\mu\pi/2)$, $0 < \mu < 1$

Table F.9 Matrices and characters of representations of the group C_{3v} of the point Λ

	1	5	9	37	41	45
Λ_1	1	1	1	1	1	1
Λ_2	1	1	1	-1	-1	-1
Λ_3	1 0	ε 0	ε^2 0	0 ε	0 ε^2	0 1
	0 1	0 ε^2	0 ε	ε^2 0	ε 0	1 0
	2	-1	-1	0	0	0

$\varepsilon = \exp(i2\pi/3)$

Table F.10 Matrices and characters of representations of the group C_{2v} of the point Σ

	1	16	28	37
Σ_1	1	φ	φ	1
Σ_2	1	φ	$-\varphi$	-1
Σ_3	1	$-\varphi$	$-\varphi$	1
Σ_4	1	$-\varphi$	φ	-1

$\varphi = \exp(-i3\mu\pi/4)$

Table F.11 Matrices and characters of representations of the group C_{2v} of the point S

	1	23	27	45
S_1	1	λ	λ	1
S_2	1	λ	$-\lambda$	-1
S_3	1	$-\lambda$	$-\lambda$	1
S_4	1	$-\lambda$	λ	-1

$\lambda = \exp(-i\mu\pi/4)$

Table F.12 Matrices and characters of representations of the group C_{2v} of the point Z

	1	2	27	28
Z_1	1 0	1 0	0 $-\nu$	0 ν
	0 1	0 -1	ν 0	$-\nu$ 0
	2	0	0	0

$\nu = \exp(-i\mu\pi/4)$

Table F.13 Matrices and characters of representations of the group C_2 of the point Q

	1	17
Q_1	1	1
Q_2	1	−1

Table F.14 Matrices and characters of representations of the group C_{1h} of the points LK and LU

	1	37
	1	45
$(LK)_1$ $(LU)_1$	1	1
$(LK)_2$ $(LU)_2$	1	−1

Table F.15 Matrices and characters of representations of the group C_{1h} of the point KW

	1	28
$(KW)_1$	1	σ
$(KW)_2$	1	$-\sigma$

$\sigma = \exp(-i3\pi/4)$

Table F.16 Matrices and characters of representations of the group C_{1h} of the point WU

	1	27
$(WU)_1$	1	κ
$(WU)_2$	1	$-\kappa$

$\kappa = \exp(-i\pi/4)$

Table F.17 Compatibility relations at the point Γ

Γ_1	Γ_2	Γ_{12}	Γ'_{15}	Γ'_{25}	Γ'_1	Γ'_2	Γ'_{12}	Γ_{15}	Γ_{25}
Δ_1	Δ_2	$\Delta_1\Delta_2$	$\Delta'_1\Delta_5$	$\Delta'_2\Delta_5$	Δ'_1	Δ'_2	$\Delta'_1\Delta'_2$	$\Delta_1\Delta_5$	$\Delta_2\Delta_5$
Λ_1	Λ_2	Λ_3	$\Lambda_2\Lambda_3$	$\Lambda_1\Lambda_3$	Λ_2	Λ_1	Λ_3	$\Lambda_1\Lambda_3$	$\Lambda_2\Lambda_3$
Σ_1	Σ_4	$\Sigma_1\Sigma_4$	$\Sigma_2\Sigma_3\Sigma_4$	$\Sigma_1\Sigma_2\Sigma_3$	Σ_2	Σ_3	$\Sigma_2\Sigma_3$	$\Sigma_1\Sigma_3\Sigma_4$	$\Sigma_1\Sigma_2\Sigma_4$

Table F.18 Compatibility relations at the point X

X_1	X_2	X_3	X_4
$\Delta_1\Delta'_2$	$\Delta_2\Delta'_1$	Δ_5	Δ_5
Z_1	Z_1	Z_1	Z_1
S_1S_3	S_2S_4	S_1S_2	S_3S_4

Table F.19 Compatibility relations at the point L

L_1	L_2	L_3	L'_1	L'_2	L'_3
Λ_1	Λ_2	Λ_3	Λ_2	Λ_1	Λ_3
Q_1	Q_2	Q_1Q_2	Q_1	Q_2	Q_1Q_2
$(LK)_1$	$(LK)_2$	$(LK)_1(LK)_2$	$(LK)_2$	$(LK)_1$	$(LK)_1(LK)_2$
$(LU)_1$	$(LU)_2$	$(LU)_1(LU)_2$	$(LU)_2$	$(LU)_1$	$(LU)_1(LU)_2$

Table F.20 Compatibility relations at the point W

W_1	W_2
Z_1	Z_1
Q_1Q_2	Q_1Q_2
$(KW)_1(KW)_2$	$(KW)_1(KW)_2$
$(WU)_1(WU)_2$	$(WU)_1(WU)_2$

Table F.21 Compatibility relations at the point K

K_1	K_2	K_3	K_4
Σ_1	Σ_2	Σ_3	Σ_4
$(LK)_1$	$(LK)_2$	$(LK)_1$	$(LK)_2$
$(KW)_1$	$(KW)_2$	$(KW)_2$	$(KW)_1$

Table F.22 Compatibility relations at the point U

U_1	U_2	U_3	U_4
S_1	S_2	S_3	S_4
$(LU)_1$	$(LU)_2$	$(LU)_1$	$(LU)_2$
$(WU)_1$	$(WU)_2$	$(WU)_2$	$(WU)_1$

References

1. B.I. Reser, *The OPW method and interpolation based on the tight binding in calculating the energy band structure of crystals* [in Russian]. Diss. for the degree of cand. of phys. and math. sciences, Inst. of Metal Phys., Ural Science Center, USSR Acad. of Sciences, Sverdlovsk (1972), 153 pp
2. O.V. Kovalev, *Irreducible representations of space groups* [in Russian] (Acad. Sci. Ukrain. SSR, Kiev, 1961). Engl. transl.: Gordon and Breach, New York, 1965

Appendix G
Factorization of the Characteristic Equation for O_h^7

At high symmetry points of the Brillouin zone, the characteristic equation in the tight-binding method (14.1) is factorized into *different* factors corresponding to the blocks of the Hamiltonian matrix. For the diamond structure, the factors are as follows [1].

Point	Rep	Block
$\Gamma(000)$	Γ_1	$H_{11} + H_{15} - \varepsilon = 0;$
	Γ_2'	$H_{11} - H_{15} - \varepsilon = 0;$
	Γ_{25}'	$H_{22} - H_{26} - \varepsilon = 0;$
	Γ_{15}	$H_{22} + H_{26} - \varepsilon = 0.$

Here

$$H_{11} = x_1 + 12x_7, \qquad H_{15} = 4x_3 + 12x_{14},$$
$$H_{22} = x_2 + 8x_{10} + 4x_{13}, \qquad H_{26} = 4x_5 + 4x_{17} + 8x_{19}.$$

Point	Rep	Block
$X(010)$	X_1	$\begin{vmatrix} H_{11} - \varepsilon & H_{17} \\ H_{17}^* & H_{33} - \varepsilon \end{vmatrix} = 0;$
	X_3	$H_{22} - iH_{28} - \varepsilon = 0;$
	X_4	$H_{22} + iH_{28} - \varepsilon = 0.$

Here

$$H_{11} = x_1 - 4x_7, \qquad H_{17} = 4i(x_4 - x_{15} - 2x_{16}),$$
$$H_{22} = x_2 - 4x_{13}, \qquad H_{28} = 4i(x_6 + 2x_{18} - x_{20}),$$
$$H_{33} = x_2 - 8x_{10} + 4x_{13}.$$

N. B. Melnikov and B. I. Reser, *Space Group Representations*,
https://doi.org/10.1007/978-3-031-13991-8

Point	Rep	Block
$L\left(\frac{1}{2}\frac{1}{2}\frac{1}{2}\right)$	L_1	$\begin{vmatrix} H_{11}+\sigma H_{15}-\varepsilon & H_{12}-\sigma H_{16} \\ 3(H_{12}-\sigma H_{16})^* & H_{22}+2H_{23}-\sigma H_{26}-2\sigma H_{27}-\varepsilon \end{vmatrix} = 0;$
	L_2'	$\begin{vmatrix} H_{11}-\sigma H_{15}-\varepsilon & H_{12}+\sigma H_{16} \\ 3(H_{12}+\sigma H_{16})^* & H_{22}+2H_{23}+\sigma H_{26}+2\sigma H_{27}-\varepsilon \end{vmatrix} = 0;$
	L_3	$H_{22}-H_{23}-\sigma H_{26}+\sigma H_{27}-\varepsilon=0;$
	L_3'	$H_{22}-H_{23}+\sigma H_{26}-\sigma H_{27}-\varepsilon=0.$

Here

$$
\begin{aligned}
&H_{11}=x_1, && H_{12}=-4x_9,\\
&H_{15}=\sqrt{2}(1-i)(x_3-3x_{14}), && H_{16}=\sqrt{2}(1-i)(-x_4-x_{15}-2x_{16}),\\
&H_{22}=x_2, && H_{23}=-4x_9,\\
&H_{26}=\sqrt{2}(1-i)(x_5-x_{17}-2x_{19}), && H_{27}=\sqrt{2}(1-i)(-x_6+2x_{18}-x_{20}),\\
&\sigma=-\tfrac{\sqrt{2}}{2}(1+i).
\end{aligned}
$$

Point	Rep	Block
$\Delta(0\frac{1}{2}0)$	Δ_1	$\begin{vmatrix} H_{11}+H_{15}-\varepsilon & H_{13}+H_{17} \\ (H_{13}+H_{17})^* & H_{33}-H_{37}-\varepsilon \end{vmatrix} = 0;$
	Δ_2'	$\begin{vmatrix} H_{11}-H_{15}-\varepsilon & H_{13}-H_{17} \\ (H_{13}-H_{17})^* & H_{33}-H_{37}-\varepsilon \end{vmatrix} = 0;$
	Δ_5	$\begin{vmatrix} H_{22}-\varepsilon & H_{26}-H_{28} \\ (H_{26}-H_{28})^* & H_{22}-\varepsilon \end{vmatrix} = 0.$

Here

$$
\begin{aligned}
&H_{11}=x_1+4x_7, && H_{13}=8ix_8,\\
&H_{15}=2\sqrt{2}(x_8+x_{14}), && H_{17}=2\sqrt{2}i(x_4+x_{15}-2x_{16}),\\
&H_{22}=x_2+4x_{10}, && H_{26}=2\sqrt{2}(x_5+x_{17}),\\
&H_{28}=2\sqrt{2}i(x_6+2x_{18}+x_{20}), && H_{33}=x_2+4x_{13},\\
&H_{37}=2\sqrt{2}(x_5-x_{17}+2x_{19}).
\end{aligned}
$$

Point	Rep	Block
$\Sigma\left(\frac{1}{2}\frac{1}{2}0\right)$	Σ_1	$\begin{vmatrix} H_{11}+H_{15}-\varepsilon & H_{12}+H_{16} & H_{14}-H_{18} \\ 2(H_{12}+H_{16})^* & H_{22}+H_{23}+H_{26}+H_{27}-\varepsilon & 2(H_{24}-H_{28}) \\ (H_{14}-H_{18})^* & (H_{24}-H_{28})^* & H_{44}-H_{48}-\varepsilon \end{vmatrix} = 0;$
	Σ_2	$H_{22}-H_{23}-H_{26}+H_{27}-\varepsilon=0;$
	Σ_3	$\begin{vmatrix} H_{11}-H_{15}-\varepsilon & H_{12}-H_{16} & H_{14}+H_{18} \\ 2(H_{12}-H_{16})^* & H_{22}+H_{23}-H_{26}-H_{27}-\varepsilon & 2(H_{24}+H_{28}) \\ (H_{14}+H_{18})^* & (H_{24}+H_{28})^* & H_{44}+H_{48}-\varepsilon \end{vmatrix} = 0;$
	Σ_4	$H_{22}-H_{23}+H_{26}-H_{27}-\varepsilon=0.$

Here

$$H_{11} = x_1, \quad H_{12} = 4ix_8,$$
$$H_{14} = -4x_9, \quad H_{15} = 2(x_3 - x_{14}),$$
$$H_{16} = 2i(x_4 + x_{15}), \quad H_{18} = -2(x_4 - x_{15} + 2x_{16}),$$
$$H_{22} = x_2, \quad H_{23} = -4x_{11},$$
$$H_{24} = 4ix_{12}, \quad H_{26} = 2(x_5 - x_{17}),$$
$$H_{27} = 2(-x_6 + 2x_{18} + x_{20}), \quad H_{28} = 2i(x_6 + x_{20}),$$
$$H_{44} = x_2, \quad H_{48} = 2(x_5 + x_{17} - 2x_{19}).$$

Reference

1. B.I. Reser, *The OPW method and interpolation based on the tight binding in calculating the energy band structure of crystals* [in Russian]. Diss. for the degree of cand. of phys. and math. sciences, Inst. of Metal Phys., Ural Science Center, USSR Acad. of Sciences, Sverdlovsk (1972), 153 pp

Index

N. B. Melnikov and B. I. Reser, *Space Group Representations*,
https://doi.org/10.1007/978-3-031-13991-8